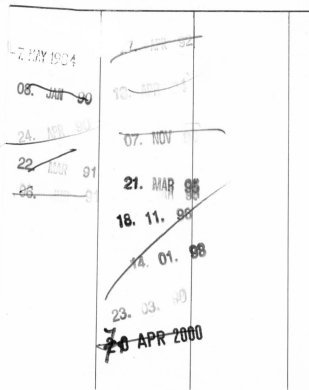

EXTINCTION

EXTINCTION

*The Causes and Consequences of the
Disappearance of Species*

Paul and Anne Ehrlich

London · Victor Gollancz Ltd · 1982

Grateful acknowledgment is made to the following for permission to reprint previously published material:

The Age: Excerpt by John Larkin from *The Age* Newspaper, June 18, 1977, is reprinted courtesy of *The Age* Newspaper.

BioScience: Excerpt by F. H. Bormann, reprinted, with permission, from the Volume 26, 1976, issue of *BioScience,* © American Institute of Biological Sciences.

The Bulletin of the Atomic Scientists: Excerpt from "Criteria for an Optimum Human Environment" by H. H. Iltis, P. Andrews and O. Loucks, from *The Bulletin of the Atomic Scientists,* Volume 26, No. 1, pp. 2–6. Reprinted by permission of *The Bulletin of the Atomic Scientists,* a magazine of science and public affairs. Copyright © 1970 by the Educational Foundation for Nuclear Science, Chicago, Illinois.

The California Native Plant Society: Excerpts from a manuscript prepared for *The California Desert: An Introduction to Natural Resources and Man's Impact,* Special Publication No. 5, 1980, by Robert Stebbins. © 1981 The California Native Plant Society, 2380 Ellsworth, Suite D, Berkeley, California 94704.

Harvard Magazine: Excerpts by E. O. Wilson from *Harvard Magazine,* January/February 1980 issue. Copyright © 1980 by Harvard Magazine. Reprinted by permission.

The Living Wilderness: Excerpt from "ORV's Threaten a Wild Canyon" by Dave Foreman is reprinted by permission from *The Living Wilderness,* September 1979. Copyright 1979 by The Wilderness Society.

The New York Times: Excerpt from "Give Me the Old-Time Darwin" by Sam Witchell, May 3, 1974, Op-ed. © 1974 by The New York Times Company. Reprinted by permission.

British Library Cataloguing in Publication Data
Ehrlich, Paul
    Extinction: the causes and consequences of the
    disappearance of species.
    1. Ecology 2. Extinct animals
    I. Title II. Ehrlich, Anne
    574.5'223   QH545
    ISBN 0-575-03114-X

Printed in Great Britain by
St Edmundsbury Press, Bury St Edmunds, Suffolk

To *Homo sapiens,*
which through the extinction of others
endangers itself

CONTENTS

The Rivet Poppers

As you walk from the terminal toward your airliner, you notice a man on a ladder busily prying rivets out of its wing. Somewhat concerned, you saunter over to the rivet popper and ask him just what the hell he's doing.

"I work for the airline—Growthmania Intercontinental," the man informs you, "and the airline has discovered that it can sell these rivets for two dollars apiece."

"But how do you know you won't fatally weaken the wing doing that?" you inquire.

"Don't worry," he assures you. "I'm certain the manufacturer made this plane much stronger than it needs to be, so no harm's done. Besides, I've taken lots of rivets from this wing and it hasn't fallen off yet. Growthmania Airlines needs the money; if we didn't pop the rivets, Growthmania wouldn't be able to continue expanding. And I need the commission they pay me—fifty cents a rivet!"

"You must be out of your mind!"

"I told you not to worry; I know what I'm doing. As a matter of fact, I'm going to fly on this flight also, so you can see there's absolutely nothing to be concerned about."

Any sane person would, of course, go back into the terminal, report the gibbering idiot and Growthmania Airlines to the FAA, and make reservations on another carrier. You never *have* to fly on an airliner. But unfortunately all of us are passengers on a very large spacecraft—one on which we have no option but to fly. And, frighteningly, it is swarming with rivet poppers behaving in ways analogous to that just described.

The rivet poppers on Spaceship Earth include such people as the President of the United States, the Chairman of the Soviet Communist Party, and most other politicians and decision makers; many big businessmen and small businessmen; and, inadvertently, most other people on the planet, including you and us. Philip Handler, the president of the United States National Academy of Sciences, is an important rivet popper, and so are industrialist Daniel Ludwig (who is energetically chopping down the Amazon rainforest), Senator Howard Baker, enemy of the Snail Darter, and Vice President George Bush, friend of nuclear war. Others prominent on the rivet-popper roster include Japanese whalers and woodchippers, many utility executives, the auto moguls of Detroit, the folks who run the AMAX corporation, almost all economists, the Brazilian government, Secretary of the Interior James Watt, the editors of *Science, Scientific American,* and the *Wall Street Journal,* the bosses of the pesticide industry, some of the top bureaucrats of the U.S. Department of Agriculture and some of those in the Department of the Interior, the officers of the Entomological Society of America, the faculties of every engineering school in the world, the Army Corps of Engineers, and the hierarchy of the Roman Catholic Church.

Now all of these people (and especially you and we) are certainly not crazy or malign. Most of them are in fact simply uninformed—which is one reason for writing a book on the processes and consequences of rivet-popping.

Rivet-popping on Spaceship Earth consists of aiding and abetting the extermination of species and populations of nonhuman organisms. The European Lion, the Passenger Pigeon, the Carolina Parakeet, and the Sthenele Brown Butterfly are some of the numerous rivets that are now irretrievably gone; the Chimpanzee, Mountain Gorilla, Siberian Tiger, Right Whale, and California Condor are prominent among the many rivets that are already loosened. The rest of the perhaps ten million species and billions of distinct populations still more or less hold firm. Some of these species supply or could supply important direct benefits to humanity, and all of them are involved in providing free public services without which society could not persist.

The natural ecological systems of Earth, which supply these vital services, are analogous to the parts of an airplane that make it a suitable vehicle for human beings. But ecosystems are much more complex than wings or engines. Ecosystems, like well-made airplanes, tend to have redundant subsystems and other "design" features that permit them to continue functioning after absorbing a certain amount of abuse. A dozen rivets, or a dozen species, might never be missed. On the other hand, a thirteenth rivet popped from a wing flap, or the extinction of a key species involved in the cycling of nitrogen, could lead to a serious accident.

In most cases an ecologist can no more predict the consequences of the extinction of a given species than an airline passenger can assess the loss of a single rivet. But both can easily foresee the long-term results of continually forcing species to extinction or of removing rivet after rivet. No sensible airline passenger today would accept a continuous loss of rivets from jet transports. Before much more time has passed, attitudes must be changed so that no sane passenger on Spaceship Earth will accept a continuous loss of populations or species of nonhuman organisms.

Over most of the several billion years during which life has flourished on this planet, its ecological systems have been under what would be described by the airline industry as "progressive maintenance." Rivets have dropped out or gradually worn out, but they were continuously being replaced; in fact, over much of the time our spacecraft was being strengthened by the insertion of more rivets than were being lost. Only since about ten thousand years ago has there been any sign that that process might be more or less permanently reversed. That was when a single species, *Homo sapiens,* began its meteoric rise to planetary dominance. And only in about the last half-century has it become clear that humanity has been forcing species and populations to extinction at a rate greatly exceeding that of natural attrition and far beyond the rate at which natural processes can replace them. In the last twenty-five years or so, the disparity between the rate of loss and the rate of replacement has become alarming; in the next twenty-five years, unless something is done, it promises to become catastrophic for humanity.

The form of the catastrophe is, unfortunately, difficult to predict. Perhaps the most likely event will be an end of civilization in T. S. Eliot's whimper. As nature is progressively impoverished, its ability to provide a moderate climate, cleanse air and water, recycle wastes, protect crops from pests, replenish soils, and so on will be increasingly degraded. The human population will be growing as the capacity of Earth to support people is shrinking. Rising death rates and a falling quality of life will lead to a crumbling of post-industrial civilization. The end may come so gradually that the hour of its arrival may not be recognizable, but the familiar world of today will disappear within the life span of many people now alive.

Of course, the "bang" is always possible. For example, it is likely that destruction of the rich complex of species in the Amazon basin would trigger rapid changes in global climatic patterns. Agriculture remains heavily dependent on stable climate, and human beings remain heavily dependent on food. By the end of the century the extinction of perhaps a million species in the

Amazon basin could have entrained famines in which a billion human beings perished. And if our species is very unlucky, the famines could lead to a thermonuclear war, which could extinguish civilization.

Fortunately, the accelerating rate of extinctions can be arrested. It will not be easy; it will require both the education of, and concerted action by, hundreds of millions of people. But no tasks are more important, because extinctions of other organisms must be stopped before the living structure of our spacecraft is so weakened that at a moment of stress it fails and civilization is destroyed.

Part I

INTRODUCTION

Chapter I

Should We Mourn the Dinosaurs?

The worst thing that can happen—will happen [in the 1980s]—is not energy depletion, economic collapse, limited nuclear war, or conquest by a totalitarian government. As terrible as these catastrophes would be for us, they can be repaired within a few generations. The one process ongoing in the 1980s that will take millions of years to correct is the loss of genetic and species diversity by the destruction of natural habitats. This is the folly our descendants are least likely to forgive us.

—E. O. WILSON,
Harvard Magazine,
January–February 1980

IN a hillside clearing in the riverine forest sloping down to Lake Tanganyika, Flo is grooming her daughter Fifi. Seated with them are Fifi's son Freud and Flo's youngest son Flint. Freud, still a baby, is in his mother's arms. Flint, half grown, is grooming his sister. The small group of Chimpanzees seems oblivious to the idyllic setting, the lake sparkling through gaps in the foliage, the flight of metallic-colored *Naja* butterflies, the calls of birds. They also seem oblivious to our presence and that of two young women from Stanford University who are making careful notes on their behavior.

Suddenly a roaring breaks the spell. The chimps scatter, scrambling up trees, as Figan—a young male soon to become the dominant individual of his troop—plunges into the clearing, running on all fours with hair erect. In an apparent fit of rage, he dashes about and flings a five-pound rock forty feet as easily as a man might do with a tennis ball. We all stand stock-still while Figan picks up a dead palm frond and thrashes first one of the students and then the other. He then runs at us—we're next in line. Figan pulls up short in front of Paul, raises the frond, and then hesitates as they stare at each other. Then Figan drops his arm and saunters off, his "anger" apparently spent. No one is injured —indeed, no human being has ever been hurt by a chimp here. But our adrenaline has flowed. The researchers tell us that the chimps seem to sense the difference between men and women and are less likely to act aggressively toward men—even though an adult male chimp could easily demolish the strongest man.

This incident occurred in the early 1970s at the Gombe Stream Reserve, established by the British behaviorist Jane Goodall a decade before as a base for studying chimp behavior. We were starting long-term work on the population biology of some of the forest butterflies, hoping that the presence of the famous chimps would guarantee protection of the site. Finding places to do long-term research in tropical forests had become increasingly difficult because those forests were (and still are) under escalating assault by the rapidly growing populations of poor tropical countries and by the commercial interests of rich temperate-zone nations. Most preserves in Africa are maintained to protect large savannah animals such as lions, elephants, giraffes, and antelopes. Having already been forced to abandon one tropical forest research site in South America, we had chosen Gombe, in Tanzania, for a second try.

Even if one is working fourteen hours a day marking numbers on butterflies, releasing them, seeing where they are recaptured, writing field notes and tabulating data, it is impossible to ignore the chimps (to say nothing of the baboons!). Before arriving at Gombe, we had forsworn imputing human characters to these animals, being determined to look at them through the "dispassionate" eyes of scientists. Our resolution lasted about ten minutes. It disappeared the first time that a frightened young chimp jumped into its mother's arms and was comforted with precisely the kinds of pats and strokes that a human mother would use in similar circumstances.

It is Jane Goodall's hope that by observing these seductively humanlike animals, light will be shed on the behavior of *Homo sapiens*. Whether that hope is fulfilled or not, the behavior of the chimps is so fascinating that most would agree that the existence of wild Chimpanzees enriches the human environment. Indeed, we found watching them irresistible, to the occasional detriment of our own research.

One incident observed by researchers while we were at Gombe made an especially deep impression on us. A strange female carrying a young infant wandered into the territory of the local Chimpanzee troop. She was attacked by local males, beaten to the ground, and then stomped on. The baby was caught beneath its mother and was badly injured. The males took the infant, killed it, and then passed it around and ate small parts of it. Then, according to the watchers, the males "seemed to know they were doing something wrong." One took the infant's body, carried it two miles through the forest, and deposited it on the front doorstep of Jane Goodall's laboratory!

Did the chimps realize they were being studied? Could the concept of "being studied" possibly pass through the mind of a chimp? Did they feel "guilty"? Did they think that Jane and her associates were some sort of gods? Was it all pure coincidence or a response to some undetected cue given by the researchers? We'll leave the interpretation of this disquieting incident to you.

What is indisputable is that chimps are among our closest living relations and have some intellectual abilities that exceed similar abilities in many human beings. Some chimps can perform better than many people on some tests of intelligence. One chimp has solved problems involving five levels of ambiguity, while many children and some adult human beings are stumped by tests presenting only three levels. (As an example, a test in which a reward was under either a square or round cover on a red or blue tray—under square if the tray is red, under round if it is blue—presents two levels of ambiguity.) If there is any species that *Homo sapiens* should have natural empathy for, it should be *Pan troglodites,* the Chimpanzee. And yet we are wiping them out.

Jane Goodall first traveled to Gombe in 1960 to start what to many seemed a hopeless task—gaining the confidence of Chimpanzees in order to study them. When she arrived, the forest habitat of the chimps stretched unbroken for sixty miles eastward from the shores of Lake Tanganyika. Ten years later Jane had become world-famous because of her success with the Chimpanzees. In the same period the exploding human population of Tanzania cleared the forest and established farms over almost all of those sixty miles, narrowing the chimp habitat to the patrolled reserve area, a slender strip along the lake, extending inland to the first ridge line less than two miles from the shore.

Africa's potential for explosive population growth is enormous. Unlike most other poor areas of the world, tropical Africa has not yet felt the full impact of Western medical technology. As a result, death rates are still relatively high, around eighteen deaths per thousand people in the population per year in East Africa, in contrast with rates of around thirteen per thousand in Southeast Asia and nine in tropical South America. If modern death-control methods take hold in Africa as they have elsewhere, there will be a substantial increase in the rate of population growth from the present level of about 3 percent per year to almost 4 percent per year—unless there is a counterbalancing decline in birthrates.

Under such great human population pressure, it is extremely unlikely that the Chimpanzees will long survive in nature; extinction is their almost certain fate. They most likely will not disappear because of deliberate hunting by human beings, but, as at Gombe, because of destruction of the ecosystem of which they are living components. They will doubtless survive in zoos and laboratories, at least for a while. But natural groups of chimps will no longer exist. Then all that is left will be movies and reports of them caring for each other, being tolerant of their rambunctious young, using twigs as tools to dig food out of narrow holes, battling over territories, helping their own wounded,[1] and behaving in many other ways that are reminiscent of their much more abundant human relatives.

WHY SAVE ENDANGERED SPECIES?

Most sensitive human beings will care, will mourn the loss. But only a relative few will realize that the coming disappearance of this prominent endangered species[2] is not just a single tragedy but symptomatic of a planetary catastrophe that is bearing down upon all of us. For along with the chimp will go the other living elements of the chimp's ecosystem—components all of Earth's crucial life-support systems.

There are four prime arguments for the preservation of our fellow travelers on Spaceship Earth. One is that simple compassion demands their preservation. This argument is based on the notion that other products of evolution also have a right to existence, that the needs and desires of human beings are not the only basis for ethical decisions.

A second argument is that other species should be preserved because of their beauty, symbolic value, or intrinsic interest: the argument from esthetics. Chimpanzees, elephants, multihued reef fishes, iridescent blue *Morpho* butterflies, and plants with beautiful flowers or bizarre shapes seem automatically to appeal to most members of Western culture, if not most human beings. And many people, especially biologists, find beauty in such unlikely places as the fine scaling on the wings of a malaria-carrying mosquito, the iridescent patches on the back of an African tick, or the delicate sculpturing of the "shell" of a microscopic single-celled diatom.

The third argument is basically economic: preserve the whales because X dollars can be made annually through harvesting them on a sustained-yield basis; save the Amazon jungle because of the immense value of the as-yet-undiscovered foods and drugs that could be extracted from Amazonian plants. In short, other species provide *direct* benefits to *Homo sapiens* and should be preserved for that reason.

The first three arguments for the preservation of other species are easily understandable, even by those who do not find them persuasive. The fourth argument is rarely heard and even less frequently understood, because it involves *indirect* benefits to humanity. This argument is that other species are living components of vital ecological systems (ecosystems) which provide humanity with indispensable free services—services whose substantial disruption would lead inevitably to a collapse of civilization. By deliberately or unknowingly forcing species to extinction, *Homo sapiens* is attacking *itself*; it is certainly endangering society and possibly even threatening our own species with extermination. This is the most important of all the arguments—the one embodied in the rivet-popping analogy of our preface.

There Have Always Been Extinctions

There are, of course, counterarguments raised by those who see nothing wrong with humanity helping to usher other species off the stage. Perhaps the commonest is that extinction is a perfectly natural evolutionary process, one that has gone on for millions of years with or without human participation. Why worry if we're just helping nature take her course?

When people think of evolution, they tend to think of new kinds of life being produced from old—of one species (kind) changing through time into another, or of two or more species evolving where previously there was only one. When Charles Darwin put forth his theory of evolution and the evidence supporting it in 1859, he not only came up with natural selection as the driving force of the evolutionary process but also recognized the inevitability of extinctions: " . . . as new forms are continually and slowly being produced, unless we believe that the number of specific forms goes on perpetually and almost indefinitely increasing, numbers inevitably must become extinct."[3]

Before Darwin, the idea of extinction had been toyed with by various naturalists and geologists, but it was a shocking concept to most people in the mid-nineteenth century. Living things were thought to have been designed by God in a sequence of increasing complexity. Species had been created once and for all, and extinction was explicitly ruled out in the "creationist" view of the origin of species. But now things seem to have come full circle. Not only is extinction no longer a shocking idea, but Darwin's name is even taken in vain by people who want to justify the extermination of other species of *Homo sapiens.*

For example, one Sam Witchell, a financial and corporate public relations consultant, published an op ed piece in the *New York Times* (May 3, 1974) entitled "Give Me the Old-Time Darwin." The message of Witchell's article was that, since Darwin showed the extinction of species to be part and parcel of the evolutionary process, there is no reason to be concerned about the disappearance of species: "The Darwin people tell us that species come and go, that this is nature's way of experimenting with life. The successful experiments survive for a time; the failures disappear to no one's detriment."

There is, of course, one thing that people who make this kind of argument overlook: that humanity has already raised the *rate* of species extinction far above the historic rates of species formation. Species now are disappearing much more rapidly than they are appearing, and the rate of disappearance promises to continue accelerating rapidly. The statement above reminds one of a man who, observing water spurting through ever-widening cracks in the face of a huge dam, says to the people downstream, "Not to worry—after all, water has always come over the spillway anyway."

The rate of extinction of bird and mammal species between 1600 and 1975 has been estimated to be between five and fifty times higher than it was through most of the eons of our evolutionary past. Furthermore, in the last decades of the twentieth century, that rate is projected to rise to some forty to four hundred times "normal."[4] To understand the significance of these estimates, one should know something about what species are, how they have come into being, and how they have disappeared—a topic we'll take up in the next chapter.

Who Misses the Dinosaurs?

One possible response to the news that extinction is rapidly outpacing species formation is so what? After all, the dinosaurs went extinct, and humanity has not suffered any loss—or so the litany goes. Indeed, when the talk turns to extinction, it often also turns to dinosaurs. An economist once said to us that anything can be had for a price. He challenged us to tell him something that couldn't be produced if one were willing to pay enough. We answered, "Produce a living *Tyrannosaurus rex.*" After claiming (incorrectly) that, given enough time and money, one of the giant carnivorous dinosaurs could be produced, he fell back on a familiar theme. The dinosaurs, after all, were nothing of value. They became extinct and he certainly did not miss them. This argument is sometimes generalized to: the dinosaurs became extinct and nobody misses them; why, then, should we worry about extinction at all?

We do not find this line of reasoning at all persuasive. First, in a sense, the dinosaurs did not all go extinct. The group that included the dinosaurs, the "ruling reptiles," has living representatives.[5] Crocodiles and alligators are members of this group, and they at least supply human beings with some fancy leather goods as well as spine-chilling stories of man-eating. The roles of crocodilians in the ecosystems of swamps and estuaries are not well understood; they may be crucially important. Alligators, for example, are vital elements of the sawgrass marsh ecosystem of the Florida Everglades. With their tails, alligators scoop out depressions ten to a hundred feet wide. Scarce water collects in these "alligator holes" in the dry season. This permits a variety of aquatic and semiaquatic organisms, including plankton, fishes, frogs, turtles, and the alligators, to survive until the rainy season refloods the Everglades.[6]

It is also quite possible, of course, that the ultimate demise of most crocodilians would affect only people with a lust for fine shoes and handbags or a feeling of compassion for these interesting, though somewhat dim-witted, reminders of an ancient era. Their concern, however, should be considered, too.

An even more important group of dinosaur relatives is the birds. They are direct descendants of dinosaurs. Indeed, some biologists think they should be considered living dinosaurs. People would surely have missed the birds had their progenitors disappeared without issue, as did most other types of dinosaurs. Colonel Sanders might never have gotten rich, innumerable groups of human beings could never have decorated themselves with feathers, pillows would have been stuffed with straw until somebody invented foam rubber, bird-watching would not amuse millions of people, and poets would have been deprived of much material for describing songs and graceful flight. Indeed, without the birds' example, airplanes might never have been invented. Much more important, insects—by many measures already the most important predators and competitors of *Homo sapiens*—probably would be even more abundant and successful. Indeed, it is conceivable that without the birds, life would have been sufficiently more rigorous for people in an insect-dominated world that the agricultural revolution, and thus the rise of civilization, would have been impossible.[7]

But what about all those dinosaurs that did die out? Should we regret their passage? Well, the answer is yes and no.

From an esthetic standpoint, some of us do regret the disappearance of dinosaurs. What a thrill it would be if in national parks people could see great lumbering brontosauruses weighing forty or fifty tons grazing across the landscape, or herds of ceratopsian dinosaurs roaming like rhinoceri with three gigantic horns! With luck one might even get to watch an attack on a grazer by that mighty predator *Tyrannosaurus*—just as on one occasion we were lucky enough to see a lioness stalk her prey in Africa. Or in Texas a gigantic pterosaur with a forty-foot wingspan—by far the largest animal ever to take to the air—might soar overhead. No human being ever saw these fascinating animals, which died out 50 million years before anything remotely resembling *Homo sapiens* appeared. But people are the poorer for it, nonetheless.

Whether anyone misses the dinosaurs is only one question, however, and a relatively minor one. The crucial point is that the dinosaurs became extinct at a time when evolutionary processes were capable of replacing them with the mammals. The huge grazing dinosaurs were eventually replaced by such grazing mammals as deer, antelopes, sheep, goats, buffalo, and cattle—some of which have been domesticated by humanity. The big carnivorous dinosaurs such as *Tyrannosaurus* were replaced by members of the cat, dog, and bear families—and by human hunters.

If, on the other hand, the dinosaurs had become extinct and the mammals had not evolved to take over the roles that dinosaurs had played, it would be a very different world indeed. Since we too are mammals, there would of course be no people; that would be the most important difference from our point of view! The principal reason, then, that people *don't* miss the dinosaurs or other groups of long-extinct organisms is because replacements for them have evolved.

Extinctions that are occurring today and that can be expected in the future are likely to have much more serious consequences than those of the distant past. First of all, unless action is taken, contemporary extinctions seem certain to delete a far greater proportion of the world's store of biological diversity than did earlier extinctions. Furthermore, the same human activities that are causing extinctions today are also beginning to shut down the process by which diversity could be regenerated. Entire new groups of organisms are unlikely to evolve as replacements for those lost if Earth's flora and fauna are decimated now.

Would We Miss the Snail Darter?

Okay, you say, everyone should be concerned about the growing imbalance between the rate of extinction and the rate of species creation. But does this mean that our concern should be lavished on *every* species? Was the Snail Darter, for example, really worth all the fuss by the conservationists? Wasn't it preposterous to try to stop the Tellico dam, a multimillion-dollar construction project in Tennessee, because it would destroy an insignificant fish, unknown even to most ichthyologists?

It wasn't. Even if the Tellico dam had not been a massive boondoggle (which a distinguished committee of Cabinet members decided it was), even if it had not threatened other values, it should have been stopped *just because* it threatened the Snail Darter. We could, and will later, raise the arguments of compassion and esthetics. We could point out that the Snail Darter is one more rivet in our spaceship—and that popping *any* rivet in this day and age is inherently stupid and potentially dangerous. But there is, we think, a still stronger argument: *the line has to be drawn somewhere.*

If the value of each endangered species or population must be compared one on one with the value of the particular development scheme that would exterminate it, we can kiss goodbye to most of Earth's plants, animals, and microorganisms. After all, dam X will be able to supply power to illuminate fifty thousand homes; freeway Y will make it possible to cut twenty minutes from the drive between Jonesville and Smith City; Sunny Acres Apartments will provide decent housing for two thousand people now condemned to existence in a slum; mine Z will create two hundred and fifty badly needed jobs. How can any organism win in the face of such arguments?

In an overdeveloped society such as the United States, questions and alternatives can be posed with respect to all such projects, however worthy they may appear. For instance, how critical is the twenty minutes' commuting time between Smith City and Jonesville? Couldn't the same result or nearly the same result be achieved by improving an existing highway—and at lower cost? Couldn't the new apartments be provided by redevelopment right on the slum site? Might there not be some less destructive way to provide jobs than by opening a new mine? Can't the ore be produced from existing mines? Is the electricity from the dam really needed in a nation that wastes enormous amounts of energy, as the United States does?

A great many development projects yield short-term benefits for a few and pass on long-term costs to society, as studies of their environmental consequences often make abundantly clear. "Development" in the United States has until recently been considered entirely beneficial. In the last decade or so, however, as conflicts over land use, water rights, and environmental values have proliferated, this view increasingly has been questioned. And the need for preserving endangered species has lately become one of the conflicting values. In our view, it may be the most crucial value; the rate at which populations and species are being eradicated has reached the point where a society like the United States would be better off doing without even the most "necessary" projects if they cannot be carried out without causing further extinctions.

The choices will be very tough ones, as you will see. Indeed, they often will be difficult to pose. In many cases, species and populations are endangered by activities far from their habitats. A new coal-burning power plant in central Indiana, for instance, may add to the acid rain that is exterminating trout populations in Maine. Against the value system of a growth-oriented Western society, defending endangered organisms on a one-by-one, place-by-place basis will be difficult even when the connections are clear. In the long run, it is a losing game. A Tellico dam will eventually be found for every population and species of nonhuman organism, and there will always be developers, politicians, and just plain people to argue that short-range economic values must take precedence over other values. For they do not understand that *their own fates* are intertwined with the Snail Darters of our planet. They are unaware of how much they would indeed miss those little fishes.

Are There Any Organisms We Would Not Miss?

Of course, not all killing of members of other species in the past or in the present is automatically bad, even if one takes a view of the world that is not centered on *Homo sapiens*. Since the time of Darwin, biologists have recognized that the success of some species is normally paid for by the reduction in population size or extinction of others. It is, for example, perfectly natural for human beings to attempt to control, or even force to extinction, populations or species that prey upon people or threaten their resources.

If, for example, some magic way were found to exterminate just the *Anopheles* mosquitoes that transmit the most important of all human diseases, malaria, it would be a tempting thing to do. But ecologists would caution that doing so would entail some small chance that the inevitable consequent changes in Earth's ecosystems would make the world less hospitable for humanity, causing worse suffering than that previously inflicted by malaria. And some demographers might warn that the sudden decline in human death rates in some developing countries could cause an acceleration in population growth that would exacerbate their already serious social and economic problems. Life is full of chances, though, and we personally would be sorely tempted to kiss the *Anopheles* goodbye and attempt to deal with the other problems if and when they arose.

But of course it is never that simple, since magic devices do not exist for selectively deleting a single species—especially an insect pest—from Earth's living complement, and since most of the techniques we use for dealing with our predators and competitors have turned out to be two-edged swords.

Even when things appear superficially simple, close investigation usually reveals unhappy complexities. In some parts of Africa, elephants have become dangerous pests—destroying the fields of poor, hungry, hard-working Africans and sometimes killing them. At the moment, either the elephants are being killed or the temporary compromise solution of moving them to other places is being attempted. But Africa is fast running out of both elephants and places to move them to.

Wiping out the massive beasts would be relatively simple—indeed, that is one of the reasons elephants are endangered. But, even from the point of view of the Africans, that is not necessarily the "right" answer. It can be argued that the future economic prosperity of large areas of Africa would not be maximized by converting all usable land to subsistence agriculture. Future African generations would almost certainly be better served by maintaining some of those areas as potential tourist meccas, with elephants being the top tourist attraction.

The basic conflict, of course, is caused by burgeoning human populations encroaching more and more on the elephants' habitat. In this case, as in so many others, efforts to control the growth and spread of the human population would inevitably be more beneficial to Africans in the long run than efforts to exterminate the elephants. With the elephants out of the way, the people could expand for a few more years, but they would soon run out of more land to convert to agriculture. Then human population growth would be halted by nature anyway. African leaders, though, are locked into a short-term decision-making process just as surely as American politicians are. Of course, from a broader ethical point of view, one could argue that the remaining herds of those gigantic and intelligent beasts should be protected absolutely.

The degree to which a strongly anthropocentric, or human-centered, approach to these problems should be taken is clearly an area on which moral and honorable people will continue to differ. For many millions of years, quite likely up until this century, an anthropocentric viewpoint was the only possible and sensible one for *Homo sapiens*. And for people living in an industrial city today, surrounded by human-made objects and supported by what seems to be an entirely manufactured system, it is very easy to believe that human beings are separate from nature and quite independent of it. Religions have reinforced this arrogant idea by teaching that dominion over Earth and other living things is humanity's God-given right. But now, with our utter domination of the planet, the time has come for a softening of such human chauvinism.

Perhaps the best way to erode human chauvinism in industrial nations would be to launch crash programs of education—to familiarize people, especially children, with our traveling companions on Spaceship Earth, and teach people not only to appreciate them for their beauty or intrinsic interest, but to understand that *we need them* as much as they need our protection to survive. Television shows on the living world have accomplished a great deal in this regard, but they are no substitute for personal experience. It is too bad that everyone can't have a guided tour, for instance, of some of the world's great coral reefs. Reefs could provide an ideal introduction to the beauty and fascination of relatively unfamiliar organisms, the intricacy of their relationships with one another, and their importance to humanity. And such appreciation of the reef ecosystem might lead to real concern about the ongoing destruction of Earth's biological resources.

THE CORAL REEF SYSTEM

An introduction to coral reefs might be given at the outer edge of the Australian Great Barrier Reef, perhaps at a spot where we have dived several times, near Lizard Island at the northern end of the reef. There a fantastic coral garden marks the seaward side of an area that contains literally thousands of different reef fish species. Squadrons of large parrot fishes swim gracefully over the outer slope, scraping coral heads with their beaklike teeth, digesting out the living parts, and excreting clouds of white limestone remains. They move over the reef like mowing machines, leaving behind them pure white sand. Schools of large jacks float by over the abyss, and gazing down the slope toward the miles-deep Coral Sea, one might occasionally glimpse the shape of a great oceanic shark. Both jacks and sharks are carnivorous; they eat smaller fish.

But it is over the shallower reefs that one can best enjoy the diversity of living things and observe their relationships. In this area we have been able to study twenty-five species of butterfly fishes, among the most beautiful denizens of the reefs. They are flattened from side to side, like the angelfishes often found in aquariums, and brightly colored with white, black, gold, blue, orange, or yellow.[8]

The butterfly fishes do not all feed on the same things. Some species are specialists in eating certain kinds of corals; others search for small invertebrates on the corals or in the sand. In more recent work around Lizard Island, we and our colleagues have seen the orderly way in which species with similar lifestyles—potential competitors—replace each other in different locations. For example, three species feeding on hard corals in the clear waters near the outer Barrier are replaced by two other hard-coral feeders in the murky waters over the reefs near the Queensland shore.[9] Not only is the unfamiliar living world of the reef beautiful and diverse, but it has a complex organization not immediately apparent to an untrained visitor to the underwater "classroom." Lesson: even though they are very similar, two species usually function differently in ecosystems. If one goes extinct, the other may be unable to take over its role.

Even to a trained biologist, the fish fauna of an area like Lizard Island presents a daily series of lessons in ecology. The entire surface of the reef is occupied by small damselfishes, each defending a little territory and feeding mostly on plants growing there. Above the reef hang other damselfishes feeding in the constant flow of plankton—tiny plants and animals floating in the water—that drifts over the reefs. Each of the two related groups has specialized in different food sources.

In tubes in the reef live saber-toothed blennies, roughly the shape of your forefinger and almost twice as long. These are "dash-and-grab" predators. When a large sluggish fish comes by, if you are lucky, you may see a blenny spring from its tube like a crossbow bolt, grab a bite from a passing fish, and vanish back into its hole so quickly that the entire motion is a blur. The larger fish just twitches and moves on. The abundance of these blennies is determined by the abundance of suitable holes. Lesson: a shortage of resources (holes) can limit population size. Question: why are there no dash-and-grab predators that live on land? It is a nice puzzle to which ecologists can give no real answer. We can just be thankful, since if they did exist, Sunday strolls could be a lot less pleasant!

At certain places on the reef, brilliantly colored four-inch-long "cleaner" wrasses have set up their stations, and one can observe the various other fishes of the reef lined up like motorists at the gas pump waiting to be groomed. When its turn comes, each customer assumes a characteristic "cleanee" pose and hangs in an apparently stunned state as the cleaner wrasse goes over it, meticulously removing parasites from the skin, gills, and inside of the mouth. It is one of the wonders of nature to see these little fishes enter the mouth of a large predator and in perfect safety pick parasites from between its daggerlike teeth! Lesson: evolution can produce some very unexpected associations and patterns of behavior.

And if you were really lucky, you might even observe the activities of the flimflam man of the reef—a cleaner mimic. Blennies, relatives of the dash-and-grab predator, have evolved a color pattern virtually identical to that of the cleaner wrasse and a behavior pattern that matches also. They advertise cleaning, but when a naïve fish comes along and goes into its cleanee pose, the mimics simply bite off a piece and proceed to chew it with extreme brazenness. The victim seems incapable of bringing itself to attack a fish in cleaner garb. Lesson: things are not always what they seem in the natural world either. And, like human victims of confidence games, even fishes find it difficult to admit they've been conned!

Fascinating and diverse as the fish fauna of the coral reef is, it is only part of the story of the living reef. The structure of the reef itself is the result of the activities of tiny coral animals, which have within their bodies symbiotic algae that, like other green plants, photosynthesize. The gigantic reefs—far larger than any human-made structures—are the product of the activities of these tiny colonial animals and the algae, being made up of their limestone skeletons deposited over untold millennia. Lesson: tiny organisms, given enough time, can produce very large geological features.

The rest of the animals of the reef, including myriad crustacea, worms, snails, and other invertebrate animals, are in many ways as fascinating as the fish fauna—although not always as obviously beautiful to behold. And there are lessons to learn from all of them, if we care to look in the right way.

The economic values supplied to *Homo sapiens* by the reef are also manifold. Reef fishes provide an important protein supplement to the diets of many peoples in tropical areas; on some islands they are a major source of food. In addition, the reef ecosystem is an esthetic resource that is important to the tourist industries of many tropical countries and to the burgeoning recreational industry of scuba diving. The entire reef complex is a focus of high productivity in the otherwise relatively unproductive ecosystems of tropical seas. Their destruction would inevitably have cascading effects on fish species throughout tropical oceans, as well as threatening the shores and harbors they now physically protect from erosion and wave action. It also has been suggested that the reefs are involved in the crucial task of regulating the salt content of the oceans. In essence, the reefs create vast evaporation lagoons between themselves and tropical shores.[10] Lesson: things that are beautiful and educational can also be economically valuable.

Homo sapiens, unfortunately, is threatening the destruction of the world's coral reefs—indeed, in a few places that destruction is already far advanced. If the destruction of coral reefs is carried to completion, many human beings will almost certainly be hungrier and will surely have lost important biological resources. All of us would lose as well an esthetic resource that could delight our descendants for countless generations. The loss of the reef system and the species that comprise it could be a much more severe one than the loss of the giant dinosaurs, for it is quite possible that nothing would evolve to replace the reef ecosystem. Certainly nothing could replace it on a time scale of any interest to humanity.

Chapter II

The Origin of Species—
and Their Extinction

When on board H.M.S. 'Beagle,' as naturalist, I was much
struck with certain facts in the distribution of the
inhabitants of South America, and in the geological
relations of the present to the past inhabitants of that
continent. These facts seemed to me to throw some light
on the origin of species—that mystery of mysteries, as it
has been called by one of our greatest philosophers.

—CHARLES DARWIN,
On the Origin of Species

Living species today, let us remember, are the end
products of twenty million centuries of evolution;
absolutely nothing can be done when the species has
finally gone, when the last pair has died out.

—SIR PETER SCOTT,
*speaking before 1972 Conference
on Breeding Endangered Species*

IN order fully to appreciate the way species of other organisms provide
humanity with pleasures, economic values, and essential life-support services, it is necessary to understand a little more about their nature, about the
processes by which they are created, and about prehistoric patterns of extinction. Such knowledge is also crucial to an appreciation of how species are
endangered and of what options are available for correcting the imbalance
between the rates of origin and extinction of species.

Species are different kinds of plants, animals, and microorganisms. The term
species or *kind* turns out, however, to be very difficult to define precisely. Every
biologist more or less understands these terms, although they may not agree
on a precise definition. What *is* agreed upon by most biologists is that if two
visibly different though similar groups of organisms live together with little or
no sign of interbreeding, they are considered to be separate species.

Dogs and foxes are separate species because no one has found a hybrid
between a dog and a fox. California Interior Live Oaks and California Black

Oaks are different species, even though they will form hybrids. And it is often very much disputed whether animals or plants that look more or less alike but do not occur together—the European Brown Bear and the American Grizzly Bear, for example—should be considered the same species. These complications have led to endless arguments in scientific literature[1] but need not concern us here. For the purposes of this book, there is no need to define species more precisely than as distinctly different kinds of organisms.

Since there are complications in producing a precise species definition, it is clear that, even if all the organisms on our planet were well known to science, there would still be disagreement on the number of species they represent. Biologists do agree, however, that there remain a great many species that have not yet even been discovered and formally named. For these reasons, the question "How many species are there?" can only be answered within an order of magnitude—somewhere between two and twenty million. Our own ballpark guess would be about ten million. Somewhere around a million and a half have been described and given scientific names.

The vast majority of species, especially of the undescribed species, are thought to be in the tropics, particularly in tropical rainforests. Even though they have a much smaller land area, the tropics probably harbor at least twice as many species as the temperate zones. As many as one million species may live in the Amazon basin alone. Not surprisingly, most of the world's larger organisms—especially fishes (about 20,000 species), amphibia (about 2,600), reptiles (6,500), birds (8,600), mammals (4,100), and higher plants (250,000) —are already known to science. Species yet to be discovered are concentrated in groups such as insects, mites, and nematode worms. For example, in 1975 alone, 786 previously unknown ("new") species of flies were described and named, joining about a million other insects that had already received latinized names.[2]

Even in the United States and Europe, new species of obscure organisms are still being turned up continually. In more prominent groups such as the butterflies, discoveries of previously unknown species in temperate zones are largely restricted to so-called sibling species—two distinct kinds so similar that they were long thought to be just one species. So when Paul and some colleagues found a strikingly different and beautiful new species of small butterfly in the Sandia Mountains of New Mexico in 1959, it was the first such discovery in the United States in about fifty years.[3] Thus, although a huge number of species remains to be discovered and described, scientists do have a clear idea of what they are, very roughly how many there might be, and where they will be found. Unless, that is, too many of them are destroyed before they are found.

A basic problem, you will recall, is the contrast between the *rates* of species formation and of species extinction. Although most of this book deals with the causes and prevention of premature *extinctions,* a complete understanding of the situation requires some knowledge of the evolutionary process of species *formation.* After all, one might argue that, if human activities can lead to an acceleration of the disappearance of species, human beings might also intervene on the other side of the equation and increase the rate of species formation, thus maintaining the diversity of the natural world.

EVOLUTION

There are two major processes in organic evolution. The first involves gradual change within a single line of descent. An example of change in a single line over the short term would be the acquisition of resistance to a pesticide in a population of mosquitoes over a period of, say, ten generations (a mosquito generation may take less than two weeks). An evolutionary change of this sort has been seen in many populations of insect pests in the past few decades. One year the insect population is very susceptible to a pesticide, almost all individuals being killed by a relatively low dose. And a year or so later the descendants of the same population treat the insecticide as an aperitif and continue to dine merrily on the crop that was to be protected—or to sip our blood. They have evolved, in a relatively few generations, from organisms sensitive to the insecticide into organisms resistant to it.

Another example of evolution within a single line, this one detectable not over a year or so but over a much longer period, was the gradual transformation over millions of years of an apelike creature called *Australopithecus* into modern human beings. *Australopithecus* was a fully erect, tool-using denizen of the plains of Africa. The descendants of these "man-apes" slowly acquired larger brains and more complex cultures, evolving into a stage called *Homo erectus,* including Java and Peking man, among others. The process continued and *Homo erectus* slowly evolved into *Homo sapiens,* the current representative of the human line.[4] In both the mosquito and the human cases, one sort of animal is transformed into another at a later date.

Speciation is the second major evolutionary process, one that accounts for the great diversity of organisms. Speciation transforms one kind of organism into two or more new kinds of organisms. For example, back in the Triassic period, about 200 million years ago, when dinosaurs were the dominant land animals, a group of reptiles began to undergo a transformation. Their normal set of reptilian cheek teeth—in which all the teeth were simple and essentially identical—began to change into a differentiated set of complex molars and premolars. The flat scales characteristic of most reptiles started to change into slender things now known as hairs, and the evolving animals began to give intensive care to their young and to produce a nutritious, whitish fluid known as milk to feed them. These and other changes did not all occur at once, and indeed the exact sequence is not clear from the fossil record, but this group of reptiles was on its way to becoming mammals.

For a very long period—well over 100 million years—the mammals were an obscure group of small animals living in terror of carnivorous dinosaurs which preyed upon them in long-ago twilights. But then at the end of the Mesozoic era, about 65 million years ago, the dinosaurs suddenly disappeared. The time of the mammals had arrived, and a relatively few species of our obscure ancestors proliferated into a group that is represented today by over 4,000 species and includes such diverse forms as kangaroos, opossums, whales, anteaters, aardvarks, bats, seals, dogs, tigers, bears, skunks, armadillos, horses, antelopes, deer, goats, cattle, mice, rabbits, platypuses, gorillas, and people.

This process involved not only changes within a single line but also, obviously, the splitting of lines—that is, *speciation*. The exact mechanisms of speciation are not fully understood, in part because speciation tends to be a very slow process.

Some aspects of the process of speciation can be observed both in present-day living systems and in the fossil record. The fossil record indicates—and observations of living species confirm—that, relative to a human life span, the process of speciation is a gradual one. In some groups of organisms and at some times, there may be "bursts" of differentiation that take place in thousands or even hundreds of years. But in most cases, speciation seems to take tens of thousands or even millions of years. It has been more than a century since Charles Darwin started biologists thinking about speciation, and the production of a new animal species in nature has yet to be documented. Biologists have not been able to observe the entire sequence of one animal species being transformed into two or more. (Note that we carefully limit these examples to animals. The situation in plants is much more complicated, but in ways not germane to this book.) Biologists *have* been able to observe innumerable examples of animal and plant species that appear to be in various stages of splitting. But in the vast majority of cases, the rate of change is so slow that it has not even been possible to detect an increase in the amount of differentiation over the decades that have been available for observation.

The basic mechanism both of evolutionary changes within lines of descent *and* of speciation is the one proposed by Charles Darwin: *natural selection*. Let us look first at the way this mechanism works within a single line of descent. Natural selection occurs when some types of individuals in a population on the average reproduce significantly more than other types, and the types are genetically different. Individuals of sexually reproducing organisms, with the exception of identical twins,[5] are never exactly alike genetically. In a given environment, some genetic types will be better able to survive and reproduce than others. Some genetic types of insects, for example, will be more resistant to a pesticide than others. When an insect population is sprayed, natural selection will occur. The resistant genetic types will be more likely to survive and thus on the average will out-reproduce the more susceptible types. The out-reproducing of the susceptible genetic types by the resistant ones leads to genetic changes in the population. The kinds of genes that are present in the better reproducers (the resistant type) become more common in the population's "gene pool," the population's collective genetic endowment, and those of the poorer reproducers (the susceptible type) become less common. Thus the entire population gradually becomes resistant to the spray.

This replacement of one genetic type for another within populations as a result of natural selection accounts quite nicely, given the billions of generations available, for the transformation of the first self-duplicating form in which life originated through branching ancestral lines to modern organisms. But how did natural selection lead to the splitting of lines so that today there are millions of different species? The place to start looking for that answer is in *geographic variation.*

Geographic Variation

Change occurs *within* lines of descent as selection produces organisms that are better able to exploit the environments in which they live or molds them to adapt to new conditions as environments change. The most important reason for the *splitting* of lines of descent—that is, for speciation—is that environments are varied. No two places are identical in topography, climate, or in the array of organisms that inhabit them. Therefore, when a species expands its distribution, its populations diverge genetically in response to the different environments in which they find themselves, each adapting through natural selection to the conditions of its own area. Every population becomes genetically different from each of the others, and those genetic differences are reflected in differences in the structure, behavior, and other characteristics of the organisms. These latter differences among populations are known as *geographic variation.* Often this variation is recognized by taxonomists, who name and describe populations or groups of populations within a species as geographic *subspecies* (or geographic *races*).

A key to how rapidly and how much natural selection can change populations in response to environmental changes is the amount of genetic variability present in the populations. The more alike individuals are genetically, the less chance there is that a population or species will be able to adapt to changed conditions. The more genetic differences there are within and between populations of a species, the greater are the chances that the species will persist in the face of a modification of the environment that affects all the populations.

Suppose there were a widespread and rapid cooling of the climate, for example. More northerly populations, already genetically adapted to more rigorous conditions, would be able to move south and survive, replacing southern populations that had gone extinct. Or perhaps rare, more cold-resistant genetic types in the southern population might survive and take over.

Thus one can think of genetic variability as a crucial resource of natural populations and species. It is the resource that gives a population or a species a chance of remaining in the evolutionary game in a world where environmental change is not the exception but the rule.

In some rare instances, it has been possible more or less directly to observe natural selection causing geographic variation. Many years ago, Paul was working with the late Dr. Joseph H. Camin at the Chicago Academy of Sciences, studying the transmission of a malarialike disease of snakes by snake mites. The stock of water snakes used in the work was obtained from a large population on the islands of Lake Erie. The snakes occurred on the flat limestone rocks of the islands' shorelines. Often when a large rock was rolled over, several four-to-five-foot-long water snakes would be uncovered. The technique was simple—you would grab one or two of the snakes, and one or two of them would grab you. (Water snakes are nonpoisonous but vicious—they will even attempt to bite you through their birth membranes as they are born!)

The water snakes showed geographic variation in color pattern. Most of the island water snakes were a genetic type that lacks the pattern of alternating light and dark bands that characterizes the same species along the mainland shores of Lake Erie and throughout much of eastern North America. Because they were bandless, the island snakes were hard to see against the limestone rocks; the banded pattern elsewhere made the snakes difficult to see against the more varied backgrounds of their normal marsh habitats.

Frequently, when the female snakes from the islands were brought back to the lab, they would produce large litters of young. The interesting thing was that the vast majority of the newborn snakes were banded. Since the snakes do not change their colors as they mature, something must have been differentially eliminating many of the young of the banded genetic type from the population, leaving the surviving adults mostly unbanded. Seagulls that fed on the young snakes were suspected of being the selective agents. It seemed that they could more readily spot the banded snakes against the plain limestone rocks of the islands, but this could not be demonstrated with certainty. It was possible, however, to show that selection was occurring by demonstrating that there was a dramatic difference in the proportion of banded snakes between the population of newborn individuals and the population of adults. The banded genetic types in the litters clearly had less chance of growing up and reproducing than the unbanded types.

An unusual circumstance allowed the detection of selection in action in this case. Snakes from the banded population around the lake shores were continually migrating out to the islands, bringing with them the genes for banding.

Thus, although these "banding genes" were being eliminated continually by selection in the islands, they were being reintroduced into the population continually by migration. This balance between migration and natural selection provided an unusual look at the way in which geographic variation is generated.

Paul's and Joe's attempts to demonstrate that seagulls were actually doing the selective predation were not crowned with success; indeed, they only succeeded in demonstrating that seagulls could outsmart scientists. They borrowed a seagull, soon christened Herman, from Marlin Perkins of *Wild Kingdom* fame. Marlin was then director of the Lincoln Zoo. A carefully designed experiment was planned in which baby water snakes of both the banded and unbanded types were to be presented to a hungry Herman on surfaces resembling the limestone rocks of the Lake Erie islands. The basic idea was to see whether Herman could more readily find and eat the banded snakes against that background rather than the unbanded ones.

Unfortunately, Herman was a long-term zoo resident, accustomed to a diet of defrosted smelt. When tested with the presentation of baby garter snakes, Herman paid no attention whatsoever to them, even when he was very hungry. The first task, then, was to convince Herman that snakes were food. This was done over a period of a week or so by feeding him baby garter snakes intertwined in his defrosted smelt.

That triumph out of the way, the time came for the first experimental run. Herman was placed in front of two identical artificial rocks, on one of which was a banded baby water snake and on the other an unbanded baby water snake. Herman managed to step on the unbanded snake, ignoring it completely on his way to grabbing and eating the banded water snake. Obviously Herman could see the banded snake more readily than the unbanded snake. But one run doesn't make a scientific paper—a long series would be necessary. On the second trial, Herman spotted the unbanded snake first, gobbled it up, and then turned to get the banded one, too. After that it was clear that he knew that there would be a snake on each rock and would go for first one and then the other without regard for pattern. Another great experiment down the drain.

In desperation, an attempt was made to show that Herman could at least discriminate between the banded and unbanded snakes. The basic design was to feed Herman the two kinds of snakes in a random sequence and to terrify him by ringing a loud gong under the feeding plate when he picked up a banded snake. The first run was with a banded snake, and Herman went for it. The gong was rung, but instead of scaring Herman witless, it didn't faze him a bit. He simply stood there and started to devour the snake. Paul rushed into the cage, grabbed Herman, and took the snake away from him. Thinking that this trauma would alert Herman to the reason for the gong, a second run was tried, again with a banded snake. Once more Herman went for the snake, and again the gong was rung. This time the results showed that Herman had truly learned. He grabbed the snake and dashed for the furthest corner of the cage before Paul could run in and take it away from him. So much for animal behavior. Herman was returned to Marlin with thanks.

So the cause of geographic variation in the water snakes (unbanded on Lake Erie islands, banded elsewhere) is differential survival of snakes of different genetic types—that is, natural selection. The identity of the selective agents on the islands remains in doubt; gulls remain a strong possibility.[6]

It is not necessary, however, to look to rather unfamiliar animals for examples of geographic variation. Our own species is one of the most geographically variable. Most human physical characteristics—height, build, skin color, hair type, blood type, eye color, and so on—exhibit geographic variation. In some cases, the sorts of pressure that have produced this variation are understood. For example, people who live or whose ancestors once lived in parts of Africa are susceptible to a disease called sickle-cell anemia. The genetics and biochemistry of this disease are now rather well understood, and it is known that the genes for sickle cell in an African population help to give protection against an especially dangerous kind of malaria that occurs in those areas of Africa. But a small fraction of the people with the sickle-cell condition develop a fatal form of anemia. The occurrence of some individuals with the fatal anemia can be thought of as the price that the population has to pay for its otherwise improved survival and reproduction in the presence of the malaria.

Geographic Speciation

But what is the exact connection between geographic variation and speciation? At the moment in human beings there is none. All people belong to one species, *Homo sapiens,* and people from all populations of modern human beings are fully capable of interbreeding. In many cases, however, geographic variation is an early stage in the process of speciation—even though, as we shall see, in *Homo sapiens* it is not. If populations changing in response to different selection pressures are long isolated from one another (for example, by a mountain range or a wide river), they may take divergent evolutionary courses. Over a great many generations, the separated populations may become so different that if they are reunited by the erosion of the mountain range or the drying up of the river they may be incapable of interbreeding. There are then two species where previously only one existed—speciation has occurred.[7]

A classic example of such geographic speciation has occurred among "Darwin's finches" on the Galápagos Islands. When we visited the Galápagos in 1979, we found these sparrowlike birds, which had first attracted Darwin's attention in 1835, still abundant. Fourteen species are recognized by ornithologists. These small birds are all very similar structurally, but vary somewhat in color, a little in size, and a great deal in the shape and size of the beak and in their feeding habits. At one extreme is a relatively large ground-feeding species with a grotesquely large triangular beak used for crushing large hard seeds. At the other is a small, warblerlike finch with a needlelike bill used to probe for insects in trees. The species are distributed unequally over the archipelago, some occurring on as many as fourteen islands, some on as few as one. Of those that do occur on more than one island, often each population is somewhat different from those on other islands.

One species behaves rather like a woodpecker, probing into bark with its bill, searching for insects to eat. But since it lacks the long tongue that a true woodpecker uses to extract the insects it uncovers, the finch has become a tool user. It employs a cactus spine to probe into holes and pry out hidden bugs. Another species feeds primarily on seeds, but also eats ticks on iguanas. On one island this species further supplements its diet by landing on the tails of large, rather stupid sea birds known as boobies. The finch nips the larger bird and eats its blood as it seeps out among the feathers, obtaining a nice protein supplement to its diet.

By the time Darwin reached the Galápagos, he had developed doubts about the creationist view of the origin of species. After all, if God had created all species at once or in a series of separate creations (which would explain the embarrassment of the fossil record that was by then emerging), why were the fossil animals Darwin found in South America clearly related to the living animals of South America? The similarities between extinct and living organisms in each part of the world was to become a cornerstone of his theory of evolution. As he wrote in 1859 in the first edition of *On the Origin of Species*:

> On the theory of descent with modification [i.e., evolution as opposed to special creation], the great law of the long enduring, but not immutable, succession of the same types within the same areas, is at once explained; for the inhabitants of each quarter of the world will obviously tend to leave in that quarter, during the next succeeding period of time, closely allied though in some degree modified descendents.[8]

Although when Darwin first encountered them, he did not grasp the significance of the Galápagos finches, he soon realized the import of this diverse group of birds occupying an archipelago far from the mainland. In 1839, twenty years before he published his magnum opus, he wrote of the finches in *Voyage of the Beagle*: " . . . seeing this gradation and diversity of structure in one small, intimately related group of birds, one might really fancy that from an original paucity of birds in this archipelago, one species had been taken and modified for different ends."[9]

Darwin was impressed by the observation that animals and plants on islands tended to have their nearest relations on the closest mainland. This made no sense under a creationist hypothesis. If it was possible simply to create species, new ones would simply be created that were suited to island habitats, with no reference to what had been created for adjacent mainlands. Island species would tend to be like one another where climatic and other conditions are similar, not like species from totally different mainland habitats.

On the other hand, similarity to mainland species was exactly what would be expected under an evolutionary hypothesis. One group of finchlike ancestor birds had reached the Galápagos archipelago centuries earlier from South America, perhaps a flock blown there by a storm of rare intensity. Isolated from the mainland and under different selective pressures on the relatively barren volcanic Galápagos, the birds began to diverge from their mainland relatives. There followed an adaptive radiation, in which speciation transformed the original immigrants into the diverse array of species now found on the islands.

Why is there not just a single species in the Galápagos archipelago? The answer is that geographic speciation has occurred within the archipelago itself. The finches rarely cross the water barriers between the islands. Consequently, populations established by occasional migrants remained largely isolated from one another. Speciation occurred in isolation, and when "daughter" species reinvaded ancestral islands, they no longer interbred with parental species. Instead, daughter and parent species had to compete with each other for food, and selection enhanced any differences that existed in beak form and diet, thus tending to reduce the competition. Darwin clearly recognized the importance of other species as a source of selection pressures. He wrote that it was a "deeply seated error [to consider] the physical conditions of a country as the most important for its inhabitants, whereas it cannot, I think, be disputed that the nature of the other inhabitants, with which each has to compete, is at least as important, and generally a far more important element of success."[10]

Speciation occurring in isolation on the different islands of an oceanic archipelago has occurred in organisms other than the Galápagos finches. Other groups of plants and animals on the Galápagos indicate at least a beginning of the same process. For example, the great tortoises isolated on the various islands, and on isolated volcanoes of Isabela (Albemarle) Island, were all differentiated from one another.

On the Hawaiian Islands, a series of speciation events strikingly similar to those that produced Darwin's fourteen finch species seems to have occurred. Twenty-three species of Hawaiian honeycreepers have evolved from a single, possibly finchlike ancestor. Like the Galápagos finches, these birds are very diverse in their beak shapes (from parrotlike to extremely slender, curved, and as long as the bird's body). But unlike the Galápagos finches, the Hawaiian honeycreepers are highly varied in color, and some are quite vividly marked.

The adaptive radiation undergone by this Hawaiian group is thus even more spectacular than that of the Galápagos finches. Yet it can be explained by the same mechanism of geographic speciation of isolation, followed by a complex pattern of reinvasions and selection to reduce competition, that explains the radiation of the Galápagos finches.

But if the circumstances are not right, geographic variation will not lead to speciation. *Homo sapiens,* for example, has not undergone geographic speciation, even though it shows abundant geographic variation. One reason is the human penchant for migration. No population of human beings has remained isolated from all others long enough for speciation to occur—that is, for groups to become so different that they cannot interbreed. Another reason is that cultural adaptations tend to replace genetic ones. For example, an Eskimo in sealskin clothing can be in as warm a "climate" as a naked Aborigine in the Australian desert. There is therefore no selective pressure for Eskimos to evolve thick coats of fur. Clothing, housing, and other cultural devices can and do compensate for differences in environments. Thus, there is a relative lack of divergent selection pressures on different human populations compared to pressures that would operate on other organisms, which are incapable of modifying their environments to the degree that people can.

Other widespread species may lack the advantage of cultural adaption but have high mobility, like that of *Homo sapiens,* which prevents speciation from occurring. Populations of such organisms—the migratory Monarch butterfly is a good example—are less likely to be completely isolated and may show little geographic variation. Or, in the case of species that have very restricted distributions, there is less opportunity for geographical variation, as the environments they occupy may not be sufficiently different to cause divergent evolution. And if the populations are not far apart, again, they may not be sufficiently isolated for speciation to occur.

THE EVOLUTIONARY BALANCE

As we have seen, geographic variation may or may not lead to speciation; when it does, the process is a slow one. This is a major reason that the recent human-induced acceleration in extinction of species should be a matter of concern. Earth's inventory of millions of species is the product of two biological processes—speciation and extinction—operating over eons. New species are created through speciation, and other species are eliminated through extinction. It is as if speciation were a faucet running new species into a sink and extinction were a drain removing others. Throughout most of Earth's history, the faucet has been running species in a little faster on the average than they have been going down the drain. As a result, the number of living species has generally increased over the ages.

Today, humanity has become a major agent of extinction, opening the drain ever wider. No longer are more species being created than are going extinct each year, and the planet's stock of biological resources—of which the number of species is a measure—is now diminishing rapidly. Some projections, based on the assumption that the extinction rate will continue to rise as it has in recent decades, indicate that perhaps *as much as one-fifth of all the species on Earth today will have vanished by the end of the century.* [11]

Humanity's pressure on species is accomplished in many ways, direct and indirect: by overexploitation and by habitat modification and destruction. By overhunting, overfishing, and overharvesting of plants, and—more importantly—by destroying or altering natural habitats, humanity is dooming populations and species to extinction in ever-increasing numbers. At the same time, human interference almost certainly is inhibiting the long-term compensating process: speciation. When a population, though not the entire species, is wiped out, some fraction of that species' genetic resources are also lost. As natural habitats of organisms are disrupted and degraded by human encroachment, and as the world in general is made more homogeneous environmentally, the opportunities for geographic speciation are generally reduced. The smaller and more uniform the overall area occupied by a species, the less likely it is there will be the isolation and environmental differences that are required for speciation. (Isolation between biological reserves is unlikely to compensate adequately for this—at least for larger animals—for reasons that are discussed in Chapter 9.) So the evolutionary processes that in the past generated a rich array of species and an even richer array of genetically distinct populations adapted to local conditions are in the process of being turned off just when they are needed most.

The most significant side of the speciation/extinction equation today, however, remains the great acceleration in the rate of extinction—estimated now to be dozens of times what it was only a few centuries ago and increasing rapidly. Even if speciation was not being inhibited by the same processes that cause extinction and was proceeding at normal rates, it could not compensate for the extinctions projected for the next few decades rapidly enough to be meaningful to humanity.

EXTINCTION

What can be said about the natural processes of extinction, which Darwin recognized and which had been going on for billions of years before industrial civilization started today's hemorrhaging of organic diversity? The fossil record itself is a testimonial to extinction. Many entire groups of organisms are gone without leaving any trace of direct descendants. For instance, some 600 million years ago, the ocean floors of the Cambrian period were the habitat of abundant trilobites, distant relatives of today's crabs and lobsters. Almost

200 million years later, predatory spiderlike animals, some over ten feet in length, hunted their prey over sea bottoms. About 100 million years after that, dragonflies with wingspreads two and a half feet wide flew through primitive forests. Those forests were eventually converted into much of the world's present supply of coal.

All of these and countless other organisms are gone, leaving only fossils to tell us of their former existence. It has been estimated that about a half-billion species have lived at one time or another, so today's stock is only perhaps 2 percent of those that have evolved over the eons. The other 98 percent have either died out without issue or evolved into something sufficiently different to be called new species.[12]

Biologists therefore know that extinction has been the fate of most species that have been generated by the prolific speciation mechanism. But there are severe problems in determining the causes of extinctions from the fossil record. It is generally accepted that natural extinctions have followed alterations in the physical environment (for instance, a climatic change) or in the flora and fauna (such as the evolution or invasion of a new predator or competitor).

A great deal of attention has been paid to periods when there appear to have been relatively rapid extinctions of entire groups. For example, about 65 million years ago at the end of the Mesozoic era—sometimes called the age of reptiles—a large number of both terrestrial and marine animals disappeared rather suddenly from the fossil record. Some were quite ordinary groups of single-celled marine animals with hard shells called Foraminifera. Others included the most spectacular of the dinosaurs.

The sudden departure of the dinosaurs over a "mere" few hundred thousand years (or even in a much shorter time—it is very difficult to tell) has long fired the interest of paleontologists. As a result, they have produced many imaginative hypotheses to explain it. At one time it was suggested that a change in the plants eaten by herbivorous dinosaurs caused them to die of constipation. The giant carnivores, deprived of their prey, quickly starved. Another plant-related suggestion has been that, as more modern plants evolved with more deadly toxins to protect themselves against herbivores, the giant beasts were poisoned. Other hypotheses have focused on rapid changes in the climate, brought on by a supernova or a buildup of carbon dioxide in the atmosphere. Could a global increase in temperature have caused, for example, a decline in the fertility of male dinosaurs? Sperm are notoriously sensitive to temperature increases. One of the most recent explanations is that an asteroid some six miles in diameter hit Earth, digging a crater a hundred miles wide. The result, it is claimed, was a curtain of dust in the atmosphere that shut down photosynthesis for about a decade, causing the extinctions.[13]

Some people think, in contrast, that it was the rise of the mammals that did in the dinosaurs, as the mammals ate their eggs or out-competed them for resources. One scientist has suggested that the dinosaurs were essentially automata, prisoners of genetically programmed behavior. He suggests that the much more flexible and intelligent mammals, the first animals able to make "conscious" choices to override their genetic programs, were quickly able to extirpate the dinosaurs from all the various ecological niches they occupied.

Objections can be raised to all of these explanations, and the "truth," if it is ever known, may combine several of them. But from what is known about the biology of present-day populations, and from evolutionary theory in general, it is clear that *change in either the physical or the biological environment is the key to extinction.*

Sometimes, perhaps often, the two subtly combine. For example, in the 1960s we were studying a population of small blue butterflies in the mountains of Colorado. The butterflies, with a name—*Glaucopsyche lygdamus*—longer than their wingspread, lay their eggs on the unopened flower buds of lupine plants. The caterpillars of *Glaucopsyche* then eat many of the flowers, preventing them from producing seeds.

In 1969, an exceptionally late snowfall and freeze destroyed all the lupine buds and exterminated the *Glaucopsyche* population. This event gave us some insight into why the lupine plants flower so early—early enough so they produce seeds long before the growing season is over. The plants live for many years. By flowering very early, they run the risk of losing their reproductive output for a season to a late freeze, as they did in 1969. But the loss of one season's seeds was a small price for the plants to pay, as the extinction of *Glaucopsyche* relieved them from the attacks of a herbivore capable of destroying a large proportion of their seed production every year. A decade later, although migrants had reestablished the butterfly population, it had not yet returned to its previous density.[14]

So the combination of evolutionary change in another population (lupines evolving earlier blooming) and a change in the weather led to the extinction of the *Glaucopsyche* population. If that had been the only existing population of *Glaucopsyche lygdamus,* then the species would have gone extinct. There is every reason to believe that events of the sort we observed in Colorado—occurring over thousands or millions of years—are the sort that have led to the demise of most of the populations and species that have ever existed. But just as scientists have been unable to observe an entire sequence of animal speciation in nature, they have also not had the time to observe the gradual extinction of a species occurring without human intervention. With human intervention, however, extinctions can all too readily be observed, from the Dodo to the Snail Darter.

Species are not all equally likely to disappear under the pressure of environmental change. The vulnerability of a species depends on a wide variety of such factors as its total population size, geographical distribution, reproductive ability, ecological relations with other species, and genetic characteristics. Much emphasis is often put on the greater vulnerability of species that reproduce slowly as contrasted with those that reproduce rapidly.[15] The situation is not simple, however, since much attention is focused on a relatively few prominent slow reproducers such as California Condors and elephants, while a great many of the rapid reproducers—insects, for example—are passing from the scene unheralded. Certain other factors, such as food-plant specialization, may actually make many fast reproducers more vulnerable than species that reproduce more slowly. No matter how fast an insect species confined to a certain plant can reproduce, it will still go extinct if that plant's habitat is paved over.

A more subtle factor affecting vulnerability is the way individuals of a population arrange themselves in space—that is, their population structure. This characteristic may have a profound influence on the susceptibility of individual populations, and thus of the entire species, to extinction. Population structure may vary quite dramatically from species to species, even in closely related groups of animals.

For example, there are five surviving species of rhinoceros, three in Asia and two in Africa. In Asia the commonest species is the Great Indian Rhino. In this species, individuals of both sexes live solitary lives on well-defined territories for most of the year. Each territory includes a small pond in a bit of open grassland and features a great central dunghill and lesser dunghills on the periphery. When the mating season comes, both sexes evidently start wandering widely. Apparently, mating takes place by random encounter, when a sexually active bull rhino happens to meet a female in heat.

The much rarer Sumatran Rhinoceros seems to wander even more widely. There are probably not more than 50 to 150 individuals scattered from Central Burma to northern Malaya and Sumatra. When we did fieldwork in Malaya in 1966, we were told of a biologist who had been studying the Burmese Rhino for a year without ever seeing one. He had, however, been able to track some and examine their droppings. The closest he came to a sighting was when he followed one rhino's trail that led him back to his own camp. His assistants informed him that it had walked through an hour before!

The third Asian species, the Javan Rhinoceros, is nearly extinct, and nothing is known of its population structure.

The two African types, the Black and White Rhinos, have very different food habits. The Black Rhino is a browser, using its overhanging upper lip to strip the leaves from bushes and trees. The White Rhino, in contrast, is a grazer on grasses and herbs. The "white" does not refer to its color; it comes from a Dutch word meaning "wide" and refers to the broad straight lips and jaw structure that make the mouths of feeding White Rhinos look like vacuum cleaners.

Black Rhinos seem to live in mated pairs or alone in unchanging territories, while White Rhinos wander in small herds. Their dispositions are quite different. The White Rhino is relatively peaceful and seemed to us quite curious. A group followed our Land-Rover closely in Wankie National Park (in Zimbabwe, then called Rhodesia), but showed no signs of aggressiveness. In contrast, it is a standard rule of African sightseeing that the motor always be running and the vehicle in gear when observing Black Rhino![16]

The rhinos, of course, like all large wild animals, are subject to the risks of having their habitats fractionated and destroyed. But rhinos also have a very special curse. The strange-looking pachyderms have become thoroughly embedded in Eastern mythology as a source of materials with great curative powers. Virtually every part of a rhino is thought to be a remedy for something—its teeth, hair, blood, internal organs, and so on. At one time, the Calcutta Zoo found a ready market for small bottles of urine from their captive Indian Rhino. But salable as they are, the value of all these parts fades to insignificance compared to that of rhino horn. The horns (not made of bone but of a compacted mass of hairs) are especially prized by Indians and Chinese, who consider them to be a powerful aphrodisiac. Small amounts of powdered horn mixed in a potion also are supposed to cure a wide range of problems from measles to diphtheria, and applied externally in a poultice it is considered good for boils and chicken pox.

But it is the sexual connection that damages the rhinos most. Because of their phallic shape, rhino horns are in great demand in the Middle East as dagger handles. Rhino horn daggers are traditional gifts at puberty rites in the area. As much as $12,000 is paid for intricately carved and decorated horn handles. Carvers in North Yemen processed more than 2,000 horns in 1975 and 1976 alone.[17]

Hunting for horn is what virtually annihilated the Javan Rhino, although human population growth would have crushed its population through habitat destruction by now if hunting hadn't done the job first. Both the Sumatran and Indian Rhinos have been severely pressed by hunters also, as well as by having their habitat fragmented and destroyed. And with their population structure, they will have great difficulty in recovering—since the lower the population density, the smaller is the chance that they will score on their extensive mating rambles. Fortunately, the Indian Rhino will reproduce in zoos—in this case, at least, artificial proximity breeds. Whether programs to maintain it in large preserves will be successful remains to be seen. We suspect the rare Sumatran Rhino is doomed.

The relatively tame White Rhino in Africa was easy to approach and was therefore almost hunted to extinction by the turn of the century. Fortunately, conservation measures were taken, and where protected, White Rhinos are making a comeback, aided by a population structure that does not require vast areas to maintain viable reproductive units. Living in herds has its advantages.

The Black Rhino, with its population structure more or less intermediate between those of the White and the Indian Rhinos, was in no difficulty until very recently. Then, with rhino horn selling for as much as $150 an ounce—making it about a quarter as valuable as gold at 1980 prices—poaching of the Black Rhino accelerated incredibly.[18] Its fate in the wild is now in doubt, but if secure reserves of reasonable size can be maintained, it should be more secure than its wandering Indian cousin.

Understanding the population structure and other characteristics of an endangered species obviously is essential to designing strategies for conserving it. But before we turn to such esoteric subjects as the optimal size and shape of preserves, there are several more basic questions to be answered in greater detail than they were in Chapter 1. For example, why should big, dumb animals such as rhinos (or small, dumb animals such as Snail Darters) be preserved anyway? Why, in short, should you *care* if many, or even most, of our fellow passengers on this spaceship disappear?

Part II

WHY SHOULD WE CARE?

Chapter III

Compassion, Esthetics, Fascination, and Ethics

Living organisms are not only means but ends.
In addition to their instrumental value to humans
and other living organisms, they have
an intrinsic worth.

—CHARLES BIRCH,
Challis Professor of Zoology,
University of Sydney, March 1979

A LMOST overcome with fear, Digit turned unarmed to face the spears and dogs of Munyarukiko and his five companions. He would have to gain time for his family to escape up the mountain slopes. It was his role, his "duty" —and although he may well have known it would mean his death, Digit stood his ground. To Munyarukiko and the other poachers, the silverback male gorilla, erect with his canines bared, was doubtless a terrifying sight, one made more terrifying by the quick demise of one of their dogs, whose frenzy at the smell of the gorilla's fear brought it too close to Digit's powerful arms. But gorillas, strong as they are, are sadly vulnerable to the weapons of their physically weaker relatives, *Homo sapiens.* Digit bought the time for his family group to flee, but he took five mortal spear thrusts in the process.

Thus on the last day of 1977 died one of the few remaining Mountain Gorillas—a death not atypical except in the detail in which it is known. Digit was one of a group of gorillas under intense study by Dian Fossey on Mount Visoke in the Parc des Volcans of Rwanda. He had, in a very real sense, become a friend of Dian's, as well as of millions of other human beings who had seen him on television. For Digit had been filmed examining Dian's pen and then her notebook, gently returning each to her, and then lying down and going to sleep by her side. This memorable scene was part of a National Geographic Society Special television program and has been reshown at least once in a collection of film clips of which the society is justly proud.

The tragedy of Digit's death was all the greater because of its motivation. In both Rwanda and Zaire, it is believed that certain portions of the body of a silverback male gorilla—testicles, tongue, ears, portions of the little fingers —have magical properties. Used in the proper potion, they are believed either to kill an enemy or make him impotent. Over the years gorillas had been killed for these parts, and the resultant distrust of human beings had to be overcome before Dian could befriend the gorilla.

But it was not this tradition of "Sumu" (poison) that led to the killing of Digit. When tourists and other Europeans began to arrive, poachers killed gorillas to make their skulls and hands into souvenirs. Digit was murdered not because the local people are meat-starved or especially impoverished, but because an African named Sebunyana-Zirimwabagabo offered Munyarukiko roughly twenty dollars for the head and hands of a silverback.[1]

SAVING SPECIES: COMPASSION

Many human beings obviously feel compassion for Digit and the other embattled Mountain Gorillas. They hope that Digit's baby, conceived before his death and christened by Dian "Mwelu" (Swahili for "a touch of brightness and light"), will have a chance for a proper gorilla existence. Others feel no such compassion; they basically ask, "What good are gorillas?" and conclude they are no good at all. In their view, Munyarukiku was right to do the beast in —the gorillas' land can be put to good use grazing cattle, and the twenty dollars can be spent for human pleasure in the form of the native beer, *pombe.*

We could counter the latter view with the standard arguments that survival of Mwelu and the other gorillas would benefit humanity far more than their extermination. For example, by studying gorillas, human beings might come to understand themselves better. Or gorillas might serve useful ends in medical research. Or they might benefit African nations as tourist attractions. We will elaborate on such arguments later in this book; here we are concerned with other reasons for caring about our fellow passengers on Spaceship Earth. The disappearance of gorillas from the Earth would be a very sad thing, apart from the real economic values they represent, simply because they are so interesting and their very evident kinship with human beings appeals to people's sense of compassion.

Compassion and curiosity have similarly been aroused with respect to many other organisms too—fortunately for them. If human beings do not *care* about the survival of other organisms, it seems unlikely that many of them will be saved from extinction. Consider what has happened to the whales, animals whose fate we will use to illustrate a number of important points about extinction.

The Whale Killers

Whales and porpoises—cetaceans—are among the most intelligent of the mammals. In complexity and size compared to body weight, their brains are comparable to human brains. Just how intelligent they are is a matter of some controversy; evaluating their intelligence accurately is extremely difficult, given that their anatomy and way of life are enormously different from those of human beings.

People who have studied and worked with these extraordinary creatures are uniformly impressed with their gentleness, cleverness, and quickness to learn. They appear to have quite effective and complex systems of communication —the rich and haunting song of the Humpback Whale being perhaps the best-known example. People who have swum with the whales report no feelings of fear, even when close to the giant creatures, which could annihilate them with the flick of a flipper. The whales clearly are aware of their presence and take care not to harm them.

Unfortunately, *Homo sapiens* has not treated whales with equal kindness. Instead, for centuries men have relentlessly hunted the larger whales until now many of them are near extinction. Weapons used against them have run the gamut from poisoned arrows and lances to harpoons and electrical charges. For the past century, the preferred weapon has been the harpoon gun, which fires a harpoon with a cast-iron explosive head controlled with a time fuse that explodes the grenade after it is inside the whale. If the first harpoon fails to kill, a second "killer harpoon" is fired.[2] Death is sometimes instantaneous, but just as often it takes from one to thirty minutes for a whale to die in fear and agony.

One observer, an Australian journalist, described what followed the striking of two whales in 1977:

> With a great swirl of blood and foam and fear the whale dived straight down, no time for fluking, thus opening the hooks on the harpoon to gouge into it further and grip it.
> It kept diving and the chaser slewed around to keep the shuddering rope in front of it.
> . . . The sea was red all around us as the whale struggled to live, but died.[3]

He then described the second one's death:

> The whale dived, and a great green cloud burst up to the surface. Blood turns green underwater at 50 feet. . . . or was this some of its intestines?
> It came up on the starboard side, its huge head, a third of its total body size, shaking itself, and then it gave out a most terrible cry, half in protest, half in pain and then it dived again.

They loaded the next harpoon, the killer, but could not get a shot at it as it twisted and turned, hurting itself all the more. Finally, the lookout in the crow's nest shouted down that it was coming up dying. Its mouth was opening.

Such horrors were tolerated—indeed, even practiced—by supposedly civilized nations long after humane methods for slaughtering livestock (which are certainly far less intelligent than whales) and laws against cruelty to animals had been established. Perhaps because the whale slaughters occurred at sea, far from the public eye, and most people knew so little about these magnificent creatures, there was scant protest against them until a few decades ago.

But the activities of conservation organizations, dramatizing the plight of the overexploited whales, drew worldwide public attention in the late 1960s and 1970s. Films and television presentations about the whales—especially those of Jacques Cousteau—have enchanted millions, as have the performances in various seaside parks of porpoises and Killer Whales. We have our doubts about the propriety of making performing pets of these agile and charming animals, but their huge cousins may owe them a debt of gratitude for educating so many people about cetaceans. If the whales are to be saved for future generations to enjoy, it will be because millions of people have learned to care about them and have insisted that the slaughtering be stopped.

ESTHETIC VALUES

Many organisms have what might be called conventional beauty. Birds, butterflies, flowers, and others are widely recognized as esthetic resources. They may also become economic resources because of their beauty, as the aquarium trade and florist business attest. There is also a second kind of beauty, a beauty of interest, which, even more than conventional beauty, develops in the eye of the beholder. The history of the opposition to whaling is in part a history of people coming to see whales as beautiful—an image of the whale born of understanding and developed side by side with compassion. People came to *know* the whales—and it's easier to make war on strangers.

There are millions of populations and species that are even more strangers to humanity than the whales, and their existence is no less threatened by human action. They also deserve our appreciation and concern. They may have conventional beauty that goes unrecognized because most people have never seen them or because they are very small. And all have the beauty of interest—the beauty of intricacy, of sophistication of design, of unusual behavior, of great antiquity—the capacity to fascinate. Both kinds of beauty are exemplified by the insects, a group that most people have been trained to shun or find repellent.

The Insect World

Any species of bug that people spray with an insecticide is "an irreplaceable marvel, equal to the works of art which we religiously preserve in museums." No one intimately familiar with the most successful group of animals on the planet could dispute this observation of the great French anthropologist Claude Lévi-Strauss.[4]

But, sadly, few people are aware of the enormous diversity, complexity, and beauty of insects. Perhaps one person in 100,000 in the United States collects or studies butterflies—which, with some 15,000 species known worldwide, are arguably the most beautiful group of organisms. New Guinea tribesmen use giant iridescent blue, green, and gold "birdwing" butterflies as hair decorations.[5] The wings of equally iridescent blue Morphos have been used so extensively in Brazil to decorate trays and other objects for the tourist trade that Brazil had to pass laws to protect those butterflies. The Morpho wings are especially valued because, like other iridescent or metallic colors of insects, they will not fade.[6]

Many insects other than butterflies also possess great beauty, although often a microscope is required to appreciate it. Certain small bees and flies, for example, look as if they were fashioned from the purest gold.

But insects can serve best as an example of the beauty of interest. They are an immense esthetic resource in the natural world that is only poorly utilized at the moment—a potential source of infinite fascination. Not only does their beauty, in our opinion, often outshine the *Mona Lisa,* their variety puts postage stamps to shame, their miniaturization far surpasses the best efforts of human engineers, and the drama of their existence can compete with those concocted by the best playwrights. For that matter, human creativity of beauty, variety, and construction has always drawn its inspiration primarily from nature. It is ironic that Western society is so frantic to find new diversions to fill its leisure time and at the same time so ready to exterminate a group of tiny animals whose study has enthralled many thousands of people and could enthrall millions more.

As one becomes acquainted with insects, they gradually transform from an amorphous mob of bugs into a highly differentiated group of rivets in the structure of Spaceship Earth—each with a unique role to play and each with a fascinating story for those interested in learning it. Let us look briefly at some vignettes from the insect world, remembering always that it is just a tiny sample of a partly explored universe.

Take beetles, as an introduction. There are probably more than a million species of them. When the famous British biologist J. B. S. Haldane was asked by a theologian what one could conclude about the nature of the Creator by studying His creations, Haldane is reputed to have replied that He must have had "an inordinate fondness for beetles." Beetles come in all shapes and sizes. The heaviest insects are giant rhinoceros beetles of the tropics, which weigh more than the smallest mammals (some species of shrews). The males use their grotesque horns in fierce lilliputian battles over females. Among the smallest insects are ptilliid beetles, which are about the size of the periods on this page —yet each one has six functional legs, a pair of wings, a complete nervous system, a complete digestive tract, and reproductive organs.

Recent research has shown the elaborate means some beetles have evolved to foil their enemies. Bombardier beetles have become specialists in chemical warfare. Cornell University biologist Thomas Eisner has taken magnificent movies of the behavior of these small insects when they are attacked by ants (or by a pair of forceps simulating ants). The hind end of the beetle contains a pair of glands and a chamber in which glandular secretions can be mixed with crystalline enzymes. When an ant grabs the beetle's leg, the flexible end of the beetle's abdomen is pivoted toward the attacker, and muscular contractions push the secretions into the chamber where they mix with the enzymes. The resulting reaction produces a chemical explosion and the ejection with a pop of a boiling hot, unpleasant spray that drives off the ant. Ultra-high-speed photography has recently revealed that the spray is not continuous, but because of the ingenious structure of the chambers, it is pulsed at a rate of hundreds of times a second—a design that keeps the reaction going properly and prevents internal overheating.[7]

Tom Eisner has become one of the most respected biologists in the world through his studies of the ways in which insects and other arthropods interact with their enemies. In many cases, he has found them, like the bombardier beetle, to be engaged in chemical warfare. But some of the discoveries of Eisner and his students convince us that, if one looks long enough and closely enough, virtually anything imaginable will be discovered in the six-legged world.

One of Eisner's former students, James Lloyd of the University of Florida, has worked with fireflies, which actually are not flies but beetles. And they don't flash their soft lights for the pure pleasure of human lovers sitting on porch swings on warm summer nights. No—it's sex for the fireflies themselves that's involved. The males of a species fly around flashing in a pattern characteristic of that species. The female remains stationary, replies in kind, and the two sexes get together. Lloyd discovered that some female fireflies mimic the signals of other species, sending out the alluring code. Then, when a love-sick male of another species arrives, they promptly catch and eat him![8]

Hungry females are common in the insect world. After mating, the large female praying mantis devours her smaller mate. In primitive dance flies— small midgelike predators—the females occasionally catch and eat courting males. In more advanced dance flies, the male gives the female a "present" to distract her and keep her from eating him during mating. The simplest form of present is simply a dead fly of another species for the female to eat. A later stage in the evolution of the ritual is found in dance-fly species in which the males form swarms to attract the females. The males add bits of silk to their dead-fly offerings to make the swarm more attractive. Other species of dance flies enclose the prey entirely in a silken balloon, and eventually the prey is so small relative to the balloon that it cannot be fed upon by the female. The female is distracted by the balloon alone. In species that have reached the final evolutionary stage, there is no prey at all, just the balloon. The female remains distracted, however, by her "toy" and the male copulates safely! We will draw no parallels to higher species.[9]

The eating habits of insects present a variety that rivals or exceeds that of all other groups of animals combined. Virtually every part of every plant that grows is attacked by some insect. They suck on roots, invade fruits, bore in stems and seeds, devour flowers, and munch on leaves. The caterpillars of some tiny moths undergo development as leaf miners, living entirely within a thin leaf, where they excavate a long serpentine mine. The young of minute wasps, studied in detail by Alfred Kinsey before he started investigating the sexual habits of larger organisms, attack leaves in a way that causes them to form galls in which the wasps develop. Leaf-cutter ants strip the foliage of large chunks of tropical forest—not to eat the leaves themselves, but to use them in their nests as a mulch on which to grow the fungi that they do eat.

Predacious insects attack everything from tiny mites and other insects to people and elephants. They may dig sandy traps in which they lie in wait for their victims (as do ant lions and the larval stage of tiger beetles) or simply run them down (as do adult tiger beetles). Like ladybird beetles, they may graze on helpless prey (mostly aphids that suck plant juices), or, like tarantula hawk wasps, they may engage giant spiders in duels to the death which the wasps virtually always win. Young blackflies live on tiny organisms strained from fast-running streams; adult blackflies live on the blood of, among other animals, *Homo sapiens.* Some parasitic insects lay their eggs on other insects or on plants where other insects will eat them. The young parasites then devour their hapless prey from the inside out, carefully avoiding vital organs so that the victim remains alive and its meat stays fresh until the last moment.

One tropical American fly, *Dermatobia hominis,* probably holds the record for weird ways of reproducing itself. *Dermatobia* captures mosquitoes and lays its eggs on its captives. Then it sets the mosquitoes free. When a mosquito carrying *Dermatobia* eggs lands on a human being and bites, the eggs hatch and the young *Dermatobia* maggots crawl down the mosquito's proboscis and onto the person's skin. Then the maggots bore into the skin and develop there, eventually getting quite large and quite painful—maintaining a hole to the outside through which they breathe. *Dermatobia* is a creature that humanity could probably do without. But one has to admire the ingenious behavior it has evolved.

Insects are decomposers, too. They help dispose of wastes—and sometimes dispose of things people don't consider wastes. Termites may decompose your house and other odds and ends of dead wood with the aid of single-celled protozoa in their guts that can digest cellulose. Dung beetles play a grotesque game of soccer as they make balls out of cowpies, roll them to a suitable place, lay an egg in the ball, and bury it—assuring a supply of food for their offspring. And unpleasant as the subject may be, a wide variety of flies, beetles, and other insects busily help dispose of the cadavers of other animals, thus playing a vital role in the functioning of ecological systems. The maggots of some species of flies discriminate between dead and living tissue so finely that they were used extensively in medicine before antibiotics were discovered to clean up wounds and prevent gangrene. They are still used occasionally for this purpose.

Insects are incredibly widely distributed in space and time. They are found deep in caves and floating high in the atmosphere, in the Arctic and the Antarctic, in rainforests and prairies, in deserts and lakes and streams. They even live in pools of oil, salt lakes, and hot springs. Only the oceans are relatively devoid of them, though we had the pleasure of seeing one of the few marine species, a water strider, in the Galápagos Islands. Cockroaches, some of which are unwelcome residents of human habitations, are so marvelously adapted that they have persisted relatively unchanged for more than 300 million years. And they may outlast *Homo sapiens* by hundreds of millions of years more, since like most insects, they are quite resistant to radiation.

Libraries have been filled with books written about insects. A volume thicker than this one could be written about how bees communicate with each other about the location and quantity of nectar sources, and about the ingenious experiments Karl von Frisch and others have used to decipher the language of the bees. The social life of insects is today one of the most active areas of biological research.[10]

The point here is that even among the insects, a group that is shunned and loathed by many (entomophobia is a well-recognized disorder), there is a wealth of esthetic values; not just beauty in the conventional sense, but the kind of beauty of intricacy and variety that captivates gun nuts, airplane and train buffs, stamp collectors, sci-fi and computer enthusiasts, bibliophiles, bird-watchers, and so on. An important related point is that one bug is not just like another—they are not interchangeable rivets. Understanding the uniqueness of species and populations is essential to the key argument for the preservation of diversity that is elaborated in Chapter 5—the maintenance of ecosystem functions.

Other Fascinating Organisms

The ability to fascinate people permeates Earth's biota. No species lacks it upon close acquaintance. It is present in both the ugly and the lovely. Flowers are familiar repositories of much conventional beauty. But they have that other kind of beauty as well in their intriguing life histories.

Some orchids, for instance, look like female bees and wasps and thus deceive male bees and wasps into attempting to copulate with them. The insects have their fun and the orchids get pollinated. Other orchids "drug" bees before attaching their pollen onto them; still others actually entrap the bee and attach pollen to it as it struggles to get out of a specially formed tunnel. Milkweeds are gentler with their insect pollinators, simply catching their legs in grooved traps. In the process of struggling free, the bug has bags of pollen attached to it, to be deposited when the same leg is entrapped in the next milkweed flower.

The vast and varied beauty of flowers, so prized by human beings, has evolved solely to permit plants, which lack mobility, to deceive other organisms into helping with their sexual functions—for that is what pollination is all about. Not all flowers are attractive by human standards, though. Those of grasses, for instance, are not even recognized by most people as flowers. Grasses are pollinated by the wind; they don't have to seduce birds or bees into carrying their pollen around. And other plants have flowers perhaps better described as bizarre than beautiful. The blooms of South African relatives of the milkweeds, for example, may mimic the eyes, ears, nostrils, anuses, and wounds of large animals—complete with appropriate smells. These flowers are pollinated by flies that normally lay their eggs in the orifices and wounds of cows and antelopes.[11]

Flowering plants, indeed, have a curiously ambivalent relationship with animals. On one hand, they wish to attract them, not only to use them as pollen transporters but often to disperse seeds. On the other hand, the plants are prey to herbivores, which they must avoid either by being inconspicuous or by developing mechanical and chemical defenses. Running away from enemies is not an option available to plants.

Beauty and intricacy are characteristics of living organisms from the smallest to the largest. One of our most interesting looks at the world of life was watching the first movies taken of living spirochetes, the tiny, undulating bacteria that cause, among other diseases, syphilis. These organisms are spiral in shape and rotate as they move. They looked for all the world as if they were continuously dissolving at one end and forming at the other. In fact, a little pond water placed under a microscope reveals a microcosm so filled with diverse single-celled organisms that it can supply at least as much entertainment as a movie. And anyone who has seen pictures of the hard shells of diatoms, single-celled plants that as fossils are much used by geologists to predict whether strata will contain oil, knows they are as varied and as lovely as snowflakes.

At the opposite end of the size scale again are the whales, whose appeal, when people have a chance to become acquainted with them, is extraordinary. Most of our own contacts with whales have been on visits to the Hawaiian island of Maui, where the town of Lahaina was once a thriving whaling port. Lahaina still thrives on whales, but now whale carvings, whale pictures, whale T-shirts, whale books, and fake scrimshaw are supporting the economy. One of the nicest resort shopping centers is the Whaler's Village, which features mounted whale skeletons and whaling memorabilia on exhibit.

All this is because, in the winter and spring, the waters off Maui are host to a population of Humpback Whales. Walking along Kaanapali Beach in the morning, we have often been treated to the arched backs and blows of a group of Humpbacks. And on lucky occasions we have watched the whales "breaching"—behavior unique to these gentle giants, who may be over fifty feet long. Suddenly the whale lunges almost entirely out of the water. It looks rather like a leaping dark salmon in slow motion. It hangs suspended for an instant, gleaming black, its characteristic long flippers bright white, and then flops back into the water on its back with a thunderous splash. The whole act may be repeated a dozen or more times, each leap looking deceptively slow as an animal the length of a sizable yacht rears out of the depths.

Why the Humpback breaches is not known. A dreary explanation is that it is attempting to dislodge parasites. We prefer to believe that these intelligent creatures do it out of sheer exuberance. They also "lobtail," lifting their flukes (tail fin) out of the water and slapping the surface with it like a gigantic beaver. Herman Melville wrote of the Humpback: "He is the most gamesome and lighthearted of all the whales, making more gay foam and white water generally than any other of them."[12]

One day Paul was snorkeling off Kaanapali Point with a volunteer assistant, winemaker George Burtness, attempting to take photographs of a surgeonfish characteristic of the "surge zone"—the disturbed water near the point where waves were breaking on the rocks. A faint high-pitched sound filled his head, and he at first thought his snorkel had developed a leak. Then he realized what it was. Paul and George dived to the bottom, and lying there holding their breaths, they were treated to an unworldly concert of squeals, moans, groans, bleeps, and other sounds from deep bass to high soprano ranging over ten octaves: the famous song of the Humpback Whale.

It is easy to understand how early whalers and other seamen were mystified and terrified by similar eerie sounds echoing through their ships. The sounds have long been an established part of sailors' lore, but they were only explained in the 1950s through the work of three scientists—W. E. Schevill, Scott McVay, and Roger Payne.[13] The ghostly, beautiful songs may last as long as thirty minutes and may be repeated verbatim indefinitely. Each individual whale seems to have its own variation of the song, whose function remains unknown. Perhaps the best guess is that it allows individuals to keep track of other members of their group—a logical conclusion in these highly social animals.[14]

Relatively little is known about the whales' social life, unfortunately, but some of what is known is intriguing. For example, cows with calves are accompanied by "guardian" females. When the trio is over deep water, the guardian swims *below* the mother and offspring—as if any danger were expected from that direction. It does not take much of a stretch of the imagination for a scuba diver or snorkeler to guess the whales' psychology. When the sea bottom is in view, the ocean seems a friendly place. But being unable to see the bottom is disturbing and disorienting—and it is easy to imagine a thirty-foot Great White Shark zooming up from the depths at twenty-five knots.

Whales not only have fascinating behavior but are physical wonders as well. The Blue Whale is the largest animal that has ever graced our planet. Such giant herbivorous dinosaurs as the brontosaurus weighed up to fifty tons. A Blue Whale weighs that much long before it reaches puberty; full grown they weigh a hundred and fifty tons, as much as three brontosauri! Blue Whales grow to more than a hundred feet—longer than any other animals. When such a whale is vertical in the water with its tail at the surface, its nose is deep enough to be subjected to the weight of three atmospheres. It is possible that they dive deep enough, more than a mile, to be subjected to more than two hundred times the atmospheric pressure experienced by people on land at sea level.

A Blue Whale's heart weighs half a ton; a child could crawl through the major artery leaving it. Its blood supply amounts to some two thousand gallons —enough, if it were gasoline, to drive a small car twice around the world. A Blue Whale's brain can weigh over a dozen pounds, quadruple that of an adult human being. The brain is highly convoluted and shows other anatomical signs of great intelligence—a conclusion supported by the observed intellectual ability of whales in captivity.[15]

The case of the whales clearly demonstrates how acquaintance with organisms can transform attitudes about them. Whales were once thought of simply as an economic resource, and, as will be described later, several species are threatened with extinction because of it. Now they are increasingly viewed as an esthetic resource, and one not without economic value.

A Biologist's View

To say that our own lives have been esthetically enriched by contact with the other organisms of our planet would be a vast understatement. To a biologist, every day is likely to bring pleasant surprises, be it the discovery that a "bird dropping" in a tropical forest is actually many bugs of two colors sitting together to simulate a dropping; or the finding, virtually in your own backyard, that an insect you thought fed exclusively on one plant actually depends on another for its survival.

Still, is making life interesting for biologists a sufficient reason to preserve organic diversity? Perhaps, for the pleasure and employment of this minority of *Homo sapiens* and for what their studies can offer to the rest of the world. Another reason would be the very large number of nonbiologists who are, or potentially could be, interested in the same diversity. Most children are extremely interested in natural history, although the interest is often killed rather than nurtured by school systems.

There are, moreover, millions of adults who enjoy nature in some way. There are some 8 million bird-watchers in the United States alone. A 1965 outdoor recreation survey found that about 20 million Americans took nature walks and almost 3 million photographed wildlife annually.[16] Keeping aquaria, an activity heavily dependent on the diversity of fishes, is the largest hobby in the country. People fascinated by succulent plants are so common that "cactus rustling" has become a crime in the southwestern United States. And there are enormous numbers of people who keep birds and reptiles as pets. So many people are interested in the wildflowers on Stanford University's Jasper Ridge Preserve that tight controls have had to be established to prevent research areas from being trampled. There are hundreds of nature and conservation centers in the United States. Similar statistics apply to many other countries, where nature hobbies are often even stronger; Winston Churchill and Vladimir Nabokov unashamedly collected butterflies.

The Lindblad *Explorer,* the luxury "nature cruise" ship, is booked years in advance for its trips to the Antarctic by people willing to pay many thousands of dollars to wade through guano to be among thousands of nesting penguins, shiver in rubber boats to watch crabeater seals sunning themselves on ice floes, or hike up difficult slopes to observe albatrosses nesting in the Falklands. The *Explorer* is so popular that another company has built a ship and set up a cruise schedule to emulate it.

Much the same can be said for the highly popular safaris to the game parks of East Africa. We will never forget the day that the lioness used our Land-Rover as cover to stalk (unsuccessfully) a wildebeest, the first time a bull elephant threatened us by making a short charge, or the day Hugo von Lawick introduced us to the wild dogs that he and Jane Goodall had been studying in the Serengeti, each one known as an individual by its distinctive color pattern. Nor will we forget the hour we watched two elephants wash each other, with what seemed to be loving care, at a waterhole in Kenya. The plains of East Africa can provide *Homo sapiens* with one of the last glimpses of what Earth was like before humanity exploded across its surface.

Of course, while the total number of people who can visit Antarctica, the plains of East Africa, tropical rainforests, or other exotic places as touring naturalists is large, it still represents a tiny fraction of humanity. But as many millions of backyard bird, bug, and flower watchers have learned, the pleasures of knowing other organisms are not restricted to the well-off. Indeed, how many ghetto children have been thrilled—or could be—by watching a female guppy give birth in a school aquarium? Or by seeing a gorgeous male Siamese fighting fish, ablaze with crimson or blue, build a bubble nest, embrace his mate beneath it, and then catch her fertilized eggs in his mouth and spit them into the nest? Not one encounter with sharks, Killer Whales, lions, rhinos, elephants, chimps, rare butterflies, or army ants in Paul's lifetime as a field biologist has had the impact of his discovery as a child that a giant silk moth —a Polyphemus—had emerged from a cocoon he had been watching. That is a thrill available to virtually any child at no expense.

It is not even necessary to have direct experience with other organisms in the wild in order to be enriched by them. A trip to the primate house at the zoo will tell a perceptive person volumes about his or her kinship with other animals, as well as about what those animals endure in captivity.

But beyond compassion for leopards and baby seals slaughtered for their hides, beyond the naturalist's delight in the millions of diverse lifestyles produced by evolution, there seems to be a deeper feeling for other life forms that runs through all societies. In the West, as naturalist Jim Fowler likes to point out, it can be seen in the use of symbols. Not just in the Mercury Cougar, Ford Falcon, and Audi Fox, but in metaphors and symbols that go far back in history and are perpetuated in such phrases as *a real tiger, lionhearted, brave as a bull, sturdy as an oak, strong as an ox, free as a bird.* It shows up too in national symbols such as the double eagle of Napoleon, the Russian bear, and the American eagle.

In many cultures, people have maintained special relationships with other living things, especially animals, even to the extent of worshiping them. These relationships survive in our own culture in ever-popular children's stories, both fiction and nonfiction. The importance of animals in particular to children would be hard to overestimate. The extinction of many animal species would deprive future generations of children of the pleasures of becoming acquainted with real versions of the animals in their books.

People of all cultures seem to feel that they are more "human" in the context of a natural world. This clearly was a factor in the post–World War II rush to the suburbs in the United States. It may well be that contact with nature is essential to human psychological well-being. Three University of Wisconsin biologists expressed this thought as follows:

Unique as we may think we are, we are nevertheless as likely to be genetically programmed to a natural habitat of clean air and a varied green landscape as any other mammal. To be relaxed and feel healthy usually means simply allowing our bodies to react in the way for which one hundred millions of years of evolution has equipped us. Physically and genetically, we appear best adapted to a tropical savanna, but as a cultural animal we utilize learned adaptations to cities and towns. For thousands of years we have tried in our houses to imitate not only the climate, but the setting of our evolutionary past: warm, humid air, green plants, and even animal companions. Today, if we can afford it, we may even build a greenhouse or swimming pool next to our living room, buy a place in the country, or at least take our children vacationing on the seashore. The specific physiological reactions to natural beauty and diversity, to the shapes and colors of nature (especially to green), to the motions and sounds of other animals, such as birds, we as yet do not comprehend. But it is evident that nature in our daily life should be thought of as a part of the biological need. It cannot be neglected in the discussions of resource policy for man.[17]

Many aspects of human behavior confirm this observation. That the color green is soothing is well known. People try to nurture plants even in the worst city slums, and city dwellers as well as suburbanites often surround themselves with animals—dogs, cats, fishes, birds—as if trying to recapture a time when animals were an everyday part of human existence. Is it any wonder that environmental concerns remain high in the polls even in times of economic distress? Could it be that most people intuitively understand that preserving nature is not just an elitest ploy, but something essential to preserving the spirit, if not the body, of a human being?

A RIGHT TO EXIST

There remains one more important argument in favor of preserving species that has nothing to do with balancing economic costs and benefits to humanity. It is essentially a matter of *ethics.* To our minds this is the first and foremost argument for the preservation of all nonhuman species. The argument is simply that our fellow passengers on Spaceship Earth, who are quite possibly our only living companions in the entire Universe, *have a right to exist.*

David Ehrenfeld, in his provocative book *The Arrogance of Humanism,*[18] called this the "Noah Principle" after the best-known practitioner of conservation in history. In Ehrenfeld's view, species and communities should be conserved "because they exist and because this existence is itself but the present expression of a continuing historical process of immense antiquity and majesty. Long-standing existence in Nature is deemed to carry with it the unimpeachable right to continued existence."

Many others—from Buddha on—of course have questioned whether humanity has the right to kill other animals, let alone push all organisms of any kind to extinction; to play God. To many of them and to Ehrenfeld, this must be the ultimate form of arrogance—that human beings believe that basically they are the only important life form and they alone should decide whether others should be permitted to live or not.

This is fundamentally a religious argument. There is no scientific way to "prove" that nonhuman organisms (or for that matter, human organisms) have a right to exist—it is rather an ethical view held by a portion of humankind that includes Ehrenfeld, the great English ecologist Charles Elton, and many others who have been concerned about conservation. Ehrenfeld describes this view as "nonhumanistic," but because of the controversy that terminology has engendered, we will simply call it nonhomocentric—not centered on the needs and desires of humanity. In other words, *Homo sapiens* is seen as just one among millions of organisms, albeit a numerous one with extraordinary power to dominate the rest of the living world.

We and others indeed believe that along with the preeminence that *Homo sapiens* has achieved goes a very great moral responsibility—a stewardship, if · you will—upon which we must not turn our backs. Perhaps especially *because* we have the power to destroy them, *we must respect the rights of our co-inhabitants of Earth.*

The exercise of this stewardship presents many complex practical and moral problems ranging from how to get most members of our species to accept responsibility for stewardship to how to deal with other species that are our frank enemies—for example, the smallpox virus and malarial parasites.[19] What principle should replace human chauvinism in choosing between preservation of a population or a species and the well-being of human society or a portion of it? The measurement of the value of everything in monetary terms, the dominant procedure today, clearly produces wrong answers much of the time. Some other yardstick needs to be developed and applied to decisions determining the fates of our fellow spaceship passengers.

There are, as you will see, some extremely powerful homocentric arguments for the preservation of organic diversity. But all of those arguments leave the existence of each population and species open to separate negotiation. If it could be shown, for instance, that the life of the gorilla Digit's baby Mwelu is of no use to humanity, or that the land required to support Mwelu and the other gorillas of his group could be used instead to support another ten or twenty people, would it be morally right to stab Mwelu to death with spears and hunt down the rest of Digit's family?

We think not, and we think the same principle applies to the Snail Darter, the Houston Toad, the Furbish Lousewort, and the myriad other threatened life forms that, since they are less akin to *Homo sapiens* than gorillas, less easily elicit compassion from people. We think they must be granted the right to exist, regardless of whether human beings find them attractive or useful.

Along with many other ecologists, we feel that the extension of the notion of "rights" to other creatures—indeed, even to such inanimate components of ecosystems as rocks and land forms—is a natural and necessary extension of the cultural evolution of *Homo sapiens*. We believe it not only to be in our immediate physical self-interest to do so, but in our moral self-interest as well. For in our view the moral concerns of a human being must extend beyond fellow *Homo sapiens* and family pets to embrace the entire system in which humanity is embedded.

The history of the ever-widening ethical concerns of our species gives real hope that the trend is in the right direction. From an original concern only with the family or immediate group there has been a steady trend toward enlarging the circle toward which ethical behavior is expected. First the entire tribe was included, then the city-state, and more recently the nation. In this century, concern has been extended in many groups to encompass all of humanity.

A little more than a century ago, many Westerners thought there was no need to behave ethically toward certain people because, as slaves or members of "inferior" races, they were excluded from the in-group. Today few Westerners—indeed, few people in any culture—would espouse such a view, even though behavior often does not measure up to professed moral standards.

As ethical concerns have widened with respect to other people, they have also widened to include nonhuman entities. The great religious traditions of the West have in some ways nurtured disrespect for nature by destroying pagan animism and have generally promoted the idea that Earth is human property to exploit as human beings see fit.[20] An appreciation of living beings in their natural context—in wilderness—though long an element of Eastern culture and religion, came late to the West. In fact, the present dominance of Western culture may very well be partly due to its rapacious attitudes toward nature and other societies. Early settlers in North America, especially, saw the wilderness as a deadly enemy to be conquered and tamed.[21] That was a reasonable viewpoint a hundred and fifty years ago, but the idea that nature is the enemy persists in some circles today. The relentless attack now being pushed on the few remaining natural areas of the nation—and its extension into other countries—can be seen as a surviving remnant of that philosophy. In the nineteenth century, as the wilderness of the eastern United States increasingly was tamed, people began to see natural areas as refuges from hectic life in society. Such American writers as Thoreau and Emerson celebrated the beauties and benefits to the soul of the wilderness, and a school of romantic landscape painting portraying America's spectacular natural scenery throve in the last century.

In the last hundred years, the ranks of those in the United States and Europe advocating compassion for, and unity with, the rest of the natural world have swollen considerably. These ideas have perhaps been best developed by that hero to all conservationists, Aldo Leopold. Leopold's development of the land ethic—that "land," his shorthand term for natural ecosystems, could be used but not abused—was the culmination of the long historic trend of extending ethics beyond the immediate family or group:

> The land ethic simply enlarges the boundaries of the community to include soils, waters, plants, and animals, or collectively: the land.
> . . . A land ethic of course cannot prevent the alteration, management, and use of these "resources," but it does affirm their right to continued existence. . . .
> In short, a land ethic changes the role of *Homo sapiens* from conqueror of the land-community to plain member and citizen of it. It implies respect for his fellow-members and also respect for the community as such.[22]

Leopold's manifesto was written in 1948; unfortunately, the prevailing attitudes toward nature that he deplored—that natural things are valueless except as something people can exploit—are still predominant in Western culture, although something like the land ethic has come to be embraced by more and more individuals. Today the spirit of Leopold marches on in people like David Ehrenfeld and Roderick Nash, who argue eloquently against a homocentric view of conservation; attorney Christopher Stone, who advocates giving natural entities "standing" in legal cases;[23] Shirley McGreal and her colleagues at the International Primate Protection League, who are battling for the rights of Digit's family; conservationist David Brower, founder of Friends of the Earth, who fights to save the canyons, the whales, wilderness areas, and all the other natural resources of our planet; and Sir Peter Scott and his colleagues in the World Wildlife Fund. It lives in all the members of those groups and the Audubon Society, the Conservation Society, Nature Conservancy, the Sierra Club, Defenders of Wildlife, the Wilderness Society, the National and International Wildlife Federations, and many other organizations struggling for the preservation of other species in the United States and elsewhere. They and millions of other individuals fervently believe that *Homo sapiens* should tread softly on this Earth.

Even those people who still take a homocentric, "what good are they?" approach toward nonhuman species and other natural objects are nevertheless somewhat infected with the notion that ethics should extend beyond *Homo sapiens*. The seeds are there, planted deep in our society. Not even real-estate developers, oil-company executives, or the Army Corps of Engineers would approve the wanton killing of a homeless kitten—even though it is no one's property and about as "useless" as it is possible to be.

That animals have intrinsic rights and should be protected against abuse is now a notion widespread in our society. It is obvious in the general revulsion at such spectacles as the Canadian slaughter of baby seals and the killing of dolphins by Japanese fishermen, and it is enshrined in humane laws. The current trend in zoos to house animals in spacious quarters with relatively "natural" surroundings, and even more the development of "zoo parks" are other manifestations of concern and compassion. The day is long gone when a horse could be beaten to death by its master with impunity because it was merely "property." Perhaps the day will soon arrive when a species like the Snail Darter also cannot be sacrificed.

It is possible that Aldo Leopold's land ethic will spread like wildfire through the human population in the next few years, much as the family-size revolution spread through the United States population in a few years in the early 1970s. Social transformations can take place very rapidly when the time is ripe. The rules governing treatment of other species today remain largely economic, and if any part of our culture is ripe for change, it is the tottering economic system. The ethic of environmental protection should be totally separated from contamination by short-term economic considerations. Rod Nash put it most colorfully: "One does not consider the price if someone threatens to rape one's daughter. With environmental ethics a reality, the same might be true of attempts to rape the land."[24]

Considering the high rates of extinction and environmental deterioration that prevail today, however, it would hardly be prudent simply to attempt to persuade environmental rapists to adopt a new land ethic, or to preserve other species because they are beautiful or interesting or worthy of compassion. While such a new ethic might be adopted widely and rapidly, it might not be adopted at all. But without a major change in attitudes and behavior toward nature, the consequences for *all species, including our own,* will be catastrophic.

We will therefore turn now to the homocentric arguments for the preservation of organic diversity, in the belief that they are powerful enough to persuade even the most dedicated human chauvinist that protecting the Gorillas and Furbish Louseworts is in his or her own direct self-interest.

Chapter IV

Direct Economic Benefits of Preserving Species

If species can prove their worth through their
contributions to agriculture, technology and other
down-to-earth activities, they can stake a strong claim to
survival space in a crowded world.

—NORMAN MYERS,
The Sinking Ark

THE direct benefits supplied to humanity by other species are often little appreciated, but nonetheless they can be very dramatic. In 1955 Paul's father died after a grim thirteen-year battle with Hodgkin's disease, a leukemia-like disorder of the lymphatic system. Just after his death, some Canadian scientists discovered that an extract of the leaves of a periwinkle plant from Madagascar caused a decrease in the white blood cell count of rats. Chemists at Eli Lilly and Company analyzed the chemistry of periwinkle leaves, and the analysis turned up a large number of alkaloids, poisonous chemicals that plants apparently have evolved to protect themselves from animals that eat them and parasites that infest them.[1] Two of these alkaloids, vincristine and vinblastine, have proven to be effective in treating Hodgkin's disease. Indeed, treatment with vincristine in combination with other chemical agents now gives a very high remission rate and long periods where no further treatment is required in patients even in the advanced stage of the disease.

Thus a chemical found in a plant species might have helped greatly to prolong Bill Ehrlich's life—and is now available to help the five to six thousand people in the United States alone who contract Hodgkin's disease annually. As some measure of its economic value, total sales of vincristine worldwide in 1979 were $35 million.[2] Vincristine also is used along with other compounds to fight a wide variety of cancers and cancerlike diseases, including one form of leukemia, breast cancer, and cancers that afflict children. Had the periwinkle plant been wiped out before 1950, humanity would have suffered a loss—even though no one would have realized it.

MEDICAL BENEFITS

Vincristine from the periwinkle is but one example of the myriad ways both plants and animals contribute to human health and medicine. Probably the best way to appreciate future medical gifts to humanity from other species is to consider what it has already received.

Plants and Microorganisms in Medicine

Turning to the plant kingdom for medical help is an ancient human tradition. The medical use of members of the periwinkle family is not new, and plants of other families have been used to treat cancers at least since the time of Hippocrates, four centuries before Christ.[3] But the wide use of plants as sources of medicines by indigenous peoples makes us suspect that the tradition is much older than that—possibly as old as humanity itself. For instance, Navajo medicine men use nearly two hundred plants in their practices. African and Asian herbalists and shamans also employ local plants intensively, as do the traditional medical systems of India and China. Records of the use of plants against disease have been left in Egyptian pictographs and the clay tablets of the Babylonians. Even the Neanderthal people apparently made medical use of plants.[4]

In recent decades, medical scientists have discovered that many of the plants used in traditional medical systems indeed had chemical constituents that were effective against the ailment being treated. There has been a general tenor of surprise at this discovery, but there should not have been, considering the intimate knowledge of their natural surroundings found in indigenous groups. For example, Peruvian Indians had a cure for that deadly scourge malaria— extracts of the bark of plants in the coffee family. The plants, trees of the genus *Cinchona,* are found in Peru, Colombia, Ecuador, and Bolivia. The cure had reached Spain by 1639 and was distributed widely by Jesuit priests, becoming known as "Jesuit powder" or "Peruvian powder."

The most important active ingredient in *Cinchona* bark is quinine, which like vincristine is an alkaloid. It was first extracted from the bark in 1820 and was available commercially in 1823. For a century, quinine was the main treatment for malaria. Demand for quinine in the middle of the last century led to the domestication of *Cinchona* trees in plantations and, around 1930, to the synthesis of chemically similar compounds. These synthetics have now replaced quinine, but their basic design was inspired by a chemical first isolated from a wild plant species.

Plants also aid human beings—especially those in the industrial countries —against another major killer, diseases of the heart and circulatory system. Reserpine, an alkaloid from *Rauwolfia* (a plant related to the periwinkle), is widely used in the control of high blood pressure. *Digitalis* (foxglove), a plant in the snapdragon family, is the source of a key medicine used in the treatment of chronic heart failure. It stimulates the heart to pump blood more effectively with a smaller expenditure of energy.

Physicians, of course, use the dried leaves of *Digitalis* and active ingredients isolated from them in tiny quantities. A typical oral dose would be about one hundred-thousandth of an ounce. Larger doses testify to the effectiveness of the chemical defenses of the foxglove plant against grazing animals: they cause nausea, vomiting, loss of appetite, diarrhea, dimming vision, and death.

Plants have evolved some remarkably clever defenses against mammals that might eat them. Some fungi, cacti, relatives of nettles, legumes, plants of the potato family, morning glories, and poppies, to name just a few, contain chemicals that have profound effects on the nervous system, causing hallucinations or depression. Think of it: a zebra eats a leaf or two of the wrong plant, starts hallucinating, tries to make love to a lion, and bothers the plant no more!

Human beings all over the world make use of psychoactive plants for religious or recreational reasons; whether such use should be counted as a benefit derived from other species is clearly a matter of debate, since, at least in the United States and some other nations, the costs of drug addiction are very high. But what is not debatable is that great benefits have flowed from the most famous of psychoactive plants, the opium poppy. Derivatives of opium, the alkaloids morphine and codeine, are very important painkillers, even though they must be used with care in cases where the development of addiction would be a problem. In some countries another opium derivative, heroin, is used to ease the pain of terminal patients.

Entire books have been written on the plant products that affect human health.[5] The examples we have given here barely scratch the surface of the anticancer drugs, diuretics, dysentery treatments, antiparasite compounds, dentifrices, toothache remedies, ulcer cures, laxatives, venereal disease treatments, and so on that *Homo sapiens* has obtained from the higher plants alone.

The overall contribution of plants to medicine today can be seen by the presence of chemicals from higher plants as major or sole ingredients in about a quarter of all prescriptions written annually in the United States. Products of lower plants and bacteria account for another 13 percent. Similar proportions of plant-derived compounds are found in nonprescription pharmaceuticals. Moreover, 69 of 76 major drugs obtained from plants can be commercially obtained from the plants themselves more cheaply and easily than they can be synthesized. The United States and other industrialized countries are dependent on imports from developing countries for most plant materials used in drug manufacture.

The other species that have come most dramatically to the aid of humanity, of course, are the lowly fungi and bacteria. Many prescientific cultures recommended that wounds be treated with infusions of moldy bread. Like many other folk cures, this turned out to be more than an old wives' tale. Starting with Pasteur in 1877, scientists increasingly noted that the presence of certain bacteria and fungi inhibited the growth of other bacteria and fungi. In 1928 Sir Alexander Fleming observed that a mold, *Penicillium notatum,* contaminating a bacterial culture dish was killing the bacterium *Staphylococcus aureus* (a species that may cause boils and carbuncles). Fleming suggested in a 1928 paper that the bactericidal substance from the mold might have therapeutic value—thus launching the career of the first and most famous antibiotic, penicillin.

The number of antibiotics now known is around a thousand. Some, like penicillin, come from fungi—cephalosporin C and griseofulvin are examples. Others, such as bacitracin, chloromycin, erythromycin, streptomycin, and tetracycline, come from bacteria. And some antibiotics have been found in higher plants, although none of these is yet used medicinally.

The impact of antibiotics on human health can hardly be appreciated by anyone born since the Second World War. The fears that used to be associated with a wide variety of injuries and diseases have largely disappeared. War wounds and surgical operations were much less likely to kill after antibiotics became available to control bacterial infections. Scourges like bubonic plague, tuberculosis, epidemic typhus, typhoid fever, scarlet fever, diphtheria, bacterial pneumonia, syphilis, and gonorrhea—the whole spectrum of diseases caused by bacteria—could be treated much more effectively.

After the war, antibiotics quickly became available globally. Indeed, we have found them more readily available in the backwoods of some poor countries than in the United States, where their distribution is more closely regulated. Along with the use of synthetic pesticides against malarial mosquitoes and other disease-causing organisms, the use of antibiotics caused dramatic declines in death rates, especially among the young. This in turn caused human populations in the poor countries, which previously had had high birthrates and high death rates, to explode. Disease control knocked the death rates way down but did not significantly affect the birthrates.

Guaranteeing Future Benefits

It is ironic that the very population explosion that was triggered in part by antibiotics may limit the future access of *Homo sapiens* to new antibiotics and other medically useful plant products. The search is always on, for as bacteria become resistant to antibiotics in present use, substitutes must be found. Some of the most promising potential sources appear to be higher plants, which themselves are subject to the attacks of bacteria and fungi and appear often to have evolved antibiotics. Earth's greatest reservoirs of higher plant species are its tropical rainforests. And these are the most endangered major habitats on the planet—endangered in part because of the explosive growth of human populations in the tropics.

Unfortunately, in spite of the enormous medical benefits *Homo sapiens* has already obtained from the botanical world, systematic investigation of the usefulness of plant species as medicines has barely begun. For example, according to conservationist Norman Myers, only about 2 percent of flowering plants —some five thousand species—have even been tested for alkaloids.[6]

The screening that has been carried out nevertheless has turned up an amazing array of useful compounds. The estimated quarter-million flowering plant species are potentially a gold mine of additional beneficial chemicals. Slightly more than 10 percent of these plant species have been crudely screened for potential anticancer drugs in a National Cancer Institute program costing about $1.5 million per year. So far fifteen species have yielded materials sufficiently promising to justify trying them on human patients.[7]

We say the plants have been "crudely screened" because the defensive chemicals of plants are distributed very unevenly. They are often present in some parts of a plant and not in others. Certain ones will, for example, be in seeds but not in leaves. Leaves of different types or ages may have different concentrations of chemicals. There also can be considerable variation in the distribution of chemicals from individual plant to individual plant. And different populations of the same species can vary dramatically in their array of chemical defenses.

Thus the thorough screening of even a single species for defensive compounds is generally a difficult and laborious process—so much so that no knowledgeable biologist would claim that even *one* relatively widespread species had been exhaustively analyzed. If civilization wishes to take full advantage of the potential bounty of the plant world, far more than the preservation of species is required: the variability *within each species* must be conserved as well. It would be foolish, indeed, not to take advantage of that bounty and to preserve it so that future generations can explore those species this generation doesn't get around to studying.

Unfortunately, humanity is not well organized at present to take proper advantage of the drug potential of the plant world. Exploration and screening are risky, arduous, and expensive. Drug companies have suffered losses because of the risks and sometimes because the people doing the work were inadequately trained.[8] They are therefore understandably reluctant to launch large programs of exploration for new plant sources. It seems likely that the kinds of programs needed—with large commitments of funds so that properly trained people could do the exploration and preliminary screening—may have to be organized by government agencies. Promising compounds could then be made available to all interested companies for a fee. Or the screening programs could be funded by a tax on drugs. The value of plant-derived medicines in the United States alone in 1980 was probably over $6 billion; 3 percent of that figure would make about $200 million annually available to investigate tropical plant species as drug sources.

Such programs would clearly be a good investment, considering that the small efforts made so far have resulted in such tremendous benefits. Moreover, since the problems faced by human beings change through time, and their ability to ask questions also changes, even a species thoroughly studied today and declared medically useless should be preserved in the hope that our grandchildren might find it useful. *Penicillium notatum*, after all, would have been useless at the time of the American Civil War, when the role of "germs" in causing disease was not yet understood. Thus an extremely important aspect of a screening program for plant sources of drugs is the preservation of natural plant communities, especially the rich tropical forests, just so they can eventually be investigated.

Medical Benefits from Animal Species

Medical gifts already received or potentially available from nature are by no means limited to plant biochemicals. Extracts from some marine animals related to those that built the giant coral reefs show some promise as anticancer agents, and broad-spectrum antibiotics have been extracted from marine sponges.[9] There appear also to be potential anticancer drugs in a wide variety of other marine animals, including sea anemones, segmented worms, clams, sea cucumbers, sea squirts, moss animals, proboscis worms, sharks, and sting-rays. Marine animals have yielded substances with a broad range of potential uses in medicine, from antiviral and antibacterial action to anticoagulants, contraceptives, and the control of ulcers and hypertension.[10] Cytarabine, derived from a sponge, is used for the treatment of leukemia and, because of its antiviral properties, against herpes infections.[11] When people use the oceans as the ultimate dump for pollutants, they are in effect attacking a vast pharmacopoeia, whose contents, like those of the tropical forests, have barely begun to be explored.

Various other drugs of animal origin have medical uses. For example, ancrod, the venom of a Malayan pit viper, a relative of rattlesnakes, is used as an anticoagulant—an agent to prevent the formation of blood clots, which, among other things, can cause heart attacks. The venom of bees is said to relieve arthritis, and a secretion from blowfly larvae is useful to promote healing.

The major medical contribution of other species of animals, however, has been their use in research and experimentation to increase the understanding of human health and disease. Rats, mice, Rhesus Monkeys, and Chimpanzees in very large numbers have served as surrogates for human beings in a wide variety of medical experiments and for safety testing of numerous possible toxins and cancer-causing agents. Research on exotic species of wild animals also has made contributions to the knowledge of human physiology. For instance, studies of elephants under stress and in different habitat types may have given clues to the environmental origins of heart disease in human beings.[12] Squid have played a major role in investigations of how nerves, including human nerves, function. The microbe that causes leprosy is most satisfactorily cultured in Armadillos.[13]

Cottontopped Marmoset monkeys have proven to be excellent experimental animals for the development of a vaccine against a cancer similar to Hodgkin's disease that is caused by a herpes virus. The work showed that in this rather close relative of *Homo sapiens* it is possible to prevent cancer by immunization. Related work has indicated that a virus thought to be implicated in causing human cancers also was carcinogenic in Cottontops and Owl Monkeys. It seems possible that experiments with these or other unusual nonhuman primates may eventually lead to development of effective vaccines against some cancers of human beings.

Fruit flies, mice, Guinea Pigs, wasps, salamanders, African Clawed Toads, sea urchins, and butterflies are just a few of the animals that have helped scientists gain a basic understanding of genetics and embryology, and thereby of human heredity and development (molds and bacteria have played central roles here, too). The animals used in research have therefore helped fight birth defects and diseases such as mongolism, Tay-Sachs disease, and sickle-cell anemia. This sort of research also holds the greatest promise of finding really satisfactory cures for cancer—which is fundamentally a breakdown in the genetic controls of cells. And it is impossible to predict what yet-to-be-studied animals may prove to be crucial experimental tools for the conquest of that dread family of diseases.

The use of other species of animals in aid of human health raises for us and for many others moral issues related to the "rights" of other species to exist. When species are useful, does humanity have the right to exploit them to extinction? Or is *Homo sapiens* obliged to husband them and treat them with utmost kindness? What kinds of experiments on what kinds of animals are morally permissible? Only the most foolish, of course, would recommend overexploiting a useful organism.

Personal positions on the care to be used in exploitation, however, vary a great deal, even among biologists. Most would agree that the more humanlike the animal, the greater is its capacity to suffer, and the more restrained and compassionate its treatment should be. We are forced to kill many butterflies and occasional reef fishes in the course of our own research. We kill as few as possible because we don't like it, but we have no deep moral compunctions about it. Toward mammals, though, we would have very different feelings. For most scientists, where mammals are concerned, the decisions are more difficult.

Sometimes the moral choices are difficult when an endangered species can supply a direct benefit to humanity. Few people indeed would not eat the last members of a species if the alternative was starvation. Nor would most people shrink from sacrificing them in medical experiments if failure to do so would cost many human lives. But rarely are the choices so stark.

Hepatitis B, for instance, is a widespread disease of human beings caused by a virus.[14] It is transmitted by blood transfusions, by contaminated needles of drug users, by sexual intercourse, and possibly by biting insects. The disease is rarely fatal, although mortality in older patients can be as high as 10 to 15 percent, and those of any age who contract it are usually sick for several months. In 1976 there were probably 150,000 cases altogether in the United States, of which about 1 percent were fatal.

In 1978 a dispute arose over a request to the U.S. Fish and Wildlife Service by Merck, Sharp and Dohme, a pharmaceutical company, to import 125 Chimpanzees to use in testing a vaccine against hepatitis B. Primatologists protested, but not because the tests would be particularly dangerous for the chimps. The animals would be used for one low-risk test each—the vaccine is thought to be very safe—and then transferred to a breeding colony. The scientists protested because of the impact the removals would have on the remaining chimps in the wild. The standard method of procuring the animals is to find a mother and child, shoot the mother, and take the child. Mortality among young chimps in transport is high. One knowledgeable primatologist, Geza Teleki, estimated that about 500 chimps would die in the process of delivering 125 healthy ones to the United States.

Merck countered that their proposed animal supplier in Sierra Leone had told them that the chimps would be captured by surrounding chimp groups with a number of people and chasing them until the juveniles tired. The story, to anyone familiar with chimps in the wild, attested only to the naïveté of the Merck officials. Jane Goodall, with commendable restraint, called the story "utterly fanciful." Teleki a little more accurately called it "pure malarkey."

Sadly closer to the truth was the claim by Merck that the chimps were doomed in the wild because of the escalating destruction of their habitat and that their best hope for survival lay in captivity. Merck also pointed out that restraint on the part of American corporations would simply give the vaccine to Polish or Japanese drug companies, which were already contracting for chimps (and whose governments are far less enlightened than that of the United States on endangered species issues).

So the dilemma seemed fairly clear—no chimps, no vaccine. How many chimps morally could be sacrificed to help cure a widespread but rarely fatal human ailment? If a species is already seriously threatened in nature, is that any justification for reducing populations still further to bring individuals into captivity? (This is an issue we'll return to in Chapter 9.) Our views at the time were summed up well by *Science* magazine writer Nicholas Wade: "The world has a growing population of 4 billion people and a dwindling population of some 50,000 chimpanzees. Since the vaccine seems unusually innocuous, and since the disease is only rarely fatal, it would perhaps be more just if the larger population could find some way of solving its problem that was not to the detriment of the smaller."[15] The permit was not granted. What would you have done?

Fortunately, it looks as though the chimps may not even be needed after all. Beechey's Ground Squirrels turn out to be susceptible to a very similar virus and may prove to be excellent animals on which to experiment with hepatitis B. Beechey's Ground Squirrel is common on the West Coast and is not endangered.[16]

In this case, what appeared to be a tough choice between human welfare and an endangered species turned out to be nothing of the sort. If hepatitis B were a really threatening disease, not controllable by public health measures, and if Chimpanzees were unquestionably the only suitable animal for making vaccines, there might be some justification for using them. (Of course, this would be an equally strong argument for preserving them very carefully.) But we know of no cases where such a situation exists. Other, unendangered organisms (like Beechey's Ground Squirrel) can be found to replace those that are threatened, and for any experimental animals, captive breeding could be used to a much fuller extent than it now is. Most of all, we think the need for using higher animals—especially endangered primates—in research generally ought to be more carefully evaluated beforehand.

Finally, among the benefits from other species in the health field, various organisms can serve as "miner's canaries" for humanity because of their sensitivity to various pollutants. Lichens have been used widely in this capacity. Curiously enough, snakes also seem to have considerable potential as indicators of environmental contamination.[17] Exterminating such organisms is a little like letting the batteries in your smoke alarm run down.

FOOD SOURCES

No greater benefit is received from other organisms than the nourishment *Homo sapiens* extracts from them. Most of our food today comes from domesticated plants and animals, but all of these, of course, were originally derived from wild species. And wild species still do make significant contributions to human diets and therefore also deserve protection.

Plants

At one time or another, human beings have used about three thousand species of plants (about 1 percent of the total existing plant species) for food. Only about a hundred and fifty or so species have been grown commercially to any extent, and the number of plant species that are really important in human nutrition is much smaller than that.[18] The principal feeding base of humanity is three species of grasses: rice, wheat, and corn. So important are these cereal grains that something over half of Earth's total cropland is devoted to growing them.[19]

Fewer than twenty other plant species serve as major sources of food for most of humanity: a few other grains such as millet and sorghum; legumes such as peas, beans, soybeans, and peanuts (which are all crucial sources of protein); root crops such as potatoes, yams, sweet potatoes, and cassava; sugarcane and sugar beets; coconuts and bananas. The U.S. National Academy of Sciences put it very succinctly: "These plants are the main bulwark between mankind and starvation. It is a very small bastion."[20] The narrowness of the range of food plants now used is not primarily due to a lack of other potential crops. But if the still unexploited species that have potential use as food are wiped out, humanity will be limited to what it is using now. The selection will hardly have been made rationally.

In tropical countries, which contain the world's great storehouse of plant diversity, the choices of species for cultivation and improvement were determined in no small part during the colonial era by consumer preferences in European countries. The needs—and the knowledge—of indigenous peoples were largely ignored.

The situation did not change significantly when the colonial era ended. There was a research and development investment in the crops already being grown, markets for those plant products were well established, and the newly liberated nations were not in a financial position to forgo the foreign exchange that would be lost in a period of experimentation and diversification. The administrators and scientists in the tropical countries had also been trained in temperate-zone countries at institutions such as the London School of Economics and had acquired the attitudes and tastes of the former colonial powers. They had no interest in searching for new tropical species to domesticate. Finally, the European influence was so pervasive that the populations of poor countries had begun to westernize their diets, so there was little indigenous demand for the traditional crops. For these reasons, the potential of tropical plant species as crops has barely been investigated, let alone exploited.

Meanwhile, the populations of tropical countries have grown extremely rapidly, especially since World War II, and in several regions are outstripping their ability to feed themselves. By the late 1970s, at least a half-billion people —one-eighth of the global population—were seriously malnourished.[21] And the majority of these chronically hungry people are in the tropics. One response to this problem will only make it worse in the long run. Expansion of agriculture in the effort to raise food production is a major cause of the disappearance of natural areas, and along with them go countless populations and species that might have high potential for development into crops.

The extinction of potential tropical crop plants is especially serious because transplantation of temperate-zone crops to the tropics is often not successful. For example, wheat does not grow well in the tropics, in part because one of its major diseases, the fungus causing wheat rust, thrives in warm, humid climates. One of the three great cereal crops is therefore excluded from some of the hungriest parts of the world. In these regions, already poised on the brink of famine, humanity can ill afford *not* to seek new sources of potential food crops. A fair number of promising species are already known and should be developed; systematic exploration for others is clearly needed.

Let us look at some of the *known* possibilities, keeping in mind that they are only a small part of the potential, though that potential is shrinking as tropical plant species and populations succumb to human activities. Note also that what are minor crops today may lose any possibility of being transformed into major crops because of population extinctions. The genetic raw material required for their development can be lost through the extermination of populations of the crop itself or of its wild relatives.

One group of tropical plants that holds promise as a source of staple crops includes several Central American species of the genus *Amaranthus* (in the cockscomb family), whose seeds have a very high quality protein. (Most plant proteins are of relatively low quality compared to animal sources.) *Amaranthus* leaves are also protein-rich and are widely eaten as a sort of tropical spinach. Grain amaranths were once widely cultivated in Central and South America, but they were largely displaced by corn, which is nutritionally inferior. Among other reasons, the cultivation of one amaranth species was suppressed by the Spanish Church because the Aztecs had centered some of their religious ceremonies on it. These species of amaranths were considered of high potential by a National Academy of Sciences committee studying utilization of tropical plants. There are some eight hundred species in the family, many growing in tropical Latin America and Africa,[22] but few of them have been explored for development as crops.

Numerous other examples can be cited. Quinoa, a plant of the spinach family, has been a staple grain in the high Andes since the time of the Incas. In many areas it was replaced by barley, which is less nutritious, after the

Spaniards came. Closely related species are also cultivated at high altitudes. Although growing in poor "tropical" countries like Peru, Bolivia, and Ecuador, these species are anything but tropical in the usual sense, since they grow in cool, montane climates. They hold promise, however, of improving the diets of impoverished mountain peoples everywhere.

The fifteen hundred species of the spinach family also show a particular ability to grow in salt-rich soils. Such soils are common in desert and semidesert areas and are becoming more common because irrigation usually causes buildups of salts in soils, often eventually forcing farmland to be abandoned. Again, a research investment in this group of plants might yield high dividends.

There is even a group of plants called eelgrasses, which grow entirely under the sea and which could become an important substitute for traditional grains in densely populated seacoast regions. The Seri Indians of the west coast of Mexico have long made flour from seeds of one of the eighteen species in the group. Here is a potential crop that would need no fresh water, pesticides, or fertilizers.[23] And in Australia one species of wild true grass (a plant in the same family as the traditional grains) shows great potential for development into an important forage and grain crop for arid regions.[24]

The story of grains is repeated among vegetables, legumes, roots and tubers, fruits, oilseeds, and forage crops. Unexploited plants with potential value as these kinds of crops are already known, even though the investment in searching for and developing such new crops has been minimal. That investment is especially small compared to the magnitude of need embodied in the world food problem.

It is important to note that the potential value of the ancestors of today's highly domesticated crops would in most cases be far from obvious to the untrained eye. For instance, in 1978 in the hills of Galilee, Israeli geneticist Eviatar Nevo showed us wild plants that were the ancestors of wheat, barley, and rye. There we were in the homeland of crops since spread around the world. But these were no amber waves of grain, just some unimpressive, scruffy grasses, which are pitifully unproductive compared to cultivated crop plants. Selective breeding by countless generations of farmers and, in this century, by trained agronomists has brought the crop varieties to their current levels of productivity. But in Israel and throughout the Middle East, these unprepossessing "weeds" are disappearing rapidly under the spread of urbanization, agriculture, grazing, and weed control. Their potential as breeding stock to improve cultivated crops is thus being lost forever.

The same is happening everywhere both with wild relatives of now-exploited crops and with wild plant species whose possible value as future crops is unknown. Among the millions of populations and species of plants now threatened with extinction there undoubtedly are many Cinderella plants potentially equivalent to the ancestors of wheat and barley, but they may be doomed to disappear without ever making it to the ball.

The human food situation, precarious already, is being made even more vulnerable to disastrous famine in two ways. First, the human population is dependent on a narrow range of crop species, and the possibilities for expanding that range are being foreclosed by extinctions. And second, society's crop species are themselves endangered by erosion of their genetic diversity and the disappearance of their wild relatives.

The Benefits of Genetic Variability

An essential ingredient of successful crop breeding, whether to improve an existing crop or to develop a new one, is genetic variability. In order to preserve this variability in nature, not only species but *populations within species* must be saved from extinction. Wild-growing plants are perpetually evolving new ways of fending off the animals and microorganisms that attack them. In turn, the pests are always attempting to evolve new ways to neutralize the plant defenses.[25] The price for each to remain a contestant in the "coevolutionary race" is to have available the genetic variability on which natural selection can act.

A domesticated plant must remain in the race also. To some extent, farmers can help defend crops with various programs of pest control, but their inherent resistance remains a central factor. The life of a new cultivated wheat variety in the American Northwest is about five years. Then rusts (fungi) adapt to the strain, and a new resistant one must be developed.[26] That development is done through artificial selection: the plant breeder carefully combines genetic types that show promise of giving resistance.

The entire enterprise of high-yield agriculture, upon which the future of civilization is utterly dependent, rests on having an adequate supply of genetic types to use in artificial selection. But unfortunately, that supply is rapidly dwindling away; as biologists would say, there is a continuing decay of the genetic variability of crops. This is possibly the most serious single environmental problem that *Homo sapiens* faces; certainly it is the least widely understood. And it is intimately related to the problems of extinction.

The very success of high-yield agriculture is a major factor behind the loss of genetic variability. Farmers quite naturally tend to plant the crops that will give them the highest yields. As a result, in the United States, for example, something like 70 percent of the acreage of the important crops is devoted to growing only a few major varieties of each crop.[27]

Sometimes a high price is paid for the planting of such extensive "monocultures"—stands of genetically almost identical crop plants. For example, in

1970 a new genetic strain of fungal blight overwhelmed the defenses of much of the corn crop in the United States. About 80 percent of the American corn crop that year was planted in strains that were especially susceptible to the disease, and the weather that year was unusually damp, which also encouraged the blight. Almost a fifth of the nation's crop was destroyed. In some areas virtually all of the corn was lost to the blight. Fortunately, the genetic resources were still available to permit development of new, more resistant strains, but unless care is taken, humanity may be out of luck the next time a fungus evolves its way around a crop's resistance.

One reason that decay of genetic variability is an increasingly serious problem is the Green Revolution. This is shorthand for a widespread attempt to modernize agriculture in poor countries in order to try to keep their food production up with their exploding populations. Among the key elements to the Green Revolution is the distribution of so-called miracle varieties of rice and wheat—strains that (if properly handled with appropriate inputs of water, fertilizer, and pesticides) are capable of producing much higher yields (more food per acre) than traditional varieties.

Naturally, farmers in poor countries, just like their counterparts in rich countries, are happy to increase their yields.[28] As a result, a few miracle varieties are replacing numerous traditional strains; genetic uniformity is taking over from genetic diversity; unique populations are going extinct. Geneticist Reuben Olembo of the United Nations' Environmental Program expressed the problem vividly: "When farmers clear a field of primitive grass varieties, they throw away the key to our future."[29]

A second potential source of genetic variability for crops is the array of wild strains and species closely related to the domestic ones. Wild relatives of cultivated grains—some inconspicuous grasses like those we saw in Israel— have already been used to introduce new and desirable characteristics into their domestic cousins. A Turkish wild wheat bred with American strains conferred on them resistance to a series of wheat diseases known as "bunts." The value to American agriculture is estimated at $50 million for the useful life of this single genetic improvement alone.

A wild relative of corn recently discovered in Mexico is perennial—that is, it lives as a plant year after year rather than surviving between growing seasons only as seeds. If through hybridization the perennial characteristic could be transferred to the crop, farmers would no longer have to plow their cornfields and sow the crop anew each year. The savings in dollars would be enormous, and the reduction of soil erosion would be tremendously important.[30]

Unhappily, population pressures, the spread of cultivation, urban sprawl, and habitat destruction in general are increasingly exterminating populations of wild and semi-wild ancestors of crop plants. In recent decades, famines in Nigeria, Ethiopia, and the Sahel have led to the loss of much diversity among natural populations of edible plants as people and their animals ate virtually everything growing. The filling of Lake Nasser in the 1970s behind the Aswan High Dam drowned out strains of grains used as feed for domestic animals, which are impossible to replace.

Many of the wild species from which important crops are derived are "weedy"—that is, they normally grow in disturbed habitats such as landslide areas or streambanks. Before agriculture was developed, these plants were natural denizens of the disturbed sites of campgrounds and early settlements. The more desirable ones came to human attention, were cultivated, and were gradually improved through selective breeding over centuries. With the advance of civilization, some of their ancestors and relatives persisted only in areas given protection for their historical interest. As a result, even the cleaning up of ancient ruins in Mediterranean tourist areas has caused serious losses of plant germ plasm. Priceless—and irreplaceable—populations go under today as people "police up" tourist attractions.

Food Sources: Animals

Human beings have domesticated far fewer species of animals than of plants —only a couple of dozen or so. Virtually 100 percent of the protein from domesticated animals consumed by human beings comes from just nine species: cattle, pigs, sheep, goats, water buffalo, chickens, ducks, geese, and turkeys. Poultry, beef, and pork in about equal amounts add up to some 90 percent of meat production. A few other animals, such as rabbits and pigeons, contribute a statistically negligible amount. Cows yield about 90 percent of the milk and milk products people consume, with water buffalo, goats, sheep, and an occasional Reindeer chipping in the rest.

Wild animal species, unlike wild plant species, do, however, make significant contributions to human nutrition. The biggest contribution by far, of course, is made by marine fishes. Since 1971 the world fisheries catch has fluctuated around 70 to 75 million tons—quite possibly the maximum that will be achieved. As a result, with the world's human population growing rapidly, the per capita share of the catch is dropping apace. Nonetheless, the ocean's fish species supply directly something like 14 percent of the animal protein in the diet of *Homo sapiens* on the average worldwide, and they make a further indirect contribution through their presence in animal feeds. In some countries they are a major source of animal protein. Not included in world statistics are the fish caught by millions of anglers whose catch is consumed by themselves and their families.

Whales are also eaten, especially in Japan, but they have been hunted primarily for other purposes. Also included in the fisheries harvests are crabs, lobsters, oysters, shrimp, clams, mussels, and so on, which represent a commercially important harvest, although their total nutritional contributions are negligible measured against needs. These are mainly luxury foods, gracing the tables of the already well fed.

Wild mammals, birds, and sometimes reptiles are also a source of food around the world for anyone willing to hunt them. Hunting is a popular form of recreation in North America, Western Europe, New Zealand, and Australia. In the late 1970s, more than 25 million hunting licenses were issued each year in the United States.[31] Still, the contribution of game animals to diets in these countries is very small. In developing countries, this is not necessarily the case. Especially in areas where forests have not been destroyed, hunted animals probably make a significant, if unknown, contribution to the diets of the poor. This is one of the seldom counted benefits to local people of preserving forests and woodlots—at the same time the game animals they harbor are preserved.

Hunting, however, is a two-edged sword. The reaping of the benefits always carries with it the danger of forcing populations and species to extinction. Hunting, as we shall see, has been a major factor in historic extinctions and is threatening many species and populations today. We must emphasize that here, as in other areas, the price of getting direct benefits from other species usually carries the cost of husbanding them.

As with plants, the possibilities for domesticating animals have hardly been scratched. For instance, aquaculture, the domestication of marine and freshwater animals, in theory has enormous potential. Some of that potential has already been realized; more than six million tons of shellfish and fishes were produced that way in 1975. Production from aquaculture doubled between 1970 and 1975 and has continued to grow rapidly. Israel now gets half of its fish from aquaculture, and it is estimated that 40 percent of all the fish and shellfish produced in China are obtained that way. Aquaculture provides 20 percent of Indonesia's fish and 6 percent of Japan's.[32]

Far Eastern nations have traditionally practiced fish culture in their flooded rice fields. The rice is fertilized by fish droppings, the small fish gobble up mosquito wrigglers and other insect pests, and the farmers get a protein supplement to their diets from eating the fishes. Among the fishes most successfully raised for food in ponds are African fishes of the genus *Tilapia*. These are cichlid fishes, related to the freshwater angelfishes familiar to tropical fish fanciers.

Tilapia yields of almost a thousand pounds per acre annually have been obtained in some African localities, but the potential of most cichlid fishes for domestication has not been investigated. Species "swarms" of cichlids occur in the great African lakes. Lake Tanganyika has a hundred and twenty-six species, all unique to that lake. Lake Victoria has more than a hundred and seventy cichlid species, all but six found nowhere else. Lake Malawi has more than two hundred species, all save four found only in that lake. The various species in the swarms differ from one another in their diets and breeding behavior.[33]

Considering the diversity of their feeding habits, it is not unlikely that an aquaculture system using combinations of lake cichlids could be designed that would have much higher protein yields per acre than any cichlid monoculture. The different species could divide up the resources of a pond much as they now divide up the food resources of the lakes, efficiently utilizing many that would not be accessible to a single species. But the basic requirement for being able to find, assemble, test, and domesticate such combinations, of course, is the preservation of the cichlid species in nature. Similarly, a prerequisite for being able to domesticate many other salt- and freshwater organisms for aquaculture is species conservation.

The potential for domesticating more wild animal species is not restricted to aquatic habitats. Many people have recognized that domesticating or semidomesticating African grazing animals could produce higher yields of meat with fewer environmental problems than are now caused by raising cattle, which are poorly suited to most African environments. It has been suggested that elands (which have already been successfully domesticated and which give highly nutritious milk), Wildebeests, and smaller antelope such as the Uganda Kob or Thomson's Gazelle could be herded where cattle are unsatisfactory.[34] David Hopcraft has established an experimental ranch in Kenya, where he is attempting to demonstrate that superior yields can be obtained from using the native animals that are physiologically adapted to the semi-arid savannah. His ranch now has fourteen species of gazelles on it, and it will soon be clear whether the biological, economic, and political problems of making it a success can be solved.[35]

More complex, and perhaps more satisfactory ecologically, would be mixed herds that could take advantage of a broader spectrum of natural vegetation. The makeup of the herds could be carefully designed for different areas. For instance, in semi-arid East Africa, much of which is not well suited for raising crops, Giraffes, elands, and hartebeest could be herded together, since they browse at high, middle, and low levels respectively on trees and shrubs. They might further be combined with oryx, which graze in open grassland, and kudu, which like thick woodland, and so forth. A different array of animals —elephants, White Rhinos, hippos, buffalo, Warthog, and two species of antelopes—would be a suitable mix for areas around rivers or with high rainfall.[36] Using these large mammals would be excellent way to harness the rich productivity of moist areas in tropical Africa. The antelopes, buffaloes, and warthogs could be a source of food; the other animals would be economically useful as tourist attractions in game parks, and if necessary could be cropped for food. But in order to use these species, humanity must first ensure their survival.

OTHER SPECIES AND BIOLOGICAL CONTROL

Great care must be taken whenever people intervene in nature. Moving animals and plants around may do a lot of good (as in the case of wheat and *Tilapia*). It can also do a lot of harm. A great deal of care and biological knowledge are necessary in order to make transfers prudently. Otherwise an undesirable population explosion may result when a species is moved away from its natural enemies. For example, *Opuntia* (prickly pear) cactus was introduced into Australia from the Americas in the last century by early settlers, who perhaps thought it could be used as livestock fences. It fenced out livestock all right—it spread all over Queensland, covering vast areas that could have been used for grazing. Almost 100,000 square miles were affected, and about half of that was so densely infested it could not be penetrated by man or beast.[37]

Entomologists went to South America and looked for natural enemies of the cactus. One they found, a small cactus moth, was introduced into Australia. It quickly multiplied on the abundant food plant and ate the *Opuntia* down to the ground, soon restricting it to small scattered clumps. It is no longer a pest; the moth stands perpetual guard at no cost to the consumers of meat and wool.

There are many similar stories in the annals of the biological control of pests. Among the benefits that humanity has repeatedly extracted from wild species are their services in combating organisms that are doing economic harm. An Australian scale insect that got into California and began to devour citrus crops was controlled by importing two natural enemies of the scale, a beetle and a fly, from Australia. Introductions of parasitic insects have been successful against such diverse pests as white grubs and Sugarcane Weevils in Hawaii, Gypsy and Browntail moths and Alfalfa Weevils in the United States, and a rhinoceros beetle in Mauritius.[38] Myxomatosis, a virus disease native to South American rabbits, has been used with great success to control rabbit plagues in Australia and Europe.

As anyone who has been there can testify, one of the most annoying pests in Australia is the bush fly. A TV show or movie genuinely shot on the island continent can always be identified by the actors continually swatting and blowing at flies in the outdoor scenes. Bush flies breed in dung, and theoretically they hang around animals (including people) because they are the sources of dung (they apparently haven't wised up to modern people's crypto-defecatory habits).

There are some 20 million cattle in Australia, a non-native population half again as large as the human population. The cattle produce a dung bonanza and a bush-fly paradise. As a result, a novel biological control program is now underway. Some 55 dung-beetle species, mostly from Africa, were imported into Australia and released in the 1970s. The imported beetles are specialists in the type of wet droppings produced by cattle. Native Australian beetles were adapted to the dry dung of kangaroos and wombats and could not dispose of cattle droppings before the bush flies reproduced in them. The hope was that the introduced beetles would bury much of the dung before the bush flies could mature—helping to suppress the plague. Early results seem promising.

The possibilities of using other species to help people out of difficulties with pests seem almost infinite. Some of the most diverse and least known groups of insects, for example, are several families of tiny parasitic wasps that attack other insects. Here the potential value is obvious, but the dung-beetle case ought to give anyone pause at making snap judgments about the potential value of *any* other species, no matter how apparently strange or obscure. The bug that is threatened today might be tomorrow's vital ally.

ADDITIONAL PRODUCTS FROM OTHER SPECIES

Finally, let us look briefly at some of the multitudinous ways in which other organisms produce or help to produce valuable products. Wood comes to mind immediately—the gift of a wide array of large plants. Although all trees produce wood, the woods of different species have very different characteristics. One would not want the spars of an airplane wing made of white pine, or a piano sounding board made of balsa.

Wood was at one time humanity's principal fuel, and in most poor nations it remains an essential source of energy. Some 90 percent of the wood used as fuel is consumed in less developed countries, and 80 percent of the wood used in those areas is burned for fuel. The harvesting of firewood thus constitutes a major reason for the destruction of tropical forests.[39]

There is, of course, also exploitation of tropical as well as temperate forests for timber. Because of a lack of appropriate husbandry, some tree species with the most desirable woods have been reduced or exterminated over wide areas. In the western United States, redwood forests have been substantially depleted. Mahogany is extinct in Honduras and in much of the lowlands of Panama.[40] And teak is being rapidly reduced in Southeast Asia.

Two factors make essential the widest possible preservation as wood sources of tree species and populations. One is that even within a tree species, as forest geneticists are well aware, different populations may have diverse growth characteristics and wood quality. Of course, countless trees, especially in the tropics, have never been investigated as wood sources even at the species level. Many tropical tree species that have woods with desirable properties thus may not yet have been discovered.

The other factor is the continual evolution of ideas about what kinds of woods are desirable. A few centuries ago, British naval surveyors valued oaks with low, spreading crowns from which pieces of the proper shape for crucial parts of wooden ships could be obtained. That shape of oak tree is exactly wrong for today's uses of oak. Pulp and paper manufacturing interests have had an evolving view of the best kinds of wood to use, and in recent decades a whole array of new pulp species has been adopted. The future will probably bring further dramatic changes in the uses of trees. For example, a major future energy source may well be biomass—plants grown to be burned in power plants or converted to other fuels. Many suitable species for biomass operations have no commercial use at present.

Trees and other plants, of course, are sources today of an enormous variety of inedible products in addition to wood. An outstanding example is rubber, the natural form of which is made principally from the sap of a tropical tree *(Hevea)* in the spurge family. Other plant products include tanning agents, a wide variety of dyes, fibers such as cotton, flax, and hemp, insecticides such as pyrethrum and rotenone, perfumes, lotions (e.g., witch hazel and aloe), waxes, gums, cosmetics, kapok, meat tenderizers, preservatives, gutta-percha (used for insulation and waterproofing), turpentine, fat for candles, soap substitutes, aromatic resins used for making incense, fertilizers, packing materials, brooms, baskets, thatching, matting, and rattan furniture.

Many plants yield valuable oils. Some, such as those from safflower, soybeans, peanuts, and olives, are used in cooking. Linseed, soybean, tung, and flax oils are used in paints and varnishes. Castor oil is extracted from the poisonous seeds of the castor plant. In addition to its well-known use as a laxative, it also is a fine lubricant. It was commonly used in the engines of World War I fighter planes and caused uncomfortable problems for the pilots, who in their open cockpits would occasionally take in too much of the oil sprayed out by the engine! Many plant substances are also used as ingredients in the manufacture of industrial products. For example, camphor is used in making plastics, lacquers, films, and explosives.

Plants are the source of all spices, herbs, and most other flavorings. It is hard to remember when enjoying a French or Mexican meal that many of the flavors one is enjoying come from chemicals evolved by plants to repel or kill their enemies. Plants also provide us with stimulating drinks such as coffee and tea and are the basis from which alcoholic beverages are fermented or distilled. Both plants (flowers and leaves) and animals (musk) are used in making perfumes and scents for many products such as detergents, soaps, shampoos, and deodorants. Finally, plants supply other economic values because people will pay considerable sums to possess their beauty.

At the risk of sounding like a broken record, we reiterate that the potential for yielding these sorts of products from plants barely appears to have been tapped. But that potential is lost when plants are forced to extinction. What is available "out there" is indicated by recent investigations of the potential of three previously obscure plants—Guayule, Jojoba, and *Leucaena.*

Guayule is a shrub that was widely grown in northern Mexico and Texas as a source of rubber early in this century. Its latex is essentially identical to that from *Hevea* trees, and in 1910 it supplied 10 percent of the world supply and half of that in the United States. The enormous upheaval of the Mexican Revolution, overexploitation of the crop, and the great depression of the 1930s all conspired to bankrupt the industry, though cultivation was temporarily resumed in California during World War II, when imports of *Hevea* rubber from Southeast Asia were cut off.

Guayule has numerous attractive qualities. It lends itself easily to genetic improvement and grows satisfactorily in poor soils in arid regions; hence it could be an important crop for Native Americans in the southwestern United States and northwestern Mexico. The plant also lives for up to fifty years, so established stands represent a sort of living stockpile of rubber. There are some questions about whether appropriately pure rubber can be extracted economically today, but pilot plant operations in Mexico have been encouraging, and increased demand (and prices) will probably stimulate solutions to the problem.[41]

Demand for Guayule will surely increase because natural rubber is superior to synthetic rubber in many uses—it is more resilient and resistant to heat, for example.

Perhaps even more important, the synthetics are made from petroleum, an increasingly scarce and costly resource. Since *Hevea* trees appear to have about reached the peak of possible genetic improvement and are threatened by a disease that wiped out rubber production in South America, the search for other plant species that yield natural rubber carries a certain urgency.

Jojoba, a shrub related to boxwood, which is often used as a hedge, produces seeds containing up to 60 percent of a liquid wax almost identical to sperm whale oil, which has a great ability to "wet" metals—that is, to cling to them and thus provide continuous lubrication. This makes it a highly desirable component of lubricating oils in particular, though it also has a variety of other uses.

Sperm oil can no longer be imported into the United States because Sperm Whales are endangered. Sadly, though, it is available on a black market at more than twice the world market price—and that price seems doomed to escalate as the whales are either protected or exterminated. Fortunately, Jojoba oil makes an excellent substitute for sperm oil in virtually all the latter's major uses. It also has additional attractive characteristics. Jojoba oil can be processed into a wax that is superior to the best now available. And once the oil has been extracted, the seed residue has a high protein content and could be developed into an animal feed.[42] Like Guayule, Jojoba grows in now marginal, arid lands, and its cultivation would not seriously compete with other agricultural products.

Leucaena is a group of species of related shrubs and trees in the pea family, which originated in Central America. A U.S. National Academy of Sciences study reported:

> Of all tropical legumes, leucaena probably offers the widest assortment of uses. Through its many varieties, leucaena can produce nutritious forage, firewood, timber, and rich organic fertilizer. Its diverse uses include revegetating tropical hillslopes and providing windbreaks, firebreaks, shade, and ornamentation. Although individual leucaena trees have yielded extraordinary amounts of wood—indeed, among the highest annual totals ever recorded—and although the plant is responsible for some of the highest weight gains measured in cattle feeding on forage, it remains a neglected crop, its full potential unrealized.[43]

Some *Leucaena* strains have dense woods and are fast-growing (harvestable every three to ten years), making them prime candidates for growing in biomass plantations. *Leucaena* plantations could be used in conjunction with windmill "farms" to supply much of Hawaii's future energy needs. Like Guayule and Jojoba, *Leucaena* does not require high-quality farmland; it is tolerant of an extraordinary range of tropical and subtropical environments.

Biomass is not the only route by which plants may help humanity with the mobilization of energy. Some tropical plant species of the spurge family produce a hydrocarbon that shows promise of providing a renewable source of low-sulfur petroleum. With appropriate genetic improvement, one day "gasoline farms" may be supplying a significant fraction of the world's portable fuels.[44] Although this may seem strange at first glance, one need only recall that *all* fossil fuels were originally plant hydrocarbons. They were "processed" by geological action over millions of years into coal, oil, and natural gas. There is no theoretical reason that *Homo sapiens* cannot short-circuit the plant-to-gasoline sequence by starting with plants especially rich in desirable hydrocarbons. No reason, except of course the possibility that the very best plant sources may be unrecognized species and populations that are being pushed toward extinction at this very moment.

Although not remotely as rich a source as plants, animals supply a wide range of materials used in commerce, including shellac from lac insects, musk from deer, wool from sheep, and glue from rendered horses. Wool and other animal fleeces (such as camel hair, cashmere, llama, and vicuña) have qualities that have proven very difficult or impossible to reproduce in synthetics. The same is true for silk and leathers. Plastics are satisfactory for luggage and handbags, but attempts to make plastic shoes have been relatively disastrous.

Down, the fine feathers underlying the main feathers of geese and other birds, is perhaps the lightest and best insulator known—witness the current demand for down jackets, vests, and sleeping bags.

Some, though certainly not all of these products, are obtained from domesticated animals, many of which are also raised for food or other purposes. These animals obviously are not endangered, and their exploitation is controlled. When wild animals are used—Sperm Whales for their oil, rhinoceri for their horns, elephants for ivory, deer for musk, and so forth—the market for their products becomes a leading cause of extinction.

While the prospects for exploiting new animal species in industrial processes may be substantial, we doubt if it is great enough to serve as a strong argument for the preservation of animal diversity in general. (Arguing on the basis of the economic value of a product to preserve a particular species, such as the endangered Vicuña, is another matter, of course.) Reasons for preserving animals are more readily found elsewhere. The principal loss to industry that will result from the reduction in the number of living species will be from the depletion of the incredible stores of different organic chemicals found in plants.

THE OPPORTUNITY COSTS OF EXTINCTION

Humanity thus takes direct advantage of many other species at a level that is largely unrecognized by the vast majority of people, especially those in rich countries; and the potential for further use, also unrecognized, clearly is enormous. The most important direct costs of today's trend toward extinction of more and more species are what economists would call opportunity costs. They are the difference between the advantages gained in the process of extermination and the value of the species or population in its best use if preserved.

The opportunity cost of wiping out the whales, in our view, would be their value as sources of interest and recreation for future generations. For others it might be their value as a perpetual supply of dog food and other products if they were harvested on a sustainable yield basis.

Presumably, the decision to wipe out or not wipe out any species will be made "rationally" (in economic terms) when the living resource has a known value. But at the moment, the opportunity costs of wiping out thousands of species, one of which—unbeknownst to humanity—may hold the secret to curing cancer or to alleviating the world food problem, are not easily measured by standard economic calculations.

We have barely touched here on the myriad direct benefits that humanity derives from other living things. From the very beginning, human beings have depended on other organisms for all sorts of products from shoes to weapons and tools, for such elemental necessities as food, clothing, fuel, and housing materials. And people still depend on living things for most such necessities, even though they have "domesticated"—that is, harnessed and adapted for their own use—many species in order to control their production. Domesticated species are, however, still subject to nature's rules, as are human beings themselves. And humanity, together with its artificial biological systems (agriculture and forestry) is supported by the larger system, known as the *biosphere* —nature, if you like—in many indirect ways as well. While most people are completely unaware of these subtler aspects of our dependency on natural ecosystems, they are no less vital for being comparatively inconspicuous.

Chapter V

Indirect Benefits:
Life-Support Systems

Man depends on wildlife for survival, and wildlife depends
equally on man. The two must find means for living
together on planet earth or there will be no life on earth.

—RAYMOND R. DASMANN,
"Wildlife and Ecosystems,"
in *Wildlife and America*

To return to the spaceship analogy with which we opened this book,
rivet-popping—forcing other species to extinction—costs humanity
dearly because of the potential value of individual rivets. The greatest concern
about the driving of populations and species to extinction, though, must be
focused on the danger of structural weakening of Spaceship Earth.

Natural ecological systems (ecosystems) are the outcome of billions of years
of evolution. Human beings too are a product of that ongoing process and, like
every other organism, are dependent on those systems for support. Most
people seem to think they have become independent of nature, but they have
not. Domesticated crops and animals are also evolutionary products—though
humanity has acted as the dominant selective agent in their recent evolution.
The ability of human beings to obtain raw materials, manage natural resources,
and control various aspects of their environment, impressive as it may be, in
no way frees them from dependence on the biosphere.

Natural ecosystems support human life through an array of absolutely
essential, free public services. Once we look closely at the nature of those
services and the systems that provide them, and at the roles that individual
species and populations play within ecosystems, then it should become clear
why any enemy of the Snail Darter is an enemy of yours and ours, too.

ECOSYSTEM STRUCTURE

Technically, an ecosystem consists of *all* the organisms—plants, animals, and microbes—that live in an area, combined with their physical environment. All forms of life, of course, are modified and constrained by their physical environment. Some modifications take place within a single generation—that is, the *development* of an individual organism is affected by its physical environment. For example, a tree will be stunted if it is planted in relatively infertile soil.

Organisms may also be affected over many generations by having their hereditary characteristics changed by natural selection—the process of *evolution*. For example, if the climate in an area grows colder, those individuals of each species that happen to be the most resistant to cold will tend to leave more offspring than those that are less resistant to cold. They may be less likely to freeze to death or more able to move around in search of food or mates or to find shelter. If individual-to-individual variation in cold-hardiness is primarily due to genetic differences, then the whole population will gradually become more cold-resistant. The abilities of individuals to develop differently in different environments and of populations to evolve in response to environmental change are of course essential for the survival of life on this inconstant planet.

While organisms are affected by their physical environment, obviously the reverse is also true. Lichens help break down rocks to produce soils; the roots of trees and grasses prevent soils from being eroded away; forests moderate their own climate; and earthworms stir up the soil and help keep it fertile. The organisms of an ecosystem and the physical parts of the system are bound together by a maze of interactions. The maze is so complex that it is not altogether unreasonable simply to say that *every living thing potentially affects every other living thing and the physical environment of this planet.*

Many of the paths of influence are extremely circuitous.[1] For example, the major influences of animals on physical factors in the environment are probably due to their impact on plants. Herbivores (plant-eating animals) such as mice may denude an area of vegetation and cause serious erosion unless their population is kept under control by a predator—such as a hawk. So the hawk may be influencing the physical environment through a chain that includes mice and plants. If a louse is important in spreading a disease from one hawk to another, it too may influence the erosion of a hillside.

A famous incident drives home the intricate relationships within ecosystems. Some years ago, large quantities of DDT were used by the World Health Organization in a program of mosquito control in Borneo. Soon the local people, spared a mosquito plague, began to suffer a plague of caterpillars, which devoured the thatched roofs of their houses, causing them to fall in. The habits of the caterpillars limited their exposure to DDT, but predatory wasps that had formerly controlled the caterpillars were devastated.

Further spraying was done indoors to get rid of houseflies. The local gecko lizards, which previously had controlled the flies, continued to gobble their corpses—now full of DDT. As a result, the geckos were poisoned, and the dying geckos were caught and eaten by house cats. The cats received massive doses of DDT, which had been concentrated as it passed from fly to gecko to cat, and the cats died. This led to another plague, now of rats. They not only devoured the people's food but also threatened them with yet another plague —this time the genuine article, bubonic plague. The government of Borneo became so concerned that cats were parachuted into the area in an attempt to restore the balance![2]

Food Chains and Webs

In spite of such complicated connections, a great deal is known about the general structure and functioning of ecosystems. They are arranged in a more or less pyramidal way, with masses of plants at the bottom supporting the whole structure, and relatively small numbers of flesh-eating organisms at the top. The arrangement is dominated by feeding sequences—say, plant to mouse to hawk to hawk louse—called *food chains.* If the plants that support the food chains are destroyed, the whole sequence will collapse. If an animal further up the chain is destroyed, that may cause population explosions below (knock off the hawks, and the mice will multiply out of control), with dominolike effects culminating in disaster.

The whole business is further complicated because the chains are intertwined with each other in *food webs,* and all the living components are influenced by their physical environments. But if you keep the basic feeding sequence characteristic of ecosystems in mind as we get into more detail, the complexity should not obscure the basic attributes of ecosystems.

With trivial exceptions, all of the energy that runs ecosystems comes from the sun. Solar energy is "captured" by green plants in the complex process of photosynthesis. That is why they are the base on which the pyramid rests. In this process, the radiant energy of the sun is used to convert carbon dioxide and water into sugars, with oxygen given off as a by-product. Both plants and animals then use the oxygen to "burn" the sugars slowly, and the energy thereby released enables them to run their life processes.

Plants that can carry on photosynthesis are normally green because they contain a green pigment known as chlorophyll, which plays a crucial role in the photosynthetic process: it is the chemical that captures the light energy. Since without that energy source, life and ecosystems would not exist, it is not surprising that ecologists view green plants as the most fundamental component of ecological systems. All flesh is indeed grass—because all animals are dependent for sustenance on the plants they eat, or on other animals that have fed on plants, or on other animals that eat plant-eating animals.

A common way of looking at the structure of ecosystems is to divide their food chains into *trophic* (feeding) levels. The first level is that of the *producers,* green plants. The second is that of the *herbivores,* animals that eat plants. At the third level are *carnivores,* animals that eat other animals. There may be several levels of carnivores. For example, the lice feeding on the hawks would be second-order carnivores. Or the hawk would be a second-order carnivore if it ate a smaller, insect-eating bird or a snake.

A very important trophic level that is connected to all the others is that of the *decomposers.* Decomposers include the myriad small insects, mites, worms, fungi, and bacteria that make their living by breaking down the waste products and dead bodies of larger organisms. In the process of extracting energy for themselves from the organic molecules formerly present in living trees, mice, whales, and your great-grandparents, they perform the crucial role of freeing up nutrient molecules essential for both plant and animal growth and returning them to the ecosystem. In terrestrial ecosystems, most of the decomposers live in the soil.

Ecosystem Functions: Cycles of Nutrients

Neither we nor any other organisms can live by sugars alone. All organisms must have access to a wide range of elements; among the more important are carbon, hydrogen, oxygen, nitrogen, phosphorus, potassium, sulfur, iron, calcium, magnesium, copper, manganese, molybdenum, boron, and zinc. As with sugars, the ultimate sources of essential elements for all animals and decomposers are the plants at the pyramid's base.

Essential nutrients tend to move in cyclical pathways through ecosystems. To take a highly simplified example, phosphorus, which is critical to the ability of all organisms to make use of energy, is taken up through the roots of plants and passed on to the animals that eat the plants. Carnivores, in turn, obtain their phosphorus from herbivores. Decomposers in soil break down organic molecules containing phosphorus and once again make the phosphorus available for plants to pick up through their roots. In outline, if not in detail, the phosphorus cycle resembles those of other nutrients.

The ways in which nutrients cycle in ecosystems can be extremely complex. The most intricate is that of nitrogen, which as a necessary component of protein is essential to all living beings. There is a huge pool of nitrogen in the atmosphere, but it cannot be used directly by higher organisms. Certain specialized microbes (mainly blue-green algae and a few groups of bacteria) are able to "fix" atmospheric nitrogen—that is, to convert it to forms in which it can be used by other organisms. The best known of the bacteria live in nodules on the roots of plants of the legume family. The legumes include peas, beans, peanuts, clover, and alfalfa. Thus these valuable plants not only provide protein-rich plant foods for human beings and forage for their animals, but they also, with their microbial companions, replenish soil with vital nitrogen. Nitrogen fixation takes place in aquatic habitats, too, mainly by the action of blue-green algae. Many of these are free-living, but an important one lives in symbiosis with a water fern in rice paddies.

Once the nitrogen has been fixed in soil or water, it is taken up by plants and passed on through many different living pathways. Eventually, in the process of decomposition, another group of microorganisms returns some of the nitrogen to the atmospheric pool.

Energy

While nutrients tend to cycle in ecosystems, energy moves through them in a one-way flow. Physicists long ago learned that energy's capacity to do work can be used but once. In the real world, moreover, part of the theoretical capacity to do work is always wasted. It is impossible, for instance, to convert all the kilocalories of chemical energy in a gallon of gasoline into the same number of kilocalories of the energy of motion in the drive shaft of an automobile. Indeed, in typical automobile engines, about 70 percent of those kilocalories either remain in the exhaust or are converted into heat, which dissipates from the engine and becomes unavailable to turn the drive shaft (that is, the engine is only about 30 percent efficient). This trend toward the unavailability of energy to do work is one of the many phenomena that physicists describe as a consequence of the second law of thermodynamics.[3]

Living organisms use the work derived from the energy they process to help them capture more energy either from the sun or from other organisms, to build tissue for growth and repairs, to generate offspring, and to protect themselves from enemies. To all of these processes, the laws of thermodynamics apply just as inexorably as they do to automobile engines. Sometimes it is a distinct blessing that no real-world processes are 100 percent efficient. For instance, it is impossible for you to convert the 2,300 kilocalories in the chemical bonds of a pound of chocolate into 2,300 kilocalories in the chemical bonds of fat in your body!

It is the loss of availability of energy to do work that dictates the one-way flow of energy through the food chains of ecosystems. Unlike the atoms of nutrients, which can be used over and over again, the ability to do work that is present in a quantity of energy can be used just once. At each trophic level in ecosystems, organisms extract some work from the energy originally received from the sun. And at each level some of the energy's capacity to do work becomes unavailable because, like cars or human beings, other organisms cannot achieve 100 percent efficiency in energy conversion. The animals that eat plants are unable to convert all the energy found in the plants to their own uses. Some of the energy remains in waste products that are never digested, and some of it becomes unavailable in the processes of digestion and assimilation. And in the very process of running their lives, all animals and plants are continually using energy to do work. When work is done, some energy always is converted to heat and becomes unavailable. All living things are therefore continually giving off heat. This is most obvious in the so-called warm-blooded animals: you continuously give off about as much heat as is produced by a 100-watt light bulb.

The most important consequence of all this for ecosystems is that at any given trophic level there is only about *10 percent* as much capacity to do work available to the organisms as there was at the trophic level below. Crudely, this means that if the weight of herbivores in an ecosystem were ten tons, one would not expect to find more than about one ton of carnivores. But there would be at least a hundred tons of plants supporting the herbivores.[4] That is, because energy's capacity to do work cannot be recycled, the weight and numbers of organisms in most ecosystems are arranged in a *trophic pyramid*: a large mass of plants supporting a much smaller mass of herbivores, which supports a much smaller mass of first-order carnivores, which supports a still smaller mass of second-order carnivores, and so on.

An obvious though simplified example would be a mountain forest/meadow system with a great variety of plants that support deer, elk, herbivorous insects, rabbits, mice, marmots, ground squirrels, and seed- and fruit-eating birds, among others. Predators on the herbivores would include hawks, coyotes, bears (which also eat plant foods), carnivorous insects, snakes—and human beings. Second-order carnivores include some of the above as well as ticks, mosquitoes, mites, and other small organisms that attack larger animals. At each level, the populations tend to be smaller and the total weight of organisms much less than those of the level below. (Humanity's population isn't small, of course, but only a small proportion are hunters who utilize the forest/ meadow ecosystem as a food source, and even they obtain only a fraction of their food this way.)

PYRAMIDAL STRUCTURE AND ECOSYSTEM DISTURBANCE

The consequences of the pyramidal structure of ecosystems must be understood in order to appreciate the way these systems respond to insult and why populations and species at some trophic levels are more susceptible to extinction than are others at different levels.

Because of the pyramidal structure of ecosystems, populations of predators usually are smaller than populations of herbivores, and all else being equal, small populations are more susceptible to extinction than larger populations. One reason is simply a matter of chance. Suppose that some catastrophic event, say an unusually late freeze, kills 99 percent of all the organisms exposed. In a population of a thousand organisms, there will be ten survivors—perhaps enough to permit eventual recovery. If the original population size were only a hundred, however, the population size would be reduced to one. If it is a sexually reproducing organism, then the result is extinction.

For sexually reproducing organisms (which are the vast majority), a reduction in population size has many hazards that are not obvious in just the brute numbers. For example, when population sizes are reduced to a few individuals, much of the genetic variability of the population is lost. If the population remains small, that variability will tend to erode further. That can have very serious consequences and in most cases makes extinction more likely. Reduced variability means that a population is less able to adapt to any change in its environment—whether it be a sudden change in climate, a reduction in food supply, pressure from a new predator, or a disease.

In many social animals, moreover, reduction in the size of the group can make the remaining individuals much more vulnerable in other ways. There is evidence, for instance, that many animals are more secure from predators and are more efficient feeders if they are in sizable groups. Examples are herds of deer and antelope, migrating flocks of birds, and fishes that feed in large schools.

Finally, if the area occupied by the population is extensive, an extreme reduction in population size may cause difficulties in finding mates. Bears, who lead solitary lives most of the year and whose populations have been drastically reduced on all continents, may be prime candidates for final extinction because of this. Some species of rhinos, you will recall, are made especially vulnerable in this way by reduced population size. Plants, too, may suffer from the mating problem if individual plants become so widely scattered that pollination cannot take place. This is particularly likely to happen in moist tropical forests, where individuals of plant populations already tend to be dispersed among numerous other plant species.

Bioconcentration

One important characteristic of biological systems is their ability to concentrate substances selectively in the environment—and this too creates differential vulnerability to extinction. Many people assume that any substance released into an ecosystem will simply disperse until it is evenly distributed in the environment into which it was released. For example, British Nobel Laureate Chemist Sir Robert Robinson made a classic blunder in 1971 when he calculated the dilution of lead as a pollutant in the oceans and announced that it would occur at such a low concentration that its impact would be biologically negligible.[5] This conclusion totally ignored the process of *bioconcentration.*

One of the simplest and most effective forms of bioconcentration is found in clams and oysters—filter feeders—which make their living by sifting minute amounts of nutrients from their environments. Sometimes they inadvertently concentrate pollutants associated with the nutrients at the same time. Oysters have been shown to accumulate up to 70,000 times the concentration of DDT-like insecticides found in their environment.

An important mechanism of bioconcentration is found in the affinity of certain kinds of chemicals in the environment for the components of living systems. For example, chlorinated hydrocarbons—the group of chemicals that includes DDT and PCBs—have a high affinity for fat. Therefore, if a chlorinated hydrocarbon has been released into the environment, the place to look for it is not in air or water but in living organisms. Human beings, for example, will concentrate DDT in their fat bodies to levels far above those normally found in the food they eat, and in the late 1960s concentrations were found in human milk that were higher than those permitted in cow's milk in interstate commerce. (Since DDT was all but banned in the United States in 1972, concentrations in body fat and human milk among Americans have declined substantially.) All else being equal, longer-lived organisms will concentrate more toxins this way than shorter-lived ones, since over time they are exposed to more.

The pyramidal feeding structure of ecosystems itself provides an additional mechanism of bioconcentration. For instance, herbivores incorporate into their bodies only a small proportion of the energy that was originally present in the plants. The rest either is excreted, used to power their activities, or dissipated as heat. But if the plants have been sprayed with DDT, a very large proportion of the DDT that was on the plants is taken up and stored in the fat of the herbivores. In turn, carnivore populations feeding upon the herbivores will capture only a small amount of the energy present in the herbivores, but most of the DDT once again will be transferred. Thus the weight of DDT moving up the food chain will decline very slowly, while the weight of organisms at successive trophic levels in most cases declines precipitously. With each link in the food chain, the amount of DDT per pound of flesh escalates. The concentration of DDT at the upper levels in food chains therefore will be very much higher than at the lower levels—resulting in, among other things, the death of cats in Borneo.

In a classic study in the 1960s, ecologists George Woodwell, Charles Wurster, and Peter Isaacson unraveled the food web of a Long Island estuary and examined the concentrations of DDT in its elements. They found that some of the water plants had DDT concentrations of less than 0.1 part per million (ppm), while at the top of the food chain predatory birds had concentrations as high as 75 ppm.[6] This is the combined result of the pyramid mechanism and of the organisms at the top of the food chain's being longer lived.

Indeed, high concentrations of DDT and other chlorinated hydrocarbons at the tops of food chains have presented a serious threat to populations of pelicans, hawks, eagles, and other predatory birds. These poisons interfere with the manufacture of eggshells, which become so thin that the eggs can be crushed by the weight of the nesting parents. Fortunately, because of the almost total ban of DDT in the United States and restrictions in other countries, the threat to predatory birds in these areas is somewhat reduced.

But the DDT experience dramatized how vulnerable predators are to poisons that humanity releases into ecosystems, both because of the predators' small population sizes (making them already more subject to chance extinctions and loss of genetic variability) and because the process of bioconcentration often exposes them to much higher concentrations of these poisonous substances than those to which herbivores are exposed.

There is still another reason, however, that predators tend to be more vulnerable to extinction than herbivores. Remember that for many millions of years herbivores have been engaged in an evolutionary war with plants. The plants have been evolving better and better ways of discouraging the herbivores, and the herbivores have evolved ways to evade the plants' defenses. This stepwise, reciprocal evolution of ecologically intimate organisms is called *coevolution.* [7] The plants have one great disadvantage in their "coevolutionary race" with the plant eaters—they cannot run away. Therefore plants have resorted to armor and, most important, chemical warfare. Virtually everyone realizes that the spines of plants like cacti are there to discourage animals from eating the plants. But few people realize that, as we outlined in the last chapter, virtually all spices, caffeine, nicotine, the active ingredients of drugs like marijuana, cocaine, opium, and heroin, and a great many medicines are defensive compounds evolved by plants to poison herbivores.

Thus herbivorous insects have had a great deal of evolutionary experience with pesticides. They are pre-adapted to being assaulted with toxins and are easily able to respond to human attempts to poison them by evolving resistant strains. In contrast, predators have not previously had as much pressure to evolve physiological defenses against poisoning and may lack systems that can quickly be modified to detoxify a new pesticide. Combined with their smaller population sizes, this puts them much more at risk than the herbivores and makes them susceptible to extinction.

In general, each population in a natural ecosystem has evolved a set of characteristics that make it a unique functional part of that system. It can tolerate the range of physical properties—temperature, humidity, salinity— that are characteristic of that ecosystem. And it has adjusted where necessary so that it is more or less compatible with the other organisms of the ecosystem. Geographic variation is a manifestation of populations "fitting into" ecosystems. It is as if each rivet in the airplane's wing, to return to our analogy, were a slightly different shape. Once a rivet has been popped, it may be difficult or impossible to find another that satisfactorily fills the hole.

ECOSYSTEM SERVICES

With some background on the structure and organization of ecological systems, let us now consider the ways in which they support human life. In this discussion it must be remembered that populations and species, endangered or not, are all vital components of natural environmental systems. Each kind of organism has its own role within its ecosystem and is to some degree—often a very great degree—essential to the continued healthy functioning of that system. And each kind of organism in turn depends on other elements of the ecosystem for its own life.

It is impossible to separate protection of species from protection of natural ecosystems; they are two aspects of the same fundamental set of resources. Any of the public-service functions of an ecosystem may theoretically be affected by the deletion of *any* species from the system, and continued extinctions in the system are *certain* to cause disruption.

It is the dependence of human civilization on the services provided by ecosystems that concerns us here, though, and we will examine each of those services separately.

Maintenance of the Quality of the Atmosphere

The mixture of gases and other substances in Earth's atmosphere to a very large degree was created by the living beings of the planet, and its life-supporting composition is maintained by those beings. The roughly 21 percent oxygen, which permits people and other terrestrial animals to breathe, is the product of the photosynthetic activities of green plants. There was no oxygen in the atmosphere, and thus there were no animals, before photosynthesis began. Most of the rest of the air is nitrogen (78 percent). The concentration of nitrogen is controlled by the nitrogen cycle, already described.

The proportion of these two gases is of no small interest to civilization. A significant drop in the oxygen concentration would create much distress, as everyone realizes. Few are aware, though, that even a small rise would be lethal. If the oxygen concentration increased from 21 to 25 percent, terrestrial life would become impossible. In an atmosphere of 25 percent oxygen, even damp tropical rainforests would burn uncontrollably after lightning struck, and all vegetation would be consumed in global fires.

The roughly 1 percent of dry air that is not oxygen or nitrogen contains a variety of gases that are important to humanity and whose concentrations are also controlled by functioning ecosystems. These include carbon dioxide and ozone, both important in regulating climate, and methane, which is involved in regulation of the oxygen concentration.

Because of the enormous amounts in the atmospheric pools, significant changes in oxygen or nitrogen concentration would probably take place only over thousands of years, even if their regulation by ecosystems were seriously disrupted. And the degree of disruption required to make substantial changes would quite likely terminate civilization through other means anyway—so you probably needn't worry about a dropping oxygen concentration.

The same is not true, however, for changes in the "minor" gaseous components or for water vapor, which is also maintained in part in the atmosphere by biological processes. Ozone, for example, which makes up only one part per hundred million of the atmosphere, is very important because it screens out wavelengths of the sun's ultraviolet radiation that are very damaging to both plants and animals and because changing the ozone concentration may affect the climate.

Another gas produced by ecosystems, nitrous oxide, influences the ozone concentration in ways that are not fully understood. What *is* understood is that the microorganisms that produce nitrous oxide do so, up to a point, at a higher rate as their environment becomes more acid. In Chapter 7, we discuss ways that human activities have been making rain more acid. In the process, humanity has probably been causing the creation of more nitrous oxide, which in turn may be damaging the crucial ozone shield.

Moreover, it should be noted that ecosystems also help to control the dust content of the atmosphere (which affects both human beings and the weather) through their soil-retaining functions. And plants, in ways that are only poorly understood, remove dust and other pollutants from the atmosphere. Forests in particular can be thought of in this context as giant air filters.

Control and Amelioration of Climate

Not only are ecosystems responsible for the present composition and quality of the atmosphere, but they profoundly influence the global patterns of air circulation that determine the weather and climate in any given place. The weather machinery of our planet is driven by the power of the sun. The sun's heat creates columns of rising air over warm areas of the surface, which are balanced by descending air over cooler areas. The result is circulation of the atmosphere. The sun's warmth also evaporates water from oceans and fresh-water sources. When the warm, moist air rises, it cools, clouds form, and then rain (or snow) may fall.

This basic heating/evaporating system is greatly complicated by the irregularities of the Earth's surface. Ranges of high mountains, for example, have great effects on the weather in their vicinity because they force air to rise on their upwind sides, cooling it and leading to precipitation on that side of the range, and producing a "rain shadow" on the other side.

Since the sun is the driving force of the weather system, it is not surprising that factors influencing the amount of the sun's energy that is absorbed by the Earth are important in determining climate. These include the reflectivity of the atmosphere (clouds reflect some of the sun's light back into outer space), the reflectivity of the surface (deserts reflect more light than forests), and entrapment within the atmosphere of solar energy—a mechanism called the greenhouse effect.[8]

The ecosystems of the planet are involved in all three of these critical factors. For example, the forests of the Amazon Basin recycle rainfall many times over. Not all of the water that falls on the forest runs directly out to sea; most of it is returned to the atmosphere by the prodigious pumping of plants. As anyone who has forgotten to water the garden knows only too well, the upright posture of grasses and herbs and the shapes of leaves on trees are related to the availability of water to the plant. In fact, the plant continually extracts water from the soil in order to stay alive and unwilted, and water passes from the plant into the atmosphere. Thus the cloudiness over the Amazon Basin is in no small part the result of the activities of the forest of that basin.

If the forest were to be cut down, the reflectivity of the atmosphere in that area would be reduced, while the reflectivity of the surface would probably be increased. The likely result is that the area's climate would become noticeably hotter and drier. Whenever the biotic community of an area is drastically altered, the reflectivity of the surface and possibly of the atmosphere will be changed, followed by some change in climate locally or even regionally. The direction and the size of the changes depend on how drastic the alteration of the biotic community is and over how large an area it has occurred. Climatic changes are seldom for the better from the perspective of the local human population.

There are other important ways in which ecosystems influence the reflectivity of Earth. For example, the plants in natural ecosystems generally anchor the soil rather firmly in place. When these systems are removed or altered by the introduction of grazing or the cultivation of crops, losses of soil tend to increase. Some losses are through wind erosion; in some parts of the world, the movement of soil into the atmosphere is so continuous that a more or less permanent haze is created—such as the Harmattan haze of Africa.[9] The presence of dust in the atmosphere changes the reflectivity of the atmosphere and thus affects the climate.

An example of the profound influence that ecosystems have on regional climate is the increasing problem in much of the world of desertification—the conversion of land suitable for farming or grazing into desert wastelands.

Overintensive use of the land starts a self-reinforcing process of deterioration. For example, grazing species naturally select and remove the most edible plant species. As vegetative cover is removed, surface water evaporates more rapidly, the climate becomes drier, and a downward spiral begins. Bare soil, compacted by the animals' hooves, becomes highly vulnerable to erosion from both rain, when it does occur, and wind. These changes in turn make the conditions for plant growth more and more harsh. Fewer and less desirable plants grow on the deteriorating soil. This in turn increases grazing pressure on the remaining desirable species; and rangeland is converted to desert.

There is considerable reason to believe that overgrazing, resulting from overpopulation of human beings and their animals, is largely responsible for the southward spread of the Sahara Desert in recent decades in the area known as the Sahel. In a period of drought from the late 1960s until the mid-1970s, the cattle and goats of nomadic people ate virtually every green thing that appeared above the soil's surface. The drought may have been a normal cyclic occurrence, but the population pressures of people and animals clearly intensified its consequences. Climatic models have indicated that the increase in reflectivity caused by the removal of the vegetation has itself further reduced the already sparse rainfall of the area.[10]

Perhaps the most dramatic way in which ecosystems can influence climate is through modification of the carbon dioxide (CO_2) concentration in the atmosphere. This potentially will have effects on a global scale. Many human activities cause CO_2 to be dumped into the atmosphere—the most prominent ones at present being the burning of fossil fuels and the clearing and burning of forests. At the same time, carbon dioxide is removed from the atmosphere by being absorbed into the oceans and by being taken up by green plants in the process of photosynthesis. There is considerable disagreement at the moment about how these processes balance one another, but there is no doubt that the atmospheric concentration of CO_2 is increasing.

Some scientists claim that a significant portion of this increase is due to the combination of a reduction in the global amount of photosynthesis being carried out because of the steady clearing of the planet's forests and an additional contribution of carbon dioxide to the atmosphere through the burning of the wood that was in those forests.[11] Thus the same widespread destruction of forest ecosystems that is a leading cause of species extinction may also be contributing to another important environmental problem. Some other scientists, however, claim that the forest-clearing and -burning effect is not important in comparison with the burning of fossil fuels, or that it is balanced by the additional photosynthesis on land brought under cultivation. There is no disagreement, however, that the ecosystems of Earth are intimately involved in maintaining the CO_2 balance of the atmosphere. It is one of the major services provided by other species of organisms.

Why does this matter? A buildup of atmospheric CO_2 could lead to an increase in the average global temperature. Even a small increase—in the vicinity of a degree or so—could result in dramatic changes in climate over much of the world. The stability of climate patterns depends on small differences between large numbers. A seemingly insignificant change in the average global temperature could cause massive alterations in the distribution of rainfall—just as an additional hundredth of an ounce pull can fire a gun whose trigger requires a total pull of six ounces.

Any significant change in climate, of course, affects ecosystems and the various species that comprise them. Rapid climate change is known to have pushed populations of many organisms to extinction. A planet-wide disruption of climate patterns, such as the CO_2 buildup may induce, will surely have profound effects on the world's flora and fauna.

Of more immediate concern to humanity, such a change in climate patterns —an increase or decrease in average rainfall or a change in the length of the growing season, for instance—would cause severe problems for agriculture. Farmers' practices are generally attuned to local climatic conditions. If the climate changes, farmers must adjust, but they tend to be conservative and to adjust slowly. In the meantime they may suffer crop failures. The CO_2 buildup may bring on climate changes in every important food-growing region on Earth, and this in turn could lead to widespread crop failures and catastrophic famines. As physicist John Holdren has written, it is conceivable that before 2020 a CO_2/climate-induced famine could kill as many as a billion people.[12] It would be a high price to pay for destroying populations and even entire species of tropical trees.

Regulation of Freshwater Supplies

Closely related to their function in moderating climates is the role that ecosystems play in regulating freshwater supplies. For example, forested slopes tend to retain and, over time, replenish their soil; and they have a great capacity for absorbing rainfall. This rainfall then is released gradually through streams and springs. Forests also have the ability to remove pollutants such as acids, heavy metals, and radioactive substances from rainwater. Thus the water in those streams and springs may be of much higher quality than that which originally fell on the forest.[13]

The value to humanity of the soil- and water-retaining and -conserving services performed by natural systems—particularly but by no means exclusively by forests—is seldom appreciated until the system is damaged or removed and the services are lost. After deforestation, rainwater runs off the slopes in torrents, quickly eroding away the soil, creating muddy floods in adjacent lowlands, and causing alternating periods of flood and drought.

When flying over some tropical countries, we have been very impressed by brown rivers carrying loads of precious soil from denuded highlands into the sea. One of the worst examples we have seen was the Dominican Republic. That country paid a terrible price in floods for its loss of good watershed when it was struck by a hurricane in 1979. Losses of life and property undoubtedly would have been smaller if more of the forests had been intact. In the June 1979 monsoon in India, over two billion dollars worth of damage was done and very many lives were lost in floods in the Ganges Valley. The floods were the result of deforestation not only in India's northern states but also in neighboring Nepal.[14]

Recently it has been reported that the height of the annual flood crest of the Amazon at Iquitos, Peru, has substantially risen since 1970—apparently due to large-scale deforestation accompanying rapid population increases in the upper parts of the watershed in Peru and Ecuador. The authors of the report conclude that, even though most of the Amazonian forest is still uncut, "the long predicted regional climatic and hydrological changes that would be the expected result of Amazonian deforestation may already be beginning."[15] The consequences of further rises in the flood crest for the population of Amazonia, concentrated in the river's floodplain, could be catastrophic.

Deforestation also can lead to changes in the local rainfall patterns themselves. In many areas, especially in the tropics, the loss of the moisture-recycling forest system results in reduced rainfall locally and downwind. Some formerly humid forested areas are now essentially deserts, much of Brazil's impoverished Northeast being a notable example.

Generation and Maintenance of Soils

Soils are being continuously produced from rocks by physical forces wearing the rocks away and by the concerted activities of myriads of plants and animals, many of them microscopic. These organisms help crumble the rock and add organic material to it. The rate of soil generation, however, is usually extremely slow—on the order of inches per thousand years. Soils moreover are not just pulverized rock, but enormously complex ecosystems in themselves. Their fertility is very largely the result of the activities of such unsung organisms as bacteria, fungi, earthworms, and mites, all of which play specific roles suited to the soil in which they are found.

The diversity of organisms in a small bit of soil is truly astonishing. For instance, in about a square yard of a Danish pasture soil, some 45,000 small earthworm relatives, 10 million roundworms, and 48,000 tiny insects and mites were found. A gram (.035 ounce) of fertile agricultural soil has yielded over 30,000 one-celled animals, 50,000 algae, 400,000 fungi, and over 2.5 *billion* bacteria.[16]

The importance of these tiny living components of the soil cannot be over-estimated. Among them are the microbes that "fix" atmospheric nitrogen, making it available for other organisms, and the decomposers so essential for recycling all nutrients. Earthworms and ants also perform the important task of "turning" the soil—that is, moving particles from below the surface to the top and vice versa, which facilitates the decomposing process and keeps the soil loose enough that air and water can percolate into it.

Some of the least obvious organisms may be among the most important within an ecosystem. For instance, certain kinds of fungi are essential to the maintenance of populations of various trees. Although to a superficial observer it may seem that a gigantic tree is the dominant organism in a forest, the most critical organism might in fact be an obscure relative of a mushroom living in the forest soil. The trees may be most conspicuous and exert an ecological dominance, controlling the conditions for most other organisms; but they could not survive without the fungus which helps them to obtain nutrients from the soil.

Such soil organisms are vitally important to agricultural productivity just as they are to natural ecosystems. Loss of their services in maintaining soil fertility would be disastrous to agriculture, and artificial fertilizing would not begin to compensate for that loss.

Disposal of Wastes and Cycling of Nutrients

Intimately related to the functions of ecosystems in generating and maintaining soil are those of waste-disposal and nutrient cycling. Decomposer organisms, which dispose of all the wastes produced by organisms and dispose of their bodies when they die, often live in the soil. The large, complex organic molecules found in droppings and dead bodies are reduced by the decomposer system to simple inorganic chemicals—mainly nutrients—which are then returned to their starting points in the system, often by circuitous pathways. Thus the cycling of nutrients and disposal of wastes are two aspects of the same biological process.

To return to the nitrogen example, a nitrogen atom in the muscle protein of a dead cow in the coastal mountains of southern California may move into a strand of the DNA (the molecule that holds the genetic code) of a California Condor after the condor dines on the cow's carcass. That nitrogen atom may be passed on in the DNA to a daughter of the condor and eventually enter the life processes of a beetle feeding on the carcass of the daughter condor when she dies. It could be excreted by the beetle after incorporation into a molecule of uric acid and in turn be converted back into elemental nitrogen in the atmosphere by a soil bacterium. The nitrogen atom may again be brought into living parts of the ecosystem through the operations of a nitrogen-fixing bacterium residing in a special nodule on a root of an alfalfa plant. Following its incorporation into the plant's protein, it could then be passed on to a cow that eats the alfalfa and be built into the muscle protein of the cow, thus returning in a sense to its starting point.

The organisms involved in the decomposer trophic level range from tiny bacteria to hyenas and vultures. These creatures are often highly specialized for the work they do and sometimes undergo intense competition for their food.[17] The work of decomposers is also crucial to disposing of the wastes produced by humanity. Certain bacteria are an essential part of the sewage disposal process. Unfortunately, the nutrients released by their action are often just dumped into rivers or the sea, rather than being returned to the soil. Dangerous microbes that may be present in sewage or other wastes are also destroyed by the decomposers either in sewage treatment or in natural systems. Thus rivers can purify themselves, making the water safe again for human consumption—assuming their aquatic ecosystems are not inundated by poisons, overfertilized by a flood of nutrients, or heated up by effluents of power plants.

Many of industry's waste products that find their way into the environment also can be degraded and decomposed by ecosystems. Among these are soaps, detergents, pesticides, spilled oil, phenols, acids, alkalis, paper, plastics, old tires, and so on. Most of these break down more or less rapidly under the right physical conditions and in the presence of an appropriate decomposer; but a few, such as DDT and some plastics, are virtually indestructible. These are true "wastes," produced only by *Homo sapiens.*

Maintaining the nutrient-cycling functions of Earth's ecosystems in good health is crucially important. Without the functioning of the myriad organisms responsible for breaking down wastes and recycling carbon, nitrogen, phosphorus, sulfur, and all the other essential elements, all life on Earth would quickly grind to a halt. And that emphatically includes human life; the cycles also operate in—and are absolutely essential to—agricultural ecosystems.

Pest and Disease Control

Natural ecosystems control the vast majority of the potential pests of crops and the transmitters of human diseases. Human beings have frequently tried to exercise control as well, but the results often have not fulfilled their hopes. When pest-control efforts have failed, it usually is because the controllers failed to understand how ecosystems work.

In 1949, for instance, the cotton farmers of the Cañete Valley of Peru thought that DDT was the answer to their prayers. A little DDT applied to their fields at first gave successful control of several important pests of cotton. The farmers naturally assumed that if a little DDT did some good, a lot would do a lot of good. They sprayed more and more DDT and its relatives, and the results by the mid-1950s were utter disaster. Yields dropped far below the pre-spraying level. Not only did the old pests become resistant to the poisons being used against them, but a whole new array of insects that had never been pests before achieved pest status.[18]

This "promotion" to pest status was well understood by population biologists even then, but it still has not been grasped by most people in the pesticide industry and many bureaucrats in departments of agriculture. Herbivores, as described earlier, are almost always more difficult to kill with poisons than are carnivores. In the process of sousing their fields with poisons, the Peruvians had decimated the populations of many predators and "released" the organisms the predators preyed upon from the forces that had previously controlled their population sizes. The results were instant population explosions of new kinds of pests—including leaf rollers, leaf worms, and bollworms.

The Cañete Valley experience has been repeated in many other systems the world over. Spider mites, now among the most important pests of crops, are essentially the creation of the DDT industry: they became pests after overuse of DDT (which had little effect on the mites) greatly reduced populations of the predatory insects that normally kept the mites in check.

These experiences highlight a very fundamental ecosystem service: the control of populations of species that, if the ecosystem is perturbed, have the ability to become important enemies of *Homo sapiens*. It seems certain that over 95 percent of the organisms capable of competing seriously with humanity for food or of doing us harm by transmitting diseases are now controlled gratis by other species in natural ecosystems.

Pollination

Pollination, which is required for the successful reproduction of a great many flowering plants, is another essential service performed by ecosystems. Some 90 crops in the United States alone depend on insects to pollinate them; 9 others benefit from insect pollination.[19]

A surprising number of organisms, mainly insects, perform this service. Carrot flowers in Utah were found in one study to be visited by 334 species of insects belonging to 37 different families. The most efficient pollinators include honeybees, solitary bees, wasps, hover flies, and a soldier fly. In contrast, many kinds of commercial figs are entirely dependent for pollination on a tiny wasp, which in turn is entirely dependent on the fig flowers for a breeding place. Indeed, it appears that each of the 900-plus species of fig has its own obligatory wasp species pollinating it and dependent on it in the same way.[20]

Some crops are adequately pollinated by honeybees, which are basically domesticated animals. But many others are partially dependent on the pollination services of wild insect species. Alfalfa, for example, is pollinated most efficiently by wild bees in the cooler humid regions where it is grown. Agriculture in many parts of the world would suffer greatly if the pollination services of bumblebees, solitary bees, assorted flies, and so on were not available free from adjacent natural ecosystems. And many thousands of wild plant species would become extinct as well.

Direct Supply of Foods

Besides their role in supporting agriculture, natural ecosystems also provide people with food directly. Food from the world's oceans and freshwater systems are the most obvious and important example. For the vast majority of the fisheries harvest, hunting alone is used—there is no herding or husbandry. Ecosystems on land also produce "free" food in the form of game animals and wild plants such as nuts, berries, and maple syrup. The game, like fishes, provides a critical protein supplement to diets in many poor countries.

Maintenance of a Genetic Library

Finally, as we discussed in Chapter 4, *Homo sapiens* derives many direct benefits from thousands of species, all of which are components of natural ecosystems. The living world thus amounts to an enormous organic "library" from which humanity has already withdrawn a vast array of useful substances, ranging from foods to drugs to lubricating oils. And natural systems undoubtedly harbor many thousands of directly useful organisms and products yet to be discovered. The maintenance of this library is among the most valuable— and most irreplaceable—of the services rendered by ecosystems to humanity.

EXTINCTIONS AND ECOSYSTEM SERVICES

To what degree will the various essential ecological services to humanity be compromised by a continuing trend of extinctions? The answer is that they *all* will be threatened if the rate of extinctions continues to increase, although the degree of their deterioration will undoubtedly vary from place to place.

Isn't it possible for technology to substitute for many or all ecosystem functions? In some cases, a partial substitution of the lost service is possible. For example, the soil-retaining and water-metering properties of the natural forest may be closely approximated, or even improved upon, by a planted forest containing many fewer species. The long-term stability of the planted forest, however, may be very much less than that of the natural one. In general, conversion of forest ecosystems from natural to managed can be done more successfully in temperate regions with deep, rich, nutrient-laden soils than in tropical rainforest regions where the soils are often very much poorer and the nutrients are largely concentrated in the plants themselves.

If the forest is partially or totally removed, flood control and reservoir projects become necessary. Again, these compensate in part for the loss of the ecosystem service and even allow the marshaling of greater water supplies for distant cities, but they do little to control soil erosion or to moderate local climates. And they have very destructive effects on riverine and streambank ecosystems.

At the other extreme, substitution for some ecosystem services is essentially impossible; it is obvious that the quality of the genetic library is automatically and continuously compromised by the decay of biological diversity. The loss of any species is an irrevocable loss of a potential resource.

For soil systems and for the waste-disposal and nutrient-recycling functions of ecosystems, it is relatively difficult to predict what the effects of losses of populations and species will be. It seems likely that they will vary enormously, depending on *which* populations and species are lost. If nitrogen-fixing organisms are affected, for instance, the results could be very serious for agriculture. Humanity can substitute artificial nitrogenous fertilizers, but there are environmental problems associated with their heavy use. Even in the United States, which is a heavy user of fertilizers now, natural systems still provide us with considerably more nitrogen than artificial fertilizers do.[21]

On the pest and disease-carrier front, losses of ecosystem services are likely to have catastrophic results, since sensible pest control is a difficult business, even with the aid of these functions. Unfortunately, the side effects of modern chemical pest control usually include damage to the pest-control service itself, thus worsening the problem it is aimed at solving. Substitution for natural control of both pests and disease carriers is very difficult at best—and often impossible.

Thus, technological substitutes for ecosystem services are no more than partially successful in most cases. Nature nearly always does it better. When society sacrifices natural services for some other gain—to expand agriculture, for instance, or to harvest timber or to obtain mineral resources—it must pay the costs of substitution. Furthermore, humanity's present behavior increasingly affects many services simultaneously, making the logistic problems of replacing them, even where possible, very much more difficult.

Predicting Ecosystem Behavior

One of the great problems of ecologists today is their inability in most cases to predict the consequences to an ecosystem of the extinction of any given population or species. The complications are immense, since each ecosystem, however its boundaries may be defined, is unique. Furthermore, what limited knowledge is available indicates that the consequences of the extermination of any given group of organisms are also likely to be unique. Some species, for instance, appear to function as what is known as "keystone" species. If a keystone species is removed from its system, its loss may be followed by a cascade of further extinctions.[22]

When a predatory starfish was removed from an intertidal community (the organisms living on the seashore between low and high tidelines), the community collapsed from fifteen to eight species in less than two years. The mussels that were the preferred prey of the starfish increased in its absence and outcompeted other species, forcing them to local extinction.[23] There is reason to believe that keystone species are a common phenomenon, but few have been identified because of the difficulty of doing the appropriate experiments and the disruption this would cause to the ecosystems under study.

On the other hand, many species, some of them superficially similar to keystone species at the same trophic level, may not play a keystone function; their loss may have relatively little consequence for the overall characteristics of the ecosystem and the services that it provides. One reason for this seeming lack of importance may be that their various functions also can be performed by other organisms. An insect that feeds only on a particular species of plant, for instance, may share that food resource with five other insect species, rabbits, and deer. In turn, the toads, lizards, and birds that eat this particular kind of insect would still have several other insect species to feed on if it disappeared. This sort of redundancy—analogous to the extra rivets on our airplane's wings—may be an important factor in the observed stability and resilience of the functional properties of many natural ecosystems.[24] But the system would be destabilized if that insect's competitors were also decimated —or if its predators were.

It would be a mistake, we emphasize, to believe that, just because ecologists cannot precisely detail their vulnerability, natural ecosystem services can readily be replaced by equivalent services from managed systems. For one thing, there is clearly a problem of scale—there are limits to how much effort *Homo sapiens* can put into such management. The substitutions for natural ecosystem functions that *are* possible require very large amounts of capital, energy, materials, and manpower. Yale University ecologist F. H. Bormann summarized the problem very well, discussing the loss of services from forest ecosystems:

> These natural functions are powered by solar energy, and, to the degree that they are lost, they must be replaced by extensive and continuing investments of fossil fuel energy and other natural resources if the quality of life is to be maintained. We must find replacements for wood products, build erosion control works, enlarge reservoirs, upgrade air pollution control technology, install flood control works, improve water purification plants, increase air conditioning, and provide new recreational facilities. These substitutes represent an enormous tax burden, a drain on the world's supply of natural resources, and increased stress on the natural system that remains. Clearly, the diminution of solar-powered natural systems and the expansion of fossil-powered human systems are currently locked in a positive feedback cycle. Increased consumption of fossil energy

means increased stress on natural systems, which in turn means still more consumption of fossil energy to replace lost natural functions if the quality of life is to be maintained.[25]

Even when substitutions for ecosystem services are possible in principle, political, social, and economic circumstances often interfere with environmentally sensible management. This has been demonstrated, for instance, by attempts to manipulate the flows of fresh water of the southwestern United States (which result in various ecological disasters) and to maintain our prairie soils (which are rapidly being eroded away). And, most important from the viewpoint of the concerns of this book, many if not most of the components of natural ecosystems are irreplaceable. In other words, *it simply may not be possible to synthesize and manage ecosystems successfully without access to the original parts.*

A dramatic demonstration of this last point can be seen in the results of introducing cattle into the drier areas of Africa. The cattle have not coevolved with the local vegetation. Native hoofed animals such as Thomson's Gazelles and Wildebeest graze in ways that actually *increase* the productivity of the grasslands, whereas the cattle tend to overgraze.[26] Furthermore, the native animals are well adapted to the arid conditions; they have mechanisms for greatly restricting their water loss and obtain much of their water from the plants they eat and from the breakdown of sugars. Many of them never drink at all. In contrast, cattle must constantly march back and forth to waterholes —compacting the surface (which leads to erosion), trampling vegetation that might otherwise be eaten, and wasting their own energy.[27]

The overall result of substituting a managed cattle ecosystem for the natural one has been the conversion of large parts of Africa to desert and the threat of similar degradation of much more. Astonishingly, the natural system—or a managed version such as we have already described—is capable of producing a higher yield of meat and hides than the cattle system. But there are powerful socioeconomic forces behind the cattle economy. Gradually, the populations of native animals are being forced to extinction. As they disappear, the chances for establishing a permanently productive system are disappearing also. Not only are the animals of the African veldt a magnificent esthetic resource for the planet, they are also working parts of an irreplaceable ecosystem of great importance to the people of Africa. They could become the foundation of a better human life-support system than the one now supplanting them.

Genetic Diversity and Ecosystem Services

It is important to note the role that genetic diversity may play in the functioning of ecosystems and thus in the provision of ecosystem services. For example, there is reason to believe that the genetic diversity *within* a plant population may help protect the plants from attacks by herbivores.

Some of the first evidence of this appeared in the work done by our group on the coevolution of lupine plants and the tiny blue butterfly, *Glaucopsyche lygdamus,* which as a caterpillar feeds on lupines. Early in our work, we discovered that this small herbivore could have an enormous impact on the seed production of the lupines and thus could be expected to be a powerful selective force on the lupine populations.[28]

We discovered, however, that lupine populations did not suffer equally under the assault of *Glaucopsyche.* Some were heavily attacked and lost most of their seeds; others lost very few. Since we had recently come to the conclusion that plants defend themselves against herbivores primarily with poisonous chemicals, we set out to discover whether or not there were differences in those chemicals in lupines from different populations. The most obvious poisons to look for were the alkaloids. Alkaloids are abundant in lupines, and we made the working assumption that the populations that were least afflicted with *Glaucopsyche* would have the highest alkaloid concentrations.

As so frequently occurs in science, our working hypothesis turned out to be dead wrong. Some of the populations with the highest concentrations of alkaloids were the ones suffering the most from *Glaucopsyche.* The connection was not with the *amount* of alkaloid in the plant but with the *variability* in alkaloidal content from plant to plant within a population. The best-defended populations contained plants that differed from one another greatly in the kinds and amounts of alkaloids present.

A little reflection uncovered the likely reason for this relationship. Caterpillars mature on a single lupine plant, and then adult butterflies lay their eggs on a variety of plants in the same population. If all the plants in the population were the same alkaloidal type, each generation of butterflies would be subjected to exactly the same poison stress. This would lead, as does the repeated application of synthetic pesticides, to the development of resistance to the poison in the butterflies, which in turn would lead to a heavier and heavier attack on the plants. If the populations were variable, the odds were that each generation of caterpillars would be subjected to a different array of poisons. The offspring of a butterfly that happened to be resistant to the combination of alkaloids in the lupine plant on which it matured would inherit that resistance. But in all probability they would be confronted with a different set of alkaloids in the lupine plants on which they hatched from eggs, and on which they were condemned to feed. Hence the development of resistance would be much more difficult.[29]

Since this work was done, similar results have been obtained by other workers, indicating, for instance, that biochemical variability is important to the resistance of Ponderosa Pine tree populations to insect attack.[30] Such protective variability has also been found in wild ancestors of wheat, a discovery that is now being exploited in some innovative plant-breeding research on cultivated wheat.[31]

Because plant defensive chemicals are highly variable, simply preserving a limited sample of a species often is not sufficient to maximize the potential for obtaining directly useful products from that species. Now it appears that the biochemical variability *between* and *within* plant populations often is also crucial to ecosystem functioning. Loss of that variability could have serious effects on the ecosystem's ability to maintain its life-support services.

More generally, loss of genetic variability from any sexually reproducing populations will limit their ability, and that of the ecosystem, to evolve in response to environmental change. Such losses are especially critical in times of rapid change, which stress the evolutionary capacity of an ecosystem to the utmost. Today is just such a time of change—change induced by one species: *Homo sapiens.* Whether ecosystems will be able to meet the challenge and continue to provide their essential services to society will depend both on their unknown degree of resilience and capacity to evolve and on humanity's unknown willingness to lessen its assault upon them.

In a sense, *Homo sapiens* is now betting that greatly increasing its assault over the next few decades will not lead to an unacceptable breakdown of the services. We believe the bet is both unnecessary and all too likely to be lost. The price of losing will most probably be paid in massive famines and resource shortages, possibly leading to global thermonuclear war.

Now, with some background on how ecosystems function, how they support civilization, and the consequences of losing that support, we turn to the ways in which humanity directly threatens populations and species of organisms. Two things must always be borne in mind in this context. The first is that assaults on individual species are also assaults on the ecosystems of which they are integral components. The second is that assaults on ecosystems (such as deforestation, conversion of grassland to farmland, damming of rivers, pollution of estuaries) in turn inevitably threaten the populations of organisms that are components of the ecosystem. Regardless of whether an entire ecosystem is attacked or some of its components are selectively removed, the outcome is likely to be similar: a loss of the services the ecosystem once provided.

Part III

HOW ARE SPECIES ENDANGERED BY HUMANITY?

Chapter VI

Direct Endangering

What an appalling indictment it is, what a disgrace to
mankind, that the road to his so-called civilization should
be built on the memories of extinct species and species on
the way to extinction.

—THE RIGHT HONORABLE EARL OF JERSEY,
*speaking before 1972 Conference
on Breeding Endangered Species*

IN 1974/75 we were privileged to spend two exciting weeks on a ship visiting
Antarctica. We walked through many colonies of Adelie, Gentoo, and Chinstrap Penguins—some numbering hundreds of thousands of individuals resplendent in their black and white "formal dress." Almost all were mated pairs
guarding eggs or young. One member of each pair was always guarding the
"nest" (usually a circle of pebbles) while the other was at sea catching krill,
the shrimplike animals that are a crucial link in most Antarctic food chains.

The cacophony in the colonies was incredible as returning penguins went
through complex and noisy greeting ceremonies with their mates and then
regurgitated, also noisily, the contents of their crops into the gaping beaks of
their waiting young. The penguins were so tame that it was possible to approach within inches of them to photograph the process—close enough to get
pictures of their spined tongues, which help the birds to grasp the krill.

The comings and goings of the penguins had a certain comic character. The
upright shuffling waddle is often interrupted by joyous tobogganing on
the belly down snowy slopes. When one couple's attention is distracted by the
reunion ceremony, a neighboring penguin may sneak in and steal pebbles for
its own nest. The most tragicomic scene is played out at the shoreline if there
is a Leopard Seal in the vicinity. Leopard seals are the third-level predator in
the plankton → krill → penguin → seal food chain. In the presence of this fast and

efficient marine carnivore, the penguins are reluctant to enter the water, and they all go through an Alphonse and Gaston routine, hoping another will be first to take the plunge. This continues until individuals crowding up from behind push those nearest the edge into the water. In the water the penguins are magnificent, swift, agile swimmers. Most get away safely, but the seal often gets a meal, skillfully skinning and swallowing the penguin with a few flicks of its head.

Watching the penguins was always interesting and often amusing, but also tinged with sadness. The adult penguins had for millennia been without enemies on land, although their eggs and young are often snatched away by Skuas, predatory relatives of seagulls. This freedom from predation is why the penguins are so fearless of people, and as a result they have on occasion been fearfully abused. In some localities they used to be slaughtered and rendered for their fat. Dogs to pull sleds have been brought to some of the Antarctic stations, and at one station we watched chained dogs attack and kill penguins that wandered within their reach. And we heard stories about the behavior of human visitors that appalled us. American servicemen once blew up a part of a penguin colony with dynamite so they "could see penguins fly." Argentine tourists have played soccer using penguins as the balls.

At the moment the penguins are a key esthetic resource for the infant Antarctic tourist industry and, in spite of such behavior, are not as yet directly threatened by it. They may well become endangered, though, if current plans by human beings to harvest the krill upon which penguins are dependent are carried out.

Fascinating though the penguins, the seals, the various sea birds, and the incredible scenery of the Antarctic were, it was a species of whale there that gave us one of the greatest thrills of our lives. Our ship was moving slowly north through the Lemaire Channel between some islands and the Antarctic Peninsula. It was a day of dreamlike, if stark, beauty. The sky was overcast, the water smooth and inky black—broken occasionally by groups of penguins "porpoising" (leaping out of the water and plunging back in). The dark, rocky shores were hung with glaciers, which showed flashes of pale blue, the only bright color visible outside the ship. We were in the very tip of the bow, watching as the ship penetrated ice floes a few feet thick.

Suddenly the captain came on the public address system from the bridge. "Killer Whales ahead!" And there they were—a pod of five or six of those magnificent black-and-white beasts, fifteen to twenty feet long, one a male with a tall dorsal fin. They were circling a small ice cake on which was ensconced a Weddell Seal. As the ship passed, the whales made repeated rushes in concert at the ice cake, pushing waves over it in an attempt to knock the seal into the water. Finally the seal was washed off, and the roiling water and growing distance hid the details of its demise. Everyone on the ship, biologists and nonbiologists, were in a state of what might be described as excited disbelief. It seemed almost as if we had only imagined seeing one of nature's greatest and most terrible dramas unfold perhaps fifty yards away in a strange and lonely theater.[1]

Whales also impressed naturalist Robert Cushman Murphy when he sailed to the Antarctic in 1912. He reported seeing them "in all directions," even though the Antarctic whaling industry had already progressed to the point that the shoreline near the whaling station in South Georgia Island was "lined for miles with the bones of whales."[2] Sixty years later, sailing the same waters, we saw *no* whales other than the Killers, a species not hunted commercially.

OVEREXPLOITATION

The history of hunting whales serves as a classic model of directly endangering commercially valuable species by overexploitation—the taking of so many individuals that the population is unable to maintain itself. It is an appropriate topic with which to begin this chapter, which deals with the endangering of populations and species because they are of direct interest to *Homo sapiens*— either as valuable resources or as pests.

In the days of wooden ships and iron men, technology was such that human beings could not seriously threaten the whales. A record three-year cruise in the middle of the last century killed fewer than a hundred whales. By the 1930s, fast catcher boats to kill the whales and giant factory ships to process them had enormously increased the pressure on whale populations. In 1933 almost 30,000 whales were killed, yielding 2.5 million barrels of whale oil. By 1967 twice as many whales were killed, but they yielded only 1.5 million barrels of oil. The reason was that the larger species, such as the Blues and Fins, had been hunted to the brink of extinction. The industry then turned to progressively smaller and smaller whales—Seis and Sperms (about sixty feet long) and finally Minke Whales (maximum length a little over thirty feet).

The story of the whaling industry is not only one of insensitivity toward these remarkable animals, it is a splendid example of the shortsightedness with which economists usually think about resources. It has been a tragedy for the whales and a tragedy for humanity as well.

For a long time, we did not understand the behavior of the whaling industry, which seemed self-destructive. It ignored the good advice of biologists, hired by the International Whaling Commission, to ease the pressure and let the whale stocks recover. Instead, space-age technology—helicopters to find whales, sonar to follow them, radio beacons on dead whales inflated with compressed air—continually increased the pace of the slaughter.

Then, in the early 1970s, a Japanese economist explained to us that we had been viewing the problem with the blinders of biologists. We had been thinking of preserving the whales so future generations, if they wished, could harvest them in perpetuity—or communicate with them, or enjoy their songs, or just leave them alone.

The whaling industry, however, has no interest in the whales themselves; it is interested in maintaining its stream of income. In economic terms, the value of the whale resource could be maximized, from the standpoint of the interests of an individual whaling company or nation, by driving it to economic extinction. That is, the best *economic* strategy would be to continue hunting the declining whale species until catches were no longer large enough to be profitable. Then the depreciated chasers and factory ships could be converted to other uses or sold for scrap, and the capital of the industry applied to the exploitation of some other resource.

Two things make the whales and many other renewable resources subject to this insane treatment. One is a temporal myopia built into human behavior. People care less about the future as it becomes more and more distant. The promise of a favor next week means more than the promise of one next year. In economics, this attitude goes under the title of "discounting the future."[3] Because the future is discounted, the value of a Blue Whale a century from now to the whaling business today is essentially zero.

The second factor working against the whales is that they are a "common property" resource. Owned by no one and desired by all, the resource tends to be plundered. Garret Hardin described the problem with his usual skill and clarity in his classic article "The Tragedy of the Commons."[4] He pointed out that in a communal pasture, open to all, each individual herder will continually add to his herd. That is the only sensible strategy from the individual point of view. Although the grass is limited, the person with the largest herd will get the largest share. Each user of the commons reasons the same way: "If my animals don't get the grass, someone else's will." All the users struggle to increase their herds, and eventually the carrying capacity of the pasture is exceeded, the grass is exhausted, and the herds starve. Everyone rationally pursuing his own self-interest has led to community behavior that results in the tragedy of the commons.

The sea very largely has been treated as a commons throughout history— both an international commons in which whaling and fishing operations of all nations competed freely and, within territorial waters, a national commons in which various fishermen or corporations operated. In the 1970s many nations, including the United States, extended their national jurisdiction over the ocean to two hundred miles from shore. With individual countries thus assuming responsibility for the marine biological resources near their shores, a tragedy of the commons may be avoided. Still, the pressures that encourage overexploitation persist everywhere. A hunter realizes that a whale or fish not caught will not necessarily live to reproduce. Instead, another hunter might catch it. Therefore each hunter (or ship or nation) tries to maximize his catch.

The fishing industry has developed technologically in parallel with the whalers. Factory ships, sonar, new kinds of nets, and other technical improvements have greatly increased the ability of commercial fishermen to catch fish. A single modern Rumanian factory ship caught in one day in the 1960s as many tons of fish in New Zealand waters as all of the 1,500 boats of the New Zealand fishing fleet. The industry has been proud of its growing ability to loot the sea of its protein today; let the future take care of itself. *Simrad Echo,* a periodical published by a Norwegian manufacturer of sonar fishing equipment, bragged in 1966 that 300 sonar-equipped Norwegian and Icelandic ships had brought industrialized fishing to the Shetland Islands. Using purse-seines, which took smaller fishes than the drift nets used by British fishermen, these ships gathered in unprecedented hauls of herring.

A *Simrad Echo* editorial asked: "Will the British fishing industry turn . . . to purse-seining as a means of reversing the decline in the herring catch?" The attitude of the industry was reflected in another quote from the magazine: "What then are the Shetlands going to do in the immediate future? Are they going to join and *gather the bonanza while the going is good*—or are they going to continue drifting and if seining is found to have an adverse effect on the herring stocks find their catches dwindling [emphasis ours]?"

The answer was not long in coming. In January 1969 newspapers in Great Britain declared that England's last coastal herring industry had been wiped out. The young herring, which swam through the large mesh of the British drift nets, were caught in the purse-seines, thus destroying the potential breeding stock of the herring. Between 1966 and 1970, herring catches dropped from 1.7 million tons to 20,000 tons—nearly a hundredfold reduction.[5]

Other marine fisheries have been wiped out in addition to those for the larger whales and the herring. One of the most famous was the California Sardine fishery, ruined by overfishing. In the 1936–37 season, three-quarters of a million tons were taken in California waters. By 1957–58, a mere twenty-one years later, the catch was seventeen tons! The fishery has never recovered; its monument is Cannery Row in Monterey, long since converted from processing fishes to processing tourists.

When an exploited marine organism reaches economic extinction—that is, the point where it is no longer profitable to harvest it—the species is not actually extinct. Some of its component populations probably are, however. Indeed, one of the problems of designing harvesting strategies for a sustainable yield is insufficient information on the number of populations being harvested. For example, the Peruvian Anchoveta fishery in 1970 yielded thirteen million tons, almost a quarter of the total worldwide harvest of marine fishes. Then in 1971 a combination of unfavorable meteorological conditions and overfishing caused the catch to plunge to about a third of that level. It has only partially recovered since then in spite of increased restrictions on the catch. A major problem in deciding on a rational strategy for catching the Anchovetas is that it is not clear whether the Peruvians are exploiting one population or two or more separate populations.

There are economic factors, too, that help push partly depleted fish stocks toward extinction. As a particular kind of fish becomes scarcer, its price rises. The higher the price, the greater the incentive for commercial fishermen to catch it. The pressure on that stock thus continues or increases even further. This process can be, and in some cases has been, carried to the point of economic extinction.[6]

Many factors in addition to future patterns of exploitation or protection will determine whether various overexploited marine species can make a comeback or will continue gradually downhill to extinction. The smaller populations, of course, are statistically more vulnerable to extinction. In some cases, the decline in population size itself can cause a change in the relations of the harvested species with an important competitor, which may in itself lead to the eventual extinction of the population.[7] Such a change is suspected by biologists to explain the failure of the California Sardines to recover after the fishery collapsed. An increase in its main competitor, the Anchovy, may have fundamentally changed the environment of the sardine.[8]

The most serious environmental challenge to overexploited marine species will be deterioration of the marine environment in general. It is important to note that the biological riches of the sea tend to be concentrated in shallow waters near land. That is where most of the powerful upwelling currents are located. These currents bring nutrients to the surface, nourishing phytoplankton, the tiny floating plants that are the foundation of oceanic food chains.

In contrast, as marine biologist J. H. Ryther put it: "The open sea—90 percent of the ocean and nearly three-fourths of the earth's surface—is essentially a biological desert. It produces a negligible fraction of the world's fish catch at present and has little or no potential for yielding more in the future."[9]

As a consequence of this distribution of productivity, exploited marine populations are located precisely in the areas of the oceans most subject to heavy pollution. Most pollutants are dumped into the sea at the shore and are thus most concentrated there. Furthermore, many of the fishes and shrimp hunted in the seas spend part of their lives in estuaries—places where fresh and salt waters mingle at the mouths of rivers and streams. It is sobering to realize that about two-thirds of the rich fisheries productivity off the East Coast of the United States is dependent on these threatened habitats.[10] And estuaries are among the most threatened habitats on Earth because of pollution, filling, "improvement," and "development." Thus many marine species directly endangered by overexploitation may take the last steps to extinction because their reduced and scattered populations are especially vulnerable to the indirect assaults of habitat degradation and destruction.

DIRECT ENDANGERING BY PREINDUSTRIAL SOCIETIES

It is only to be expected that a species that is evolutionarily successful will have negative impacts on other species. Darwin himself noted that ". . . as natural selection acts solely by the preservation of profitable modifications, each new form will tend in a fully-stocked country to take the place of and finally to exterminate, its own less improved parent or other less-favoured forms with which it comes into competition. This extinction and natural selection will . . . go hand in hand."[11]

Early on, however, there were signs that *Homo sapiens* would carry this natural process to an extreme. For example, at the end of the Pleistocene epoch some 12,000 years ago, roughly two-thirds of the large mammal species in North America went extinct in a short time. Included were several species of elephant-like mammoths, a tall relative of the Dromedary, giant ground sloths, and saber-toothed cats. Similar losses of large animals occurred all over the planet at roughly the same time. As Alfred Russel Wallace, co-developer with Darwin of the theory of evolution, wrote in 1875: "It is clear . . . that we are now in an altogether exceptional period of the earth's history. We live in a zoologically impoverished world, from which all the hugest, and fiercest, and strangest forms have recently disappeared . . . yet it is surely a marvellous fact, and one that has hardly been sufficiently dwelt upon, this sudden dying out of so many large mammalia, not in one place only, but over half the surface of the globe."[12]

The dates of these extinctions more or less coincide with the arrival of the first *Homo sapiens* in the Western Hemisphere. This, along with less well-defined patterns of extinction associated with the spread of prehistoric peoples and with the improvement of human hunting skills (as seen, for example, in the evolution of stone arrowheads), has led to the theory of Pleistocene overkill. The idea is that the relative poverty today of Earth's megafauna (animals weighing more than a hundred pounds each) is due primarily to overexploitation of relatively defenseless large animals by ever-increasing numbers of more skillful hunters. Wallace, who had originally thought that ice age glaciations had caused the extinctions, later recognized that many of the large animals had outlived the glaciers only to succumb later. By 1911 he had accepted *Homo sapiens* as the principal agent of extinction, writing: "What we are seeking for is a cause which has been in action over the whole earth during the period in question, and which was adequate to produce the observed result. When the problem is stated in this way, the answer is very obvious. . . . the rapidity of the extinction of so many large Mammalia is actually due to man's agency. . . ."[13]

That "obvious" answer is still controversial, but there are various lines of evidence, ably marshaled by modern paleontologists, particularly Paul S. Martin, that make it seem highly likely that humanity got an early start in the business of extinction. The timing of the North American extinctions is quite critical in this regard. The Old World was occupied over a million years ago by *Homo erectus*, early ancestors of *Homo sapiens* who only very gradually evolved the large brains and effective weapons that allowed them to be deadly hunters. By contrast, the first human groups that much more recently invaded the Western Hemisphere across the Bering Strait (which was then dry) were *Homo sapiens*, who were fully capable of efficient hunting. They spread southward and eastward at a pace that coincides in time rather closely with the extinctions of large animals in the New World.[14]

In other parts of the world, occupations of islands by human groups have led to the rapid demise of large animals. *Homo sapiens* first reached both New Zealand and Madagascar about a thousand years ago. In New Zealand there were no terrestrial mammals, and bird evolution had gone berserk. An impressive array of giant flightless moas awaited the human invaders, including the spectacular *Dinornis*, which towered more than eleven feet tall and resembled a giant Rhea. People promptly started hunting them, rendering them extinct in a period of a few hundred years and leaving behind broken and burned bones in the kitchen middens of their hunting camps. In some places in New Zealand, these bones are piled in large quantities, perhaps indicating favored slaughtering grounds. It seems unlikely, however, that hunting pressure alone caused the rapid demise of the moas. The rats and dogs that came along with the Maoris undoubtedly savaged the eggs and young of these birds, which had evolved in the absence of mammalian predators.

In Madagascar, the megafauna, including a giant flightless bird and a dwarf hippopotamus, all disappeared soon after the island was occupied. In neither New Zealand nor Madagascar is there any sign of climatic or other changes that could logically explain the extinctions otherwise. The connection with human occupation in both cases is too close to be mere coincidence.

A few hundred years ago, a combination of direct exploitation and predation by mammals introduced by *Homo sapiens*, along with some habitat destruction, wrote finis to the large flightless bird that had dominated Mauritius Island in the Indian Ocean for millions of years. That bird, whose name is derived from a Dutch word describing its sluggish behavior, has become symbolic of anything and everything that is irretrievably gone: "dead as a Dodo."

Prehistoric Extinctions in Eurasia

Pressure on populations of large animals from human hunters began early. There is no doubt that human hunters at the end of the great glaciations ten thousand or more years ago were capable of killing large animals in very large numbers. The world is littered with Pleistocene "boneyards," the locations of vast slaughters by prehistoric hunters. At a ravine near Amvrosievka in the southern Ukraine of Russia, remains indicate that almost a thousand bison were killed with flint- and bone-tipped spears.

The vicinity of Předmost in Czechoslovakia was known in the middle ages as an abode of giant people—the bones of "giants" were found there periodically. In the middle of the last century, a local farmer, Josef Chromeček, digging limestone on his land, uncovered a veritable graveyard of giants— extensive deposits of bones of huge size mixed with equally large teeth. Even though the bones were quarried and used as fertilizer on local fields and sent in wagonloads to be used as ballast in construction of the Přerov–Leipnik railway, the deposit did not come to the attention of the scientific community until around 1880.

Investigating scientists found a dark layer about thirty inches thick in the yellow-gray of the fine-grained, wind-deposited soil of the area. The dark color was caused by the ashes of innumerable ancient fires. The ashes were mixed with many chipped-flint artifacts of the Aurignacian culture and incredible numbers of bones of aurochs (an extinct progenitor of domestic cattle), Musk Oxen, bison, wild horses, Reindeer, and above all, Woolly Mammoths. The bones of mammoths made up about three-quarters of the deposit, and many of them were of young animals.

The discovery caused a scientific sensation. Even a hundred years ago, it was easily interpreted. Here was a hunting camp of a culture from the late period of deposit of a wind-blown geological formation—a period twenty thousand years ago when the enormous glaciers of the last ice age which had covered much of northern Europe were just beginning to retreat under warming conditions. Here was clear evidence that ice-age people were not the "poor savages" of nineteenth-century thought, but courageous, skilled, well-equipped, and well-organized hunters. Not only were they capable of successfully hunting the giant woolly elephants, but they were able to make them a major item in their diets![15]

That mammoths were generally important to ice-age hunters is attested not only by numerous large deposits of their bones in central Europe but also by the frequency with which they are featured in the cave art of the Paleolithic era—the Old Stone Age. It seems highly likely that the increasing pressure put on the herds of mammoths by human hunters, especially by killing the young, was a major factor—if not *the* major factor—in starting these great pachyderms on the road to extinction. There is some evidence, however, that the final extinction of the mammoth, as well as the Woolly Rhinoceros, occurred in its last strongholds in Siberia without human intervention.[16] Climatic changes, including increased depth of snow cover, may have done them in. Boneyards exist in those places, but they do not show the remains of weapons, bones shattered for the extraction of marrow, or signs of fire that are characteristic of the sites of kills by *Homo sapiens*.

There are people who claim that expanding human populations were not a prime factor in the Pleistocene extinctions. Some contend that the rate and extent of climatic change at that time was unprecedented, and that the large herbivores of the time, as well as the predators dependent upon them, could not maintain sufficiently large populations in the pockets of remaining suitable habitat and thus died out.[17] Others point out that there was an extraordinary burst of evolution of different mammalian types in the early Pleistocene and see the late Pleistocene extinctions as the inevitable shake-out caused by competition in an overdiversified fauna. In this view, it is the high rate of divergent evolution in the earlier period that requires explanation, not the later extinctions, which are attributed to too many kinds of large animals competing for inadequate resources.[18]

Sorting Out the Human Role

All these explanations are not mutually exclusive by any means. Anyone familiar with the hunting peoples that have persisted into modern times cannot help being impressed with two things. The first is their skill at killing large terrestrial herbivores when they were available. The techniques seem to have been both widespread and ancient. For example, when Paul was living and hunting with Aivilingmiut Eskimos in northern Hudson Bay in 1952, his Eskimo companions described to him the use of inukshuks in the hunting of Caribou.

Inukshuks are piles of rocks erected to look something like standing hunters. The word comes from the same root as *inuk* (an Eskimo) and *Inuit* (the People)—which is what the Eskimos call themselves. A large V of inukshuks would be erected, with its apex at a lake shore or cliff. Then hunters would attempt to drive Caribou into the mouth of the V. Once in the V, the Caribou would be funneled toward their doom by the stone cairns which they mistook for people, especially since women and children would be stationed behind some of them, waving their arms and shouting. The animals would either fall to their deaths or into the water, where they were speared from kayaks. The inukshuks helped the Eskimos compensate for their relatively low population density and their consequent lack of manpower to organize large-scale animal drives.

Presumably, Pleistocene hunters used similar techniques. For example, in the Crimea there are boneyards indicating that early hunters took advantage of hundred-foot cliffs to drive herds of Wild Asses to their death.

The second characteristic of hunting people, at least if the Eskimos are a guide (as they seem to be), is the lack of a conservation ethic.[19] The Eskimos explained the erratic availability of game in religious terms. Some years the spirits were pleased and game was abundant; in others they were displeased and game was scarce. The souls of Caribou, Polar Bears, and seals were especially important to the Eskimos and had to be appeased by observing appropriate death taboos. For example, a newly killed seal must never be placed on a dirty igloo floor—the animal's soul would take offense at lying on a surface on which a woman had been walking. Therefore, fresh snow was brought in on which the dead seal could lie.[20] Similarly, scraping Caribou skins was forbidden while doing Caribou hunting—it could wound the Caribou's soul and make other Caribou less willing to be caught.

Observing or breaking the taboos were thus, in the Eskimo's eyes, major factors in the success of the group—and the area in which success was most essential was the hunt. To assure a supply of game, one did not treat an animal population properly; one simply treated individual animals and their souls properly. Such a world view worked out fine for hunter and hunted—until some environmental change made the animals more vulnerable or some technological change made the hunter more deadly.

By the 1950s, such a technological change had occurred in the world of the Aivilingmiut. The prized source of sustenance for these "People of the Walrus" was actually several kinds of seals. Walrus were hunted for food for the Eskimo sled dogs, and their tough hides were used for the making of harness. But the meat of Harbor and Bearded Seals was much tastier, and their skins could be fashioned into soft, pliable, comfortable boots.

The classic method of hunting seals after the sea ice had broken up in the spring was to harpoon them from skin boats. Then, in the 1940s the Eskimos trapped Arctic Foxes, which were much in demand for their white pelts. Successful hunters made small fortunes, which they spent on, among other things, powerboats and rifles. This changed the entire pattern of seal hunting. Now the Eskimos took potshots at the seals from the decks of their boats, trying to wound them so that they could then motor up close and spear the wounded animals. The technique was much less grueling than the traditional method, but it had an unhappy side effect. In the spring a layer of fresh water from melted snow and ice floated on the ocean's surface. In this less dense water, seals that were killed by the shots sank and could not be recovered. With the old ways, almost every seal killed provided meat for the Eskimo's larder. In 1952, after the advent of rifles, it was estimated that nineteen out of twenty kills were lost. The result, not surprisingly, was a precipitous decline in the seal populations.[21]

It seems fair, then, to conclude that Pleistocene hunters similarly were able to put considerable hunting pressure on the large mammal populations and were also unlikely to have been constrained by notions of conservation. Like the Eskimos, they would have killed when the killing was good, and like the Eskimos they would have cached their surplus for lean times ahead. If climatic changes were also reducing the herds, things would have been tougher for the hunters, and their proportional impact would have increased.

If the fauna was "overdiversified," there still is nothing in the evolutionary rulebook that says one species cannot have been a major factor in its rapid simplification. Especially in northern America, where *Homo sapiens* appeared suddenly and was already brave, clever, and well-armed, many of the large animals had precious little time to adjust by behavioral or evolutionary change to the presence of a new, deadly, and relentless predator.

It is important, however, to keep the hunting activities of prehistoric peoples in perspective. Whatever the precise human role in the Pleistocene mammal extinctions, those extinctions were relatively minor events compared with the extinctions of all sorts of flora and fauna now threatened through habitat destruction by civilized human beings. Hunting and food gathering is by far the most successful, long-enduring (at least 50,000 years), and least destructive mode of life developed by *Homo sapiens*; it has been the lifestyle of some 90 percent of all the people who have ever lived.[22] In the view of some social scientists, it was also the most leisure-filled and satisfactory lifestyle.

As distinguished anthropologists Richard B. Lee and Irwin DeVore wrote in 1968, that evaluation did not exclude "the present precarious existence under the threat of nuclear annihilation and the population explosion." They continued:

> It is still an open question whether man will be able to survive the exceedingly complex and unstable ecological conditions he has created for himself. If he fails in this task, interplanetary archeologists of the future will classify our planet as one in which a very long and stable period of small-scale hunting and gathering was followed by an apparently instantaneous efflorescence of technology and society leading rapidly to extinction. "Stratigraphically" the origin of agriculture and thermonuclear destruction will appear as essentially simultaneous.[23]

Recent Extinctions

Historic extinctions of exploited species seem to have followed a pattern resembling at least some of the prehistoric ones. Hunting has played an important role in reducing numbers, as has environmental change, but the final push over the brink has come from a reduction in some way of the species' ability to adapt. The complexity that may characterize the path leading to extinction is well illustrated by the most famous historic extinction of all—that of the Passenger Pigeon in North America.

The Passenger Pigeon was a fascinating creature. A pretty, graceful pigeon with a slate-blue back and deep pink breast, it didn't coo like a dove, but produced "shrieks and chatters and clucks."[24] Its greatest claim to fame was the gigantic size of its populations; it may have been the most abundant bird ever to exist. Audubon observed a flock of passenger pigeons passing over a period of three days. Sometimes, he estimated, they went by at a rate of over 300 million birds an hour. The passage of large flocks created a roar of wings that could be heard a half-dozen miles away.[25] Alexander Wilson, who with Audubon founded American ornithology, estimated another flock to contain 2 billion birds. The pigeons nested in long narrow colonies that could be forty miles long and several miles across. Their droppings in favorite roosting areas piled inches thick, killing all herbs and shrubs and eventually the trees themselves.

The birds occurred throughout eastern North America, where they fed on the fruits of forest trees—especially acorns and beechnuts. The reason for their flocking behavior is not known for certain. It may have helped them to find food; it may also have been a predator defense.

Early settlers in the United States, though, had no trouble adding the Passenger Pigeon to their diets. The nesting grounds were so crowded that the adults were always being injured or killed and the succulent squabs knocked out of the nests. All that was required was to wander through the colony picking up dinner. As the human population increased, however, two things began to happen. Railroads pushed through the wilderness, opening avenues for market hunters to ship the birds to centers like New York, and the great oak and beech forests in which the birds nested began to be cleared.

The market hunters devised ingenious ways of killing large numbers of the birds. The pigeons were suffocated by burning grass or sulfur below their roosts; they were fed grain soaked in alcohol and picked up dead drunk, batted down with long sticks, blasted with shotguns, or netted (after which their heads were crushed with a pair of pincers). One ingenious trapping device depended on a decoy pigeon with its eyes sewn shut, tied to a perch called a stool. "Stool pigeon" thereby became part of the language.

The demise of the pigeons was startlingly rapid. After the Civil War, many millions were shipped from the Midwest to New York—so many that live birds were used as targets in shooting galleries. But the huge flocks were by then gone from the coastal states, and by the 1880s they were dwindling everywhere. In 1878 one hunter shipped some three million birds from Michigan, the Passenger Pigeon's last stronghold. The last wild bird was seen in that state just eleven years later, and the last captive bird died in the Cincinnati Zoo in 1914.[26] Her name was Martha.

Economic extinction preceded biological extinction. The last birds in the wild were not killed by hunting, which became unprofitable as soon as the great flocks were gone. And there are still large areas of forest extant in the eastern United States that would serve as suitable habitat. But apparently the ability to form huge flocks was essential to the survival of the pigeons. When their population sizes became too small to maintain sufficiently large breeding colonies, nesting failures, inbreeding, and mortality from predation must have escalated and pushed the species to extinction.[27]

The fate of the Passenger Pigeon illustrates very clearly that enormous numbers do not guarantee the safety of a species. Under the right circumstances, species can move from superabundance to extinction with astonishing speed. The fate of the American Bison (inaccurately called the buffalo) is another example. The eastern U.S. populations, sometimes considered a separate race, were hunted to extinction by the early 1830s, and the Oregon race by midcentury. The northern Wood Bison still lives in relatively large numbers in the forests of Alberta and the Northwest Territories of Canada.

The prairie populations of bison were huge almost beyond belief. In vast numbers they blackened the plains, an estimated 30 to 40 million individuals. They showed clearly that at least part of the megafauna could thrive in the presence of skilled hunters. Native Americans made little use of bison until they obtained horses from the Spaniards. Once mounted, some tribes based their economies on the shaggy beasts—eating their meat and making multiple use of their hides. But they made no discernible dent in the bison population; apparently the number they took each year was less than the annual production.[28]

The arrival of the settlers from Europe, and especially of the railroads in the 1860s, signaled the start of the great bison slaughter. Professional hunters shot the animals primarily for their tongues and hides, leaving the carcasses to rot. Later, others collected the bleached bones that whitened the plains and shipped them east for use as fertilizer. Perhaps 2.5 million bison were killed annually between 1870 and 1875 by white hunters, and in 1883 the last significant herd, numbering perhaps 10,000 bison, was slaughtered. At the turn of the century only about 500 Plains Bison remained—finally under legal protection.

The bison was luckier than the Passenger Pigeon—it was pulled back from the brink. Today there are perhaps 25,000 in North America, scattered through parks and in private herds, but no prairie bison exist "in the wild." Humanity may have been lucky, too. A fertile hybrid has now been produced between cattle and bison by a California rancher. The hybrids, called "beefalo," are reported to be very tasty, leaner, and more productive than beef. Beefalo are easier to raise than cattle, grow faster, and require no grain feed.

At best, if accepted, beefalo could make meat cheaper and healthier, with less fat; at the least it could add variety to human diets.[29]

ENDANGERING FOR FOOD TODAY

Wild species of animals are still hunted by human beings for food on land, just as whales and fishes are hunted in the sea. Deer hunters in Pennsylvania, Bushmen stalking gazelles in Namibia, or people in western China hunting the Chinese Giant Salamander are simply carrying on an ancient human tradition. Throughout much of the world, the level of predation from hunting is low and has little or no effect on the populations cropped. But in some cases, for instance if the prey is rare (as is the salamander) or the hunting turns to slaughter, populations and species go under.

Sometimes wildlife suffers overexploitation as a result of unusual political or economic circumstances. One serious and grotesque incident occurred in Uganda in early 1979 when Tanzanian troops massacred wildlife in what had been one of Africa's most bountiful game reserves—Ruwenzori National Park. Troops out of control of their commanders butchered the wildlife, and the meat was purchased by Ugandan businessmen. American biologist Karl Van Orsdal was an eyewitness to some of the murders at Lake Edward: "Two Tanzanian soldiers stood laughing while a third, lying on the ground, fired off rapid bursts at a large group of hippos out in the water . . . a group of seven or eight Ugandan civilians [were] butchering a dead hippo with axes and machetes a few hundred feet farther down the shore."[30]

There was a lot of money to be made. A dead hippopotamus can yield as much as 1,875 pounds of meat worth over a dollar a pound. At the end of three and a half months, when Van Orsdal left, he estimated that about 30 percent of the park's 46,500 large animals had been killed—6,000 hippos, 5,000 Uganda Kob, 2,000 buffalo, 400 Topi, 100 elephants, and 70 lions. If the slaughter stopped soon thereafter, most of the species would probably recover, although some concern has been expressed about the kob.

Events following the establishment of the Islamic Republic in Iran have taken a course not unlike that in Uganda. Wildlife has been exterminated indiscriminately. Poachers on motorcycles have machine-gunned gazelles that were once protected; sturgeon have been dynamited in the Caspian Sea. Thousands of acres have been cleared of hardwoods for conversion to grazing or farming. Many animals that had been carefully protected during the Shah's regime are now so tame that they quickly fall prey to hunters with automatic weapons. Some of the most seriously threatened mammals in the world— including the Caspian Tiger, the Wild Ass, and the Persian Fallow Deer—are now in even greater jeopardy.[31]

Such slaughters are not restricted to developing nations in turmoil. In the tightly controlled Soviet Union society, meat is in such short supply that wildlife has been attacked at an unprecedented level. In the spring of 1976, after three hundred young ducks were banded on a Siberian lake, all three hundred bands were returned to the ornithologists responsible. Hunters had bagged every bird. Poaching on Soviet reserves is rampant. The animals of the Kyzyl-Agach Reserve on the Caspian Sea have been subjected to periodic assaults by groups of army officers operating from helicopters, all-terrain vehicles, and even tanks. Not surprisingly, little wildlife remains. Difficult as it is to believe, a Soviet division stationed near Lake Baikal for years reportedly has been using heat-seeking missiles to hunt deer![32]

Perhaps the most repugnant and compassionless hunting practiced on our planet recently was not in the wilds of Africa or on the steppes of Russia, but in Australia. Australian graziers have long killed every kangaroo they could because they compete with their sheep for grass. As early as 1863, the great naturalist and artist John Gould feared that the Red Kangaroo and some other "fine species" of marsupial would be exterminated by the stockmen.[33] He was wrong—the Red Kangaroo remained common in drier areas where the sheep could not thrive.

Then in the late 1950s, a market was discovered for kangaroo meat as pet food, substandard sausage, and kangaroo-tail soup. The result was a stampede to hunt the kangaroos. The standard technique was to "spotlight" them from cars at night. The kangaroos would freeze in the light and were shot with rifles. Some were killed immediately, but some hunters purposely just wounded them —sometimes leaving them to suffer for hours or days so that their meat would remain fresh until they could be collected. The night hunts were treated as "sporting events," even though neither courage nor skill on the part of the hunters was required. In 1980 a new hunting method became popular: two people chase them on a motorcycle, one steering, the other gunning down the fleeing animals.

Since the founding of the country, about one million kangaroos have been slaughtered annually in Australia. The killing continues today, although fortunately the large kangaroos that are hunted seem to be holding their own. In contrast, some of the smaller species are succumbing to habitat destruction.

Many excuses for killing the kangaroos have been made, especially by graziers, and are related to their misconceptions about the impact of the kangaroos on pastures that the stockmen themselves have often ruined by overgrazing with sheep. But the main reason once more is greed mixed with a lack of compassion. Australian conservationists fear that, since the United States has lifted its ban on importing kangaroo skin products, the slaughter will escalate and begin endangering the kangaroo populations.[34]

The Ugandan, Iranian, Russian, and Australian slaughters of wildlife clearly are extreme examples of contemporary uncontrolled hunting. Perhaps the most distressing aspect of these affairs—and similar ones such as the much-publicized annual slaughter of baby seals in Canada, which is controlled —is that they bring home just how little compassion there is for animals in much of the human population. People may kill out of what they consider to be economic necessity or sport, but either kind of killing can be accompanied by a certain sympathy for the animals killed. Indeed, hunters and anglers often are also ardent conservationists—something that should be recognized even by people who find hunting morally reprehensible. Yet, obviously, many human beings still can commit mayhem on other species without a qualm.

THE WILDLIFE TRADE

Many species are under direct attack from humanity for other reasons than to provide food. Pressure toward extinction of such species indeed continues —and in many cases is even rising—in spite of much-heightened public awareness of endangered species and in spite of a proliferation of protective laws, particularly in developed countries. International trade in wildlife, for example, goes on at a level unsuspected by most people. Great numbers of animals are collected and shipped around the world for scientific and medical research alone. Both animals and plants are collected for display in zoos and botanic gardens, for the pleasure of private collectors, and for products that can be made from them.

As an example of how far-flung the trade for research animals has become, a few years ago we received a totally unsolicited offer from Nigeria to sell us a variety of animals for "research purposes." The accompanying list included ostriches, two kinds of geese, Marabou Storks, foxes, Crowned Cranes, two kinds of monkeys, baboons, and Chimpanzees.

Primates in particular often suffer depredations in the name of research. Collecting for zoos and laboratories has helped push Gorillas toward extinction—especially since large numbers have been killed in the process of capture or died in captivity before they could be displayed or experimented upon.[35]

One of the more preposterous and tragic examples involves the recent establishment, with the help of a French oil company, of the International Center for Medical Research in Franceville, Gabon. The center was established to study and help cure human infertility, deemed to be a serious problem in Gabon, where the prevailing view is that the country is underpopulated. The rate of natural population increase in Gabon in 1979 was 1.1 percent annually —a rate that, if continued, would double its population in 63 years. Its population density is very low by the standards of this overpopulated world, but it is not at all clear that further population increase could do anything but reduce its relatively high standard of living. Average per capita income in Gabon is almost as high as that of England—based on large resources of iron, manganese, uranium, and oil.[36]

In order to rescue Gabon from a "problem" that much of the world wishes it had, the new research center has erected a large primate facility. At the center, Gorillas and Chimpanzees will be studied to find an answer to the problem, although President Bong of Gabon has reportedly admitted that the infertility in the human population is due to an epidemic of gonorrhea. The primate facility is viewed as an outlet for baby Gorillas that become "available" as their mothers are shot by the Gabonese for food. According to one observer, in late 1979 six baby Gorillas had gone into the facility, and five had died because of the inexperienced staff.

As Dr. Shirley McGreal of the International Primate Protection League (IPPL) aptly pointed out: ". . . rabbits would have been better 'animal models' of human fertility, as gorillas and chimps breed so badly that they are getting close to extinction and can't compensate reproductively for human predation."[37] The situation is especially ironic since it is the pressure of expanding human populations that threatens Gorillas everywhere. Henry Heymann of the IPPL noted that in Gabon: ". . . the gorillas are being compelled to contribute their lives, health, freedom, and sanity to the expediting of their own demise. There is a resemblance to concentration camp prisoners being forced to dig their own graves before being murdered."[38]

So the direct pressures on the Gorilla persist in Gabon. They are openly hunted for food in a relatively rich country, and the hunting is aided and abetted by a crackpot "scientific" scheme.

The demand for great apes is high in medical research everywhere because of their close similarity to human beings, but their use today can be justified under only the tightest controls. Unfortunately, the quality of much medical research is poor, and many of the projects for which primates are imprisoned and sacrificed are without merit. Gabon's project does not stand alone in this regard. It is sad that a portion of the scientific community remains insensitive to the plight of endangered species, and even sadder that they are apparently without empathy for humanity's nearest relatives.

"Scientific" pressures are also put on endangered species by zoos, which all too often buy animals from unscrupulous animal dealers. The conditions under which the animals are obtained and shipped are often horrendous. For example, in August 1978, three Malay Tapirs, three Leopard Cats, fifty Stumptail Macaques, one Pileated Gibbon, one White-crested Gibbon, and thirty-eight White-handed Gibbons arrived at the Bangkok airport in six crowded cages. They were held for several days in "intense heat and insufferably cramped conditions" before being shipped to Belgium. The International Union for the Conservation of Nature and Natural Resources (IUCN) estimated that, because of the way these animals were captured, the forty captive gibbons, all young, represented the destruction of at least a hundred breeding groups.[39] The shipment was certainly inexcusable and probably illegal, and the animals were almost certainly bound for zoos.

There is also a substantial attrition of wild-animal species for collections outside of zoos. Very large numbers of freshwater and coral-reef fishes are collected for the aquarium trade. The numbers are not accurately known, but the magnitude of the flow can be guessed from a few statistics. In 1970 nearly 84 million living fishes were imported into the United States, and by 1979 the number had probably increased to about 250 million.[40]

Over 2 million reptiles were legally imported into the United States in 1970, and that number had doubled by 1979. Some of those reptiles were bound for private collections, but zoos probably still made up most of the trade. There is an additional flow of unknown dimensions of illegal imports, especially of rare snakes. Eight of the nation's top zoos were identified in 1977 as buyers of illegally imported reptiles. Dealers' catalogues list protected species for sale, and the poaching of rare snakes such as the Arizona Ridge-nosed Rattlesnake is becoming something of a cottage industry in the southwestern United States.[41]

There is also extensive commercial trading and collecting of reptiles and amphibians in Italy. Each spring many tree frogs, tortoises, lizards, and snakes are collected in Italy and the Balkans and shipped into Central Europe for display in zoos or to be kept as pets. Wild populations of European Tortoises have been put in great jeopardy by collecting for the pet trade, and some European lizard and snake populations may also be in trouble. The Smooth Snake in England is already endangered, but it is still collected and offered for sale in pet shops. Tens of thousands of turtles and tortoises are imported by Great Britain annually for resale in the pet trade. Between 1967 and 1972, the United Kingdom received over 1.2 million specimens of the vulnerable Mediterranean Spur-thighed Tortoise from Morocco alone—and similar numbers are believed to go to continental Europe. It is thought that 80 percent die in the first year of captivity.[42]

The birds flowing into the United States and Europe to be caged and kept as pets number in the millions, and this doubtless constitutes a serious drain on many populations. Most wild birds do not thrive in captivity, and countless numbers die in the processes of capture and transport. For example, one of the most highly valued birds in the trade is the brilliant red Cock of the Rock, an inhabitant of the northern Andes. It is thought that fifty are killed for every one that arrives to grace a zoo's display.[43]

Plant species, too, have suffered from hunting pressures. In that center of snake poaching, the southwestern United States, another illegal activity has arisen—cactus rustling. Keeping of cactus and other succulent plants is now so popular in the United States that, between October 1977 and September

1978, almost 7 million were imported from over fifty countries.[44] But many collectors are now turning to United States deserts for their specimens. These plants usually adorn the collectors' houses and gardens until they die; they cannot reproduce and maintain their populations in isolation and usually in an unsuitable habitat.

Arizona now has seven "cactus cops" attempting to stop the looting of that state's flora of ornamental native plants, mostly cacti. Arizona law prohibits the digging up of any of 222 protected plants and provides stiff penalties—up to a year in jail and fines as large as $1,000—for violators. But the problem of enforcing the law, when it requires patrolling almost 100,000 square miles, are insurmountable for such a small force in the face of the current cactus craze.

The cactus craze has swept not only the United States, but Europe, Japan, and the Soviet Union. In the U.S.S.R. there are now 114 cactus collectors' clubs. In Japan, giant Saguaro Cacti, native to Arizona, sell for about $40 a foot. Recently an expedition sponsored by Japanese dealers removed *all* of the native cacti and other fleshy plants from an island off the Baja California coast. Several of the species taken were known only from that island and are now presumably extinct in the wild.

One entire population of a species of cactus, out of two known populations, was taken into Germany in 1978 in fifteen suitcases. In 1979 customs officials at Frankfurt airport seized 3,600 individuals of some of the rarest cactus species from the suitcases of a group of collectors who had been off on a "cactus study" tour set up by a Stuttgart tourist agency.

The magnitude of the looting of cactus populations is mind-boggling. In one area of Texas near Big Bend National Park, some 25,000 to 50,000 cactuses per month are being harvested—many of which, like the caged birds, die before reaching the market. Even cacti in the national park are not safe: they are now quite rare near roads. In 1977 something on the order of 10 million cacti were shipped out of Texas alone.

In short, cacti have become big business. The more spectacular specimens sell for hundreds of dollars, and smaller ones are being huckstered through supermarkets. Rustling in Arizona is now a million-dollar-a-year enterprise. As a result, the Smithsonian Institution recently estimated that seventy-two species and varieties of cacti (representing about a quarter of the cactus family) are now in danger of extinction, or quickly becoming so. And many cacti are keystone species, providing nourishment and living space for many desert animals.

Cacti are not the only plants endangered by collectors. Orchids are extremely popular, more than a quarter-million having been imported into the United States in recent years.[45] Collecting pressure is undoubtedly endangering some populations and species of the diverse orchid family, which contains some eighteen thousand species. The rarest of British orchids, the Ghost

Orchids, cannot be molested in any way without breaking the law, and fines of up to a hundred pounds can be imposed. The last British colony of Ghost Orchids produced only five flowers in 1974. Two were stolen, and a third was trampled by collectors and sightseers. Other rare British orchids are being preyed upon by collectors who sell them on the European continent for as much as four hundred dollars each.[46]

In the Alps, a rare saxifrage plant is threatened by skilled climbers who scale north-facing vertical cliffs to collect it. At the opposite climatic extreme, in the jungles of Sumatra, the largest flowers in the world are threatened. *Rafflesia arnoldii,* a plant that parasitizes other plants, has blooms a yard across— making it an apparently irresistible magnet for collectors.[47]

The Fur Business

Perhaps the most widely known direct threat to terrestrial mammals comes from the hunting of them for one of their principal mammalian characteristics —their hair. Trade in animal pelts is a much bigger business than that of cacti, and it has proven even harder to suppress—not surprisingly, since many of the original owners of the skins are species that inhabit poor countries where the pressure to exploit them is understandably high.

The use of furs for clothing, rugs, tents, and the like is, of course, a tradition probably as old as *Homo sapiens* itself. Human beings have extracted a living from their immediate environs for most of the species' history, and any useful qualities made another species quite properly fair game. But the indiscriminate killing of a large number of mammals for their hides alone is a development of rather recent times—chiefly of the last century or two when economic conditions could support widespread trade in pelts.

Virtually no common mammal with a usable skin has escaped remorseless exploitation, and in the process common mammals have often been converted into uncommon mammals. Consider the cuddly Australian Koala—a creature that looks like a teddy bear and that many Americans think is called the "Qantas," thanks to the advertising campaigns of Australia's international airline.

In a total of about two years spent in Australia, we never saw a Koala outside a zoo or a reserve. It was not always so: the animals used to be abundant. But their skins were valuable, and they were mercilessly hunted from the earliest days of European invasion. By 1900 Koala numbers had been greatly decreased everywhere outside of the northeastern state of Queensland, although in the year 1908 it was still possible to ship almost 60,000 pelts through Sydney markets. In south-central Australia the Koala was exterminated soon after the end of World War I. Before then, one to two million skins were shipped out annually, frequently disguised with labels like "beaver," "skunk," "silver-gray possum," and "Adelaide chinchilla."

Queensland is the frontier state of Australia. It is noted for its cheerful and independent people, its unhappy aborigines, and its conservative and parochial politicians—sort of a down-under Texas. The state refuses to go on daylight saving time, and a standard quip from airline pilots is "We've just crossed the Queensland border—set your watch back one hour and ten years."

Queensland by 1927 was the last stronghold of the Koala. That year, in an event that Australian biologist A. J. Marshall called the most sordid episode in the history of the state,[48] Queensland declared open season on the Koala, even though the species' precarious position was well known. The state government licensed no fewer than ten thousand trappers and thus made itself accessory to the slaughter of over a half-million Koalas in a few months. Why did the Queensland government do it? For the same reasons that politicians often permit such atrocities to take place: votes and money. As Marshall put it: "Small landholders and farm workers wanted the money. And the government wanted their votes. Rural votes are often vital votes. These would have been alienated had the cabinet failed to proclaim the open season that a single hillbilly pressure group so eagerly wanted."

The Koala is unusual in that its numbers have been reduced much more by direct persecution than by habitat destruction. (There has been plenty of the latter, but suitable eucalyptus forests are still plentiful.)

Many other fur bearers have suffered both from being hunted for their skins and from the destruction of their habitats. And some of the most beautiful fur bearers, the great cats, have suffered additional assaults because of their own predatory habits.

The Snow Leopard, for example, used to range widely through the highlands of central Asia. Although human destruction of the Snow Leopard's habitat (and that of boar, deer, gazelles, wild goats, and its other prey) have undoubtedly had a negative effect, it is mainly threatened by hunters seeking its gorgeous pelt: the fur has a background color of pale gray tinged with cream and is marked with black rosettes. And the animal's protection is not made any easier when the cats include domestic animals in their diets.

The magnificent tigers are in deep trouble for the same combination of reasons. Moreover, like African Lions, tigers have the audacity to supplement their diets on occasion with the great exterminator itself: *Homo sapiens.* There are probably a few thousand Bengal Tigers left in its former range in India and adjacent countries, where they are especially hard-pressed by habitat destruction, concentrating the remaining individuals in ever smaller areas, ever closer to the people who fear them and want to kill them. Individuals are still poached, and as recently as 1979 skins were illegally imported into Great Britain.

The Caspian Tiger appears to be extinct, a victim of the destruction of the vegetation growing along rivers that were its habitat, in the course of the development of large-scale irrigation and agricultural schemes in the Soviet Union. Its demise was hastened by extermination squads of soldiers employed to remove the tigers as threats to people and domestic animals.

The Bali Tiger is also gone, and the Javan Tiger is at best barely hanging on by the tips of its claws. The Sumatran and Corbett's Tigers are doing better, the latter especially on reserves. There are perhaps three hundred wild individuals of the Siberian Tiger left. Its long-haired pelt is the most valued of tiger skins, and the Chinese greatly value the medicinal properties of various parts of its body. It has been heavily hunted, but its decline is blamed chiefly on the widespread destruction of the vast forests, especially in Manchuria, that were its home—and the destruction along with the forests of its natural prey. The status of the Chinese Tiger is unknown, but the species classically has been persecuted intensively, and little natural habitat remains. It seems likely that neither the Siberian nor the Chinese Tiger is long for this world outside of zoos.[49]

The Cheetah and the Lion were both once widely distributed in Asia, but few of either are left there now.[50] The Lion, for the moment, is relatively secure in Africa, but the Cheetah is in a precarious situation. The Cheetah is still poached for its fur, and it seems unlikely that the hunting will be totally controlled in the foreseeable future. But even if its fur were useless, it would still be endangered; the ecology of the Cheetah tends to make it vulnerable to extinction. This swift hunter, which can accelerate from zero to forty miles an hour in a few strides, exists naturally in the African savannahs at low densities —about one for every forty or fifty square miles. The daylight attacks in which it runs down its prey are conspicuous, and it is thus subject to loss of its captures to stronger predators, such as Spotted Hyenas, Leopards, and Lions.

Cheetah young are quite vulnerable. They follow their mothers on long cross-country treks where they fall prey to other predators. They also cannot benefit from a "baby-sitting" system of the sort that helps protect Lion cubs. The Cheetah mother is usually solitary when she hunts, whereas a lioness living with a pride made up mostly of other females often has a surrogate available to ward off predators while she hunts.

The Cheetah's hunting habits and its relative weakness make it easy for herders to detect and kill it when it turns its attention to domestic animals. And as the savannah's game gives way more and more to herds of domestic animals, that shift occurs more and more frequently. The result is an increase in the hunting of Cheetahs as livestock predators.

At the same time, there is an increasing migration of farmers onto the grasslands. This leads to a habitat fragmentation that serves the widely dispersed Cheetahs badly.[51] Even large reserves are generally not capable of holding populations large enough to be safe from random extinctions and, in all likelihood, from loss of genetic variability. And natural migration between parks to permit restocking may become impossible, forcing heavy management responsibilities on *Homo sapiens* even if the parks themselves survive.

The future of the Cheetah therefore depends on whether ways can be found both to alleviate the direct endangering from the predator control activities of herders and to arrest the fragmentation of the savannahs by cultivators—two very big tasks.

Endangering for Other Products

In addition to fur bearers, many animals have suffered, and continue to suffer, human predation for the nonedible products they can yield. In the Orient and in South America, butterflies are used to make decorative objects. Many a crocodile, alligator, and snake has been killed so that its skin could be made into shoes and handbags. The Cuban Crocodile, like so many of its relatives persecuted for its hide, persists in only two small swamps.[52] Several species of giant sea turtles are now endangered in part because their shells are used to make tortoiseshell products. Millions of birds, from ostriches to birds of paradise, have given their lives so stylish women could adorn themselves with their feathers.

Much of the pressure on elephants is generated by the continued depredations of ivory poachers. Zaire and some other African countries are hubs of an ivory trade that threatens elephants over the entire continent. The elephant populations of Kenya alone are estimated to have been reduced by two-thirds in only eight years. Poaching in Uganda accelerated during Idi Amin's reign and has continued unabated. Since 1972 the elephant population in Ruwenzori National Park has fallen from 3,000 to just 150 individuals.[53] The ivory trade is believed to account for the deaths of 50,000 to 150,000 elephants each year —up to 12 percent of the total African population. Each month C-130 cargo planes carry ivory shipments to South Africa, and ivory also leaves Africa through Burundi, the Congo, and the Central African Republic. But it is difficult to blame African villagers for being active as poachers when the sale of a large pair of tusks can bring in the equivalent of ten years' income![54]

The animal that has suffered most severely in recent years from killing for a product is that great browser of sub-Saharan Africa, the Black Rhinoceros. In recent years, that trade, in particular demand for the rhino's horn, has led to a catastrophic level of rhino poaching. Around 1970 about 20,000 Black Rhinos lived in Kenya. By 1980 the population was less than 10 percent of that —perhaps as few as 1,000. The sight of a Black Rhino on the Serengeti plain, wandering majestically, surmounted by tick birds, or wallowing like a huge pig in a mudhole, has been one of the most popular tourist attractions of Africa. Soon the memories and photographs of travelers may be the only places wild rhinos exist; the real ones will have been wiped out to "cure" the impotence of the ignorant rich.

PREDATOR CONTROL

A depressing amount of direct endangering of species is connected with predator control. Wherever human beings or their domestic livestock have become the prey of carnivores, *Homo sapiens* has quite reasonably tried to strike back. In early encounters, human beings faced animals like cave bears, lions, and tigers with clubs, spears, bows and arrows, large rocks, and pure courage. For a long time the battle was more or less even, but the invention of firearms changed the situation.

Wherever human beings had muskets and then rifles, the large predators were pushed back. The Grizzly Bear has been wiped out over much of the United States, including the state of California, where it ironically—and perhaps symbolically—is the state animal. The Wolf, victim of the most unjustly bad press of virtually any animal, has been exterminated over much of Europe and North America. That intelligent animal is anything but the vicious, treacherous beast it was once pictured—a point brought home to many in Farley Mowat's classic book *Never Cry Wolf.*[55] In Tasmania, the Wolf's marsupial namesake, the Thylacene Wolf, was similarly persecuted and driven back to inaccessible marginal areas for similar reasons.

Predator control programs, official or unofficial, contribute to the jeopardy of many species. The Bald Eagle, although strictly protected, is still gunned down by hunters in the United States. Eagles have even been pursued and shot down by men in helicopters.

Much effort in the United States goes into attempts to control the Coyote, a species in no danger whatever of extinction. Indeed, the Coyote thrives in the presence of *Homo sapiens* and has increased its range and its numbers. It seems to have evolved into a bigger, tougher, and smarter animal under the selection pressures of human attempts to suppress it. The biology of Coyotes, including an ability to increase their reproductive rate under duress, allows their populations to sustain enormous mortality and still persist.

In some areas, Coyotes may cause significant losses of sheep and lambs, but not to anything like the degree implied by popular western bumper stickers. Pressures from sheep-grazing interests once promoted broadcast use of chemical poisons against Coyotes—programs that resulted in great mortality in a wide variety of other wildlife until they were halted by executive order of the President in 1972. The effectiveness of these control programs against the Coyotes, however, is problematical. Very often they resulted in a larger Coyote population. Where they *were* successful, it sometimes was necessary to open ground squirrel control programs to keep the rodents from eating too much of the sheep's grass! The ground squirrels were previously controlled by—you guessed it—the Coyotes.

A sensible control program would involve attacking depredating individuals, not attempting to suppress all Coyotes over wide areas. It would also involve teaching sheepmen and others, especially those whose livestock graze on public land, where preserving wildlife is supposed to be one of its "many uses," that a certain level of predation is an expected cost of doing business. There certainly are ways to protect herds from predation, too, that do not require attempts to exterminate the predators. Maintaining one or two guard dogs with each flock is one simple and apparently effective method. The mere presence of such dogs often seems to be enough to discourage Coyotes from attacking a flock. But many sheepmen today think they should be free to graze their herds without protecting them in any way.[56]

Humanity has also attempted to exterminate populations and species of herbivores that attack domesticated plants. A major source of mortality for African Elephants has been control programs implemented to keep them from molesting farms. Entire populations have been exterminated in those programs. A large animal on an overpopulated island, the Ceylon Elephant has been pushed into the endangered category by hunting both for sport and for predator control. The elephants' depredations of plantations led the government to institute a bounty program in 1831. One celebrated hunter of the time, a Major Rogers, promptly killed more than fourteen hundred elephants, and the number of kills in general reached the point where the government had to cut the bounty from ten to seven shillings to save money. Today a couple of thousand individual elephants remaining in Sri Lanka are dependent on a few inadequate reserves for survival—which seems unlikely in the face of the expanding human population and expanding agriculture.[57]

In summary, *Homo sapiens* has a very long history of direct attacks on other species, some of which have resulted in extinction of the species under attack. People have hunted animals for food and other products for millennia, and probably at least contributed to the extinction of many large mammals well before the agricultural revolution. People have also killed—and still do kill—animals to prevent real or imagined threats to themselves or their domestic animals and crops.

Direct pressures against other species thus are obviously an important factor in extinctions. In many cases, however, such as the big cats, elephants, and rhinos, the direct hunting pressure has been augmented by damage or destruction of the ecosystem in which the animal lives—its habitat. Indeed, the *indirect* method of habitat destruction is by far the deadliest means by which humanity has pushed other organisms to extinction. And it is that indirect attack that holds the greatest threat to other forms of life in the future.

Chapter VII

Indirect Endangering

The whales, the rhinos, the tigers, the elephants, these are
the visible tip of the iceberg. But what we're really talking
about is the biological impoverishment of this planet.

—RUSSELL TRAIN,
Earth Day '80 Press Conference,
Washington, D.C., January 18, 1980

ONE of the prettiest commuter drives in the United States starts in the
smoggy city of Honolulu and goes northeast through the steep green
mountains of the Koolau Range. When one is disgorged from the tunnels that
penetrate the range, the view of the steep pali—cliffs plunging down to the
Pacific—is spectacular. Separated from Kailua Bay by the Mokapu Peninsula
is Kaneohe Bay, home of the University of Hawaii's Marine Laboratory. At
one time virtually all of Kaneohe Bay was a coral wonderland. Local people
caught and ate parrot fishes and other denizens of the reef. There was a
thriving small industry that took tourists on glass-bottom boat rides out to
view the coral gardens with their gorgeous fishes.

Although the Hawaiian reef-fish fauna is not nearly as diverse as that of the
Australian Great Barrier Reef, it is highly varied, and many of the fishes are
exceptionally beautiful. In our work on butterfly fishes in Hawaii, we noticed
that the schools often were feeding on plankton high in the water, far from the
protection of the coral. Such behavior is not found in Australian waters;
feeding on the plankton is largely the business of the damselfishes there and
is observed only occasionally in one butterfly-fish species, which still remains
close to the reef. But there are many fewer damselfishes in the waters off
Hawaii than over the Barrier Reef, so, with no competition from damselfishes,
the butterfly fishes have moved upward. The result is often a dazzling, multi-

hued display. In the crater of Molokini, a half-submerged volcanic cone off the southwest coast of Maui, we have often photographed schools of bright-yellow Lemon Butterfly Fishes flashing against the incredible deep blue of crystal-clear ocean waters. Molokini fortunately is a marine reserve and far from the sites of population growth and unplanned development that have desecrated so much of Hawaii's beauty.

Kaneohe Bay was not quite so lucky. As people flocked to Oahu, the once sleepy town of Kaneohe burgeoned into a genuine suburb. Rows of houses marched up the hillsides in the wake of bulldozers mowing down the vegetation. Replanting for erosion control was far from the minds of the developers, and the laws requiring it were not enforced. And as Kaneohe grew, so did the volume of sewage that had to be disposed of.

In Hawaii, as in many other tropical areas, there is a lot of rain, and it tends to arrive in cloudbursts. When the cloudbursts struck the subdivisions above Kaneohe Bay, the result was torrents of mud gushing down the cleared hillsides. Silt-laden waters poured into the bay. The corals were inundated not only by enormous amounts of silt but also by the increasing volume of un-treated sewage discharged into the waters off Kaneohe.

Corals cannot survive being covered by muck. Their tiny tentacles must be free to capture the minute animals they eat. And the light must reach the algae living within them so that photosynthesis is possible. Corals do have self-cleaning mechanisms, but they were inadequate to the heedless assault launched by humanity from the shore of Kaneohe Bay. The coral gardens in much of the bay died. The great diversity of life, the great beauty, and the edible fishes all disappeared. The bay became a coral graveyard occupied by "garbage" organisms, especially hordes of a most unattractive sea cucumber noted for its ability to resist heavy pollution.[1]

Hawaii presents many other textbook examples of the indirect impact of human activities on organic diversity. For example, of sixty-eight species of birds unique to the Hawaiian Islands when Europeans first arrived, forty-one are now extinct or virtually extinct.[2] One entire family of birds, the Hawaiian honeycreepers, has suffered greatly. Deforestation has been one major cause of the extinctions. Enormous areas, including virtually all of the island of Lanai, have been cleared for sugarcane fields, pineapple cultivation, cities, and resorts.

And, like the biota of most islands, that of Hawaii has proven extremely vulnerable to imported organisms. More plants and animals are known to have gone extinct in Hawaii than in all of North America. Domesticated animals —cattle, goats, and pigs—have played havoc with the islands' vegetation. Rats were introduced accidentally, and they attacked both ground-nesting and tree-nesting birds. Mongooses, imported to control the rats, enthusiastically added the birds to their diet. Various exotic birds such as the Common Mynah and House Sparrow have been introduced and have proven tough competitors for the natives.

As if these weren't threats enough, the accidental introduction in 1826 of a mosquito has greatly contributed to the decimation of Hawaii's birds. A denizen of the lowlands, the mosquito carried diseases against which the native birds had no resistance. Its introduction led to the extinction of several species and the restriction of others to altitudes above two thousand feet, where the mosquito cannot live.

The fate of the Hawaiian birds illustrates an important point. No one overexploited them for food, for their beautiful feathers, for use in the laboratory, or because they thought the birds had aphrodisiac qualities. Similarly, the reef fishes of Kaneohe Bay were not decimated by hungry fishermen or commercial collectors for the saltwater aquarium trade. Although many plant and animal populations have been driven to extremis or to extinction by deliberate acts of exploitation, such acts do not constitute the greatest threat to organic diversity. It is rather the unthinking endangering, the consequences of human acts not directly aimed at the endangered organism, that are the major factors in the impoverishment of Earth's biota.

Many people realize the dangers of overexploitation, but the much more general threat of habitat destruction is lost on most—including those who should know better. They are simply unaware of basic principles of ecology and thus of the various ways in which populations and species may be endangered by changes in their environment. It is these diverse, often circuitous, and usually unintentional pathways to extinction that concern us here.

PAVING OVER

"Woodside's gone. This is the last sample we'll ever get there." Stuart Weiss, one of the Stanford undergraduates in our research group, was dejected. It was April 9, 1980, and he had personally observed the bulldozing of the habitat of one of the Edith's Checkerspot Butterfly populations that our group had been studying for two decades. He wasn't alone in his depression. It was not just the loss of an important population of our experimental animals that we all found disturbing. That was bad enough. It would make more difficult our attempts to understand such things as how to protect crops from insects without overuse of pesticides.

The sadness was also caused by the opportunity to observe directly a one-way process that is going on continuously all over the planet—the extermination of multitudinous populations through the paving over of their habitats. As the bulldozer swept through the Woodside location, we knew very well that many populations of insects, mites, herbs, and other organisms that made their homes on that piece of serpentine grassland were also going to be replaced by one more housing development for *Homo sapiens.*

Down the spine of the San Francisco Peninsula, serpentine soil occurs in islandlike patches. It is a soil type that is unusually poor in nutrients, and for that reason it supports an unusual flora and fauna. We knew that the destruction of any of these patches reduced the chances for the unique serpentine plants and animals on the peninsula to be maintained. The serpentine "islands" tend to be small and the populations on them subject to occasional random extinctions. When this happens, they can be reestablished by migrating organisms from other serpentine patches. But as the number of patches is reduced, the chances of successful migration decline, and the possibility of extinction in the entire area increases.

Three subspecies of Edith's Checkerspot Butterfly live in the San Francisco Bay Area. Baron's Checkerspot in the north part and Luesther's Checkerspot to the south and east seem to be holding their own—often wisely having established their populations on chaparral slopes too steep to build on. The metropolis of the third subspecies, the Bay Checkerspot, is the San Francisco Peninsula—once one of the most beautiful areas of the United States and still a relatively pleasant place to live.

After World War II, many American servicemen returning to the United States from the Pacific Theater saw no reason to keep on going after they got back to the city by the Golden Gate. The great Bay Area population and building boom was on. One after another, the islands of serpentine disappeared under tract houses and shopping centers, and populations of the Bay Checkerspot began to disappear.

Our research group started studying the Bay Checkerspot in 1960. It proved to be a very nearly ideal tool for investigating broad questions in population biology, and in two decades that research has thrown light on important principles ranging from why plants produce all those useful chemicals to how population sizes of beneficial and harmful animals might be most effectively manipulated. Colonies of the Bay Checkerspot are now by far the best-known natural populations of any invertebrates—possibly of any nonhuman organism.[3]

For a number of reasons, butterflies are very convenient organisms for studying population dynamics and evolutionary problems. They have short life cycles and often large populations, and most species are easily identified even on the wing. In recent decades, butterflies accordingly have become more and more prominent in studies of the ecology and evolution of natural populations.[4] The great naturalist Henry Walter Bates wrote prophetically in 1864, just five years after *The Origin of Species* was published: "As the laws of Nature must be the same for all beings, the conclusions furnished by this group of insects must be applicable to the whole organic world; therefore, the study of butterflies—creatures selected as the types of airiness and frivolity—instead of being despised, will some day be valued as one of the most important branches of biological science."[5]

Naturally, we have become increasingly distressed by the threats to the experimental system into which we and many students and colleagues have put so much effort and from which so much has been learned. But the jeopardy of the Bay Checkerspot is also symbolic of the little-noticed process of the extinction of millions of populations due to urban sprawl. The losses of these populations usually go unheralded, since most of the victims—bacteria, herbaceous plants, worms, mites, insects, frogs, lizards, small mammals, and the like—are among the more obscure rivets in our spacecraft.

Butterflies play a unique role as an indicator of losses among less conspicuous organisms. They tend to have quite tight population structures, with relatively limited movements. And they are the one group of insects treasured by a large group of amateur naturalists, who keep track of the location and status of many populations.

Butterflies have long been in retreat in developed countries. As long ago as 1880, the Sthenele Brown disappeared under the spreading city of San Francisco, and in 1943 the last individuals of the small blue Xerces Butterfly were taken. Then it, too, went extinct as San Francisco spread over the Xerces' sand-dune habitat. This butterfly is remembered in the name of the Xerces Society, an organization dedicated to the preservation of endangered insects and other invertebrates.

Years ago, in the Department of Insects and Spiders of the American Museum of Natural History in New York, there was a map of the favorite butterfly hunting grounds of the well-known California collector of the early twentieth century, J. D. Gunder. The sites were almost all in the Los Angeles basin; now these are under concrete. During the late 1940s and early 1950s, when Paul was collecting butterflies in New Jersey, one after another of his favorite hunting grounds disappeared under subdivisions.

All over the United States, butterfly populations are endangered by urbanization. In rapidly growing southern Florida, the gorgeous iridescent green Atala Hairstreak is on the verge of extinction because of the construction of hotels, condominiums, houses, and the like. In New York, the habitat of the Karner Blue is threatened by a proposed shopping mall. It was named by a famed novelist, the late Vladimir Nabokov, who was an amateur butterfly collector. The remaining range of the tiny El Segundo Blue, which once occupied thirty-seven square miles of sand hills along the Pacific shores of Los Angeles County, is two small plots. One, about two acres in extent, is surrounded by the asphalt, oil tanks, and research labs of a Standard Oil refinery, which has fenced in the tiny habitat island in an attempt to save the butterfly population. The other plot, somewhat larger, is at the west end of the Los Angeles International Airport.

The protective action of Standard Oil of California was taken in response to initiatives from the Xerces Society. The firm not only erected the fence to protect the area from dune buggies but also removed imported ice plant, which was threatening to crowd out the food plant of the El Segundo Blue's caterpillars. Standard Oil has promised to monitor the site, keeping out intruders and permitting access by researchers. The behavior of this giant corporation in aid of a minute butterfly is, in the words of biologist Robert Pyle (founder of the Xerces Society), "an important milestone in American conservation."[6]

Northward along the coast on San Bruno Mountain on the southern fringe of San Francisco, other butterflies are now threatened with the fate that long ago overtook other denizens of that area—engulfment by expansion of the city ($300 million worth of housing development).[7] In Oregon, coastal salt-spray meadows are being destroyed by condominium development; along with them goes a beautiful fritillary butterfly. The Oregon Silverspot seemed to have little chance of survival because of economic pressures to develop the land. Fortunately, Senator Mark Hatfield helped rally public sentiment to preserve some of the meadows, and the situation has brightened somewhat.

Remember, though, that the butterflies are only *indicators* of the much greater losses taking place in urban areas. In general, their disappearance is most directly related to the destruction of the plants their caterpillars eat. Young butterflies are very particular about their diets; adults have much more catholic tastes, sipping nectar from a wide variety of flowers. Plants don't grow well under roads and apartment houses, and plant populations suffer severely around cities—even though the losses may less often be noted than those of their pretty pests. For example, in the San Francisco area, a beautiful white-flowered Fritillaria is now making its last stand on the serpentine islands that also house Edith's Checkerspot. Halfway around the world on the coast of Russia's Crimea, Stankevicz's Pine is threatened by the same kind of development as endangered the Oregon Silverspot.[8] And all the way around the world, a small Eucalyptus tree, the Plunkett Mallee, is slowly disappearing under the suburban sprawl of Brisbane, Queensland, Australia.[9]

Urbanization

Urbanization is proceeding rapidly in most of the world. Only a fifth of the human population lived in urban areas in 1925, but the proportion increased to two-fifths by 1975. If past urbanization trends continue, two-thirds of the world's population are projected to be city folk by 2025. In rapidly growing less developed countries, urban areas are growing even faster than the populations in general. East Asia is projected to be 63 percent urbanized by 2025; Latin America 85 percent, and Africa 54 percent.[10] Urbanization in developed countries, however, seems to be leveling off, but all are at least 60 percent urban today.

These urbanization figures have both a bad and a good side. On the bad side, it is clear that, unless appropriate steps are taken to control suburban sprawl, many more populations and species are going to be paved over in the next half-century. Species like the Attwater Prairie Chicken, which inhabited the vicinity of that center of unplanned rapid growth, Houston, Texas, will be sacrificed to further expansion. Not only is urbanization extremely destructive, since it removes entire ecosystems, but it tends to occur in areas that are both biologically rich and agriculturally productive. Areas where plants and animals thrive are usually well watered and have moderate climates. People generally have tended to settle and found their cities in such places also. The damage done by urbanization therefore tends to be concentrated in species-rich areas. In the United States, some three million acres per year are paved over or otherwise destroyed for urban development, highways, airports, and water projects. Both good agricultural land and reservoirs of natural diversity are lost in these activities, much of it needlessly.

From the standpoint of plants, one of the world's most species-rich areas is the Cape Province of South Africa. That nation is undergoing a population explosion that increases its population by 60,000 each month. And as its cities expand, its flora contracts. For example, the brilliant, sparkling flowers of the Golden Gladiolus once dappled a valley and hillside near Cape Town. Then alien plants, introduced for a nearby dune-reclamation scheme, escaped and smothered the lowland populations of gladioli.

The inevitable housing project soon followed, paving over most of the rest of the species' habitat. The surviving plants were confined to a strip ten yards wide and forty yards long—surrounded and partly invaded by smothering shrubs from the dune project. Then bulldozing for gravel almost exterminated the Golden Gladiolus; the survivors were saved by being sandwiched between two rocky banks. Finally came the pressures from the housing development: a pathway, three picnic sites, a child's swing, beer cans, broken glass, and trampling. Children picked the beautiful yellow flowers.

In 1979 there were 113 Gladiolus plants; in 1980 only 45 remained. Only two of those managed to flower: one was picked and dropped; the other had flowered early and produced two seed pods. All the rest were seedlings, poking frail leaves through the trampled earth. Their future is dim—and the future of their species depends on attempts to propagate it in "captivity."[11]

On the good side of urbanization, however, the total area of the planet occupied by cities is still quite small. For example, land devoted to cities and highways amounts to only about 3 percent of the United States, which is, however, slightly more than is in national parks and national wildlife refuges combined. Even with a substantial increase in urbanization or development such as highways and airports, the percentage of total area under concrete will not be large.[12] But on city fringes from San Francisco to Cape Town, species will suffer.

Nonurban development does have its effects on plants and wildlife, too. The network of roads that connects urban areas has an impact on animals after construction is completed. Roads can serve as barriers to movement, fractioning herds of large mammals or denying them the ability to make necessary seasonal migrations. There was a great deal of concern, for example, over the influence of the Alaskan oil pipeline and its associated road on the movements of Caribou and other wildlife. Under pressure from environmental groups, changes were made in routing and construction of the pipeline to reduce its impact—changes that the oil companies now admit were improvements from their point of view as well. The wildlife migrations nevertheless have been affected, but it is not clear that the changes are necessarily deleterious.

Smaller animals may also suffer fragmentation of their populations by highways, railways, canals, etc., changing population structures and making the remaining populations smaller and more subject to random extinction. One study has indicated that a four-lane divided highway may be a barrier to the movement of small forest mammals equivalent to a river twice as wide.[13]

Roads also produce a steady attrition in animals of all sizes that attempt to cross them. The overall impact of road kills is difficult to assess, but it is thought to be serious for various European frogs, toads, and salamanders. These animals cross roads in central Europe in great numbers on their annual spring migrations to the lakes and streams where they breed, and they are squashed by the millions. In some places, fences have been constructed to keep them off the roads; they are then collected and safely transported to their spawning grounds.

The problem is sufficiently well known for European highway authorities to be concerned, and fence-tunnel systems are being considered to funnel the hopping and crawling amphibian hordes safely under highways in critical areas. Recently the Swiss Association of Civil Engineers published detailed instructions for constructing amphibian underpasses—surely a landmark in conservation history! Apparently the Swiss have learned to value their amphibian populations, perhaps for their beauty and intrinsic interest, perhaps for the mating calls of frogs that are so much a hallmark of spring, or perhaps because of the volume of mosquitoes that frogs and toads do away with.[14]

AGRICULTURE: PLOWING UNDER

The development and spread of agriculture has always been, and continues to be, a much more serious source of habitat destruction than urbanization. Entire natural ecosystems are converted into stands of one or a few plants—

and efforts are made to exclude all herbivores. Diversity of populations and species is automatically lost, and the plants are usually the first to go.

When forests in New Zealand were cleared, in part for agriculture, Adams' Mistletoe, a shrub partly parasitic on trees, moved to the missing-in-action-and-presumed-dead list. In Swaziland in South Africa, one of the last populations of a beautiful lily species with the unattractive name of *Kniphofia umbrina* was replaced by a field of corn. In 1978 only a few thousand individuals of the entire species remained, some on land also soon to be cultivated. In Ecuador the Caoba tree has been reduced to no more than a dozen individuals. This valuable fast-growing timber species has been all but wiped out by clearance of rainforest for banana and oil palm plantations. Clearing land for bananas has had a similar impact on the Vuleito Palm of Fiji.[15] And when species of plants disappear, other organisms virtually always follow.

Butterflies as Indicators

As in urban areas, butterfly populations and species have been decimated in agricultural areas around the world. And here too, butterflies can serve as more conspicuous indicators of what is happening to other less conspicuous species. The decline of butterflies can be attributed in no small part to habitat destruction in connection with agriculture, primarily by destroying the plants they depend on.

In 1966 we visited the highlands of New Guinea to study the reproductive strategies of butterflies, but we found the highland centers in New Guinea, such as Mount Hagen, surrounded by huge areas cleared for agriculture. Though we drove great distances in a Land-Rover, we were unable to find any place where there were anything but a few widespread "weedy" butterflies— species characteristic of disturbed areas. Agriculture had displaced myriad populations of the rich New Guinea butterfly fauna.

Similarly, populations of a passion-vine butterfly *(Heliconius)* that we have studied in Trinidad[16] are threatened by agriculture. They live in a supposedly protected forest watershed in the northern mountains. Regardless of the protection, people have tried to cut down and burn the forest to bring more land under cultivation.

But one does not have to go to the jungles of Trinidad or the highlands of Papua New Guinea to find butterflies exterminated or endangered by agriculture. In conservation-conscious England, the Large Blue butterfly, which feeds on the flowers of wild thyme, has just gone extinct. The disappearance of the Large Blue, and quite likely of the Mazarine Blue, which went extinct in England in the 1870s, can be traced chiefly to the enclosure and plowing of chalk and limestone grasslands that had existed since the Bronze Age. The butterfly populations were fragmented, and then the fragments, one after another, went under—presumably through processes similar to those now threatening Edith's Checkerspot. The few remaining suitable areas of habitat were far enough apart that reestablishment after local extinctions was impossible.[17]

Ironically, the last few colonies of the Large Blue in England survived where grazing pressures kept the habitat similar to areas where the butterflies originally thrived. The fields with the wild thyme were kept closely cropped, making conditions favorable for one species of ant that "herded" the caterpillars. The butterfly's relationship with the ants that protect them is obligatory. Early in their development, the sluglike caterpillars develop honey glands. When the ants stroke the caterpillars with their antennae and legs, the glands secrete droplets of sugary fluid, which are the ant's reward. Meanwhile, the caterpillars eat not only the thyme but each other; cannibalism is rife, with larger caterpillars eating smaller ones.

When the caterpillars have reached the fifth and last stage of growth,[18] the ants carry them off the wild thyme and down into the ant nests—where they live as social parasites. At that stage, the young Blues resemble the ant's own young in size, color, and skin texture; and they use begging behavior much like that of the ant's young. This elicits a feeding response from the worker ants, which feed the caterpillars. In the meantime, the caterpillar is also busily eating the ant's own young. Eventually, the caterpillar molts into a chrysalis (pupa)—a resting stage of the life cycle in which the grublike structure of the caterpillar is transformed into the delicately beautiful butterfly. When the transformation is complete, the butterfly emerges, crawls out of the ant nest, expands and dries its wings, and flies off to mate.[19]

The grazing pressure that favored the ants came from the activities of sheep and rabbits, but economic circumstances made sheep grazing unprofitable in the last areas occupied by the Large Blue, and the sheep were withdrawn. When the disease myxomatosis decimated the rabbit populations, the result was plant growth that reduced the ant populations below the level at which they could support the butterfly, and the butterfly vanished from the English fauna.[20] The species remains widely distributed in central Europe, where we hope it will persist as a dramatic example of the complex interactions that are often required to keep the rivets of our spaceship in place. The British, on the other hand, have now lost one of only fifty-five butterfly species resident in their country—and one of the most beautiful. The situation is all the more tragic in that thirty-one of the remaining fifty-four species have declined significantly in the past few decades, signaling a general deterioration of the British environment.[21]

Plant Losses

When humanity embarked on the agricultural revolution ten thousand years ago, it also embarked upon a slaughter of Earth's natural flora that continues to this day. Undocumented billions of plant populations and countless plant species have been plowed or disced under or gobbled by domesticated herbivores. The seriousness of this loss of plants is greatly multiplied because of the foundation position of plants in food chains. Peter H. Raven, director of the Missouri Botanical Garden and a leading plant scientist, has estimated that, because of the specialized feeding habits of most organisms that attack plants, every plant species that goes extinct takes an average of ten to thirty species of other organisms with it. As he wrote: ". . . the diversity of plants is the underlying factor controlling the diversity of other organisms and thus the stability of the world ecosystem. On these grounds alone, the conservation of the plant world is ultimately a matter of survival for the human race."[22]

That agriculture is a root cause of the extinction of enormous numbers of plant populations and species is obvious. In the Cape Province of South Africa, for example, native plants have been hard-pressed by intensive agriculture as well as by urban expansion. One of the marvels of the world, the rich Cape flora boasts over 6,000 kinds of plants. There are over 600 species of beautiful heathers, weird giant proteas, lovely bulbs, and wonderful orchids. But the southwestern tip of Africa also holds a world record for endangered plants. Of the great Cape flora, over 1,200 species are threatened, several hundred are at the brink of extinction, and 36 recently have disappeared entirely.

Natural climatic changes set the stage for this massive loss of plant life. For most of the past two million years, the Cape has been cool and wet, washed by storms that have now migrated south to form the "roaring forties." This southern movement of the storms seems to have occurred for periods of 5,000 to 15,000 years every 100,000 years or so. At present the Cape has been in one of the periodic spells of warmth and summer dryness for about 10,000 years.

Previously, the cold-wet adapted plants survived the warm-dry periods well —retreating to misty mountaintops and moist valleys to wait out the unfavorable conditions. Indeed, the periodic isolation in these retreats may have been responsible for bursts of speciation producing the high species diversity. The situation, however, has changed in the twinkling of an eye in evolutionary time; more than 60 percent of the area previously occupied by the Cape flora has been replaced by the farms, plantations, dams, towns, roads, and other paraphernalia needed to support an expanding population of *Homo sapiens*. What relatively natural areas remain have been assaulted with grazing, frequent burning, and the encroachment of exotic weeds. And, as a final insult, the Cape now supports a cut-flower export industry that extracts some $4 million worth of its products (80 percent) directly from the wild flora.[23]

As one example of the results of agricultural pressures on Cape plants, the Jasmine-flowered Heather survives in just a couple of fenced acres surrounded by farmland that is frequently burned over. Its survival, as is so often the case, may depend primarily on other organisms—in this case, insects to pollinate it.[24] It is the farmland equivalent of the Golden Gladiolus, and together they symbolize the paving over and plowing under of the Cape flora.

The grazing of domestic animals is an aspect of agriculture that threatens plant populations everywhere and the herbivores dependent on them as well. A Russian orchid is on the threatened list in part because cattle eat it and compact the soil so it cannot grow.[25] Cattle are also driving a beautiful Swiss alpine relative of the daisy toward the brink.[26] In the Balearic Islands, a glorious peony of great horticultural value, whose roots may have medicinal value as a treatment for epilepsy, is being devoured by goats.[27] Around the Horn of Africa, Cameron's Euphorbia is virtually extinct because of overgrazing. When it goes, the world will have lost a valuable succulent food for livestock in arid lands. And in New Zealand the beautiful Godley's Buttercup is being destroyed by introduced grazers, especially the Chamois.[28]

But these are less than the tip of the iceberg. Most of the Mediterranean basin is already a "goatscape," largely stripped of its native vegetation by deforestation and overgrazing centuries ago. In the midwestern United States today, the few remaining relatively undisturbed native prairie plant communities occupy only tiny threatened enclaves; and in the overgrazed hills of California's Inner Coast Range east of Oakland, almost all the visible plants are introduced weeds. Indeed, California flora was so completely changed by the grazing of cattle and by competition from Mediterranean species introduced by the Spanish that botanists today are not sure what the original flora was really like. China's native flora was already so reduced by agricultural development fifty years ago that, when famed entomologist Gordon Floyd Ferris searched for scale insects on their native hosts, his labors were virtually confined to temple courtyards where a few stragglers of once-abundant plant species persisted. Much of Australia's native flora has long since disappeared down the gullets of sheep, leading to a general degradation and loss of economic values over much of the continent.[29]

But the real iceberg is the destruction of plant diversity in the tropics, and especially in tropical rainforests. Remember, there are probably at least *twice as many* species in the tropics as in temperate regions, even though the land area of the tropics is much smaller. An acre of virgin tropical forest, for instance, may contain a hundred species of trees alone. Agricultural pressures, generated by population growth locally and by the demands of rich countries abroad, are the primary causes of so many species being exterminated at such a frightening rate.

Some people in temperate-zone countries have the opportunity, interest, and knowledge to be concerned about Golden Gladioli, but the poor people of the tropics generally do not. Temperate-zone conservationists therefore tend to be preoccupied with fighting rear-guard actions around a few prominent endangered species and populations in their countries while Earth's main treasure house of diversity is being looted wholesale.

Animal Losses

Of course, the spread of agriculture over the globe has also affected the distribution and abundance of animal species over vast territories. Many animal populations, like those of plants, have simply disappeared unheralded. No one made a list of the animals of the rich Tigris and Euphrates valleys before they were destroyed by agriculture millennia ago. The vast majority of animals, from Elk to the San Joaquin Whipsnake to insects, that once thrived in California's Central Valley have been gone since it was converted into one of the richest—and most pesticide-soaked—agricultural areas in the nation. Today overgrazing and conversion of land to truck and cotton farming in Arizona are helping to push the grotesque and fascinating Gila Monster toward the brink. And across the Pacific on a small Japanese island not far from Okinawa, the unique Iriomote Cat is being extirpated by farmers. The subtropical forest habitat of this small cat, first discovered and named in the 1960s, is being cleared for pineapple, sugarcane, and other crops. The Japanese government's attempts at conservation are inadequate, and it is possible the Iriomote Cat will set a modern record for mammals—less than forty years from description to extinction![30]

As the human population continues to explode, the process of conversion to agriculture goes on. Everywhere more and more marginal land is being brought under cultivation, with a continuous loss of natural diversity. Again, the problem is especially severe in the less-developed tropical countries, which have both the most rapidly expanding human populations *and* the richest remaining reservoirs of plant and animal species.

Kenya illustrates the process speeded up as if by time-lapse photography. Its population growth rate, thanks to success in public health measures and failure in its family planning program, is now the fastest ever recorded for any nation. Exploding at a rate of just over 4 percent a year—a rate that may reach even higher levels—Kenya's population will double from 16 to 32 million in about seventeen years (barring a birth-control miracle, which seems unlikely, or a catastrophic rise in the death rates, which is unfortunately more likely). This unprecedented growth rate will obviously stress all of Kenya's resources, including its wildlife—which accounts through tourism for an important chunk of the country's foreign exchange.[31]

Kenya's economy cannot support the burgeoning population that is migrating into its cities. With its arid climate, less than 10 percent of Kenya's land is planted with crops, and per capita food production is declining. This means there is more and more pressure to bring marginal and submarginal lands—

areas especially vulnerable to irreversible damage—under cultivation and grazing. Kenya's game parks, occupying precisely those areas, are already severely threatened. There have been demands, for example, that sectors of Tsavo National Park be turned over to landless peasants.

We can pray that projects like David Hopcraft's game farm may be able to thrive and preserve at least some of Kenya's spectacular diversity of antelopes. But the handwriting seems to be on the wall for Kenya's elephants, hippos, rhinos, buffalos, and giraffes; they may all be extinct before the year 2000, largely because of attempts to expand areas under cultivation. And if Kenya's demographic patterns foretell the general trend throughout sub-Saharan Africa, then humanity may kiss goodbye all of those species early in the twenty-first century, and all of the economic, recreational, ecological, esthetic, and compassionate values that they embody. Africa, where until recently population pressures in general have been less severe than in many parts of the world, seems doomed to go from poverty to hope and directly back to poverty again.

Drying Up

Desertification, one of the most serious forms of habitat destruction, as we have already seen, is often associated with overgrazing. The amount of desert in the world has been steadily increasing over the years. About 6 percent of Earth's ice-free land surface is already barren desert, and another 28 percent is under moderate to high risk of being converted to desert. That means that slightly more than a third of the planet's ice-free land area could become desert, and one can imagine the impact on biotic diversity if the present area of severe desert were almost *quintupled*. [32]

In many areas, desertification is going on at a frightening pace. It is estimated that worldwide an area equivalent to two Belgiums is being converted to desert annually. On the southern fringe of the Sahara in the Sudan, for instance, over the last seventeen years the edge of the desert has moved southward over sixty miles.

The final insult along desert edges is often the destruction of every surviving tree and bush in the desperate search for firewood by destitute people. How often the last representative of a plant species or population has been used to stoke a fire on which a hungry nomad cooked a meager meal will never be known. Cooking the meal is obviously essential, but the circumstances are tragic for both the land and its people. At least one of the plant species pushed nearer to the edge by woodcutting Tuaregs during the Sahel drought is Lapérrine's Olive, which might prove an important genetic resource for improvement of the commercial olive. [33]

One does not, however, have to travel to North Africa or the densely populated desert of India's Rajasthan to see the impact of desertification of large areas on other species. Much of the western United States is overgrazed, and much of it is arid or semi-arid. Considerable land has already been converted to desert, some of it by Navajo grazers long before European settlers became a significant factor. And the signs of more to come are there. On the other hand, grazing properly done is one of the most sensible uses of our western land—and one of the least destructive to other species. The problem is assuring good management. Dominance of undesirable plant species, erosion, and declining rainfalls and groundwater supplies are common where grazing has been badly managed.[34] The scramble for water to supply agriculture and people in the grossly overpopulated southwestern United States has led to the drying up of some springs and rivers, and the damming, rechanneling, and general development of most others.

Overpopulation, of course, is not usually a simple matter of too many people per unit of space, which is the naïve view often adopted by politicians and developers. Population pressures must be measured relative to the *resources* necessary to support a given population size. That the American population in the Southwest cannot come close to being supported permanently on the water resources of the area in itself justifies the verdict that the area is grossly overpopulated. Human overpopulation, moreover, leads to underpopulation of most other species. For instance, the modification and destruction of aquatic environments in the dry areas of the United States has already caused a general decline in the diversity of fishes and other freshwater species.[35]

The diversion of water to thirsty population centers can have profound effects on terrestrial as well as aquatic systems. Piping operations to supply water and power to Los Angeles have reduced the flow into beautiful Mono Lake, nestled against the eastern face of the Sierra Nevada, not far as the crow flies from Yosemite National Park. In recent years the level of the lake has been dropping rapidly. One consequence is that an island, the largest known breeding ground of the California Gull, has become a peninsula. The new land bridge has provided access to the island for Coyotes, which have forced the gulls out, and the gulls have not found a safe substitute breeding area. Furthermore, Mono Lake's rich productivity of brine shrimp and brine flies is threatened. These arthropod species in turn are an important food resource for many migratory birds. And alkaline dust storms rising off large exposed areas of lakeshore are causing air pollution over the nearby White Mountains—possibly endangering populations there, including those of the famous Bristlecone Pines, perhaps the oldest of all living organisms.[36]

Spraying

One of the agricultural activities that have an enormous impact on other species is the use of insecticides and herbicides. Volumes have been written on the impact of biocides on nontarget organisms, the first and most famous being

Rachel Carson's classic *Silent Spring*. That book is credited by many as being the starting point for the environmental movement; although somewhat out of date, it is still a worthwhile read.[37]

Two of the many problems with pesticides is that they depart the areas where they are applied and assault populations and species that are not their intended victims. That is, they are mobile and nonselective. Their nonselectivity is exemplified by the "promotion" of herbivorous species to pest status by the killing of their predators, as occurred in the Cañete Valley disaster. Their mobility is exemplified by the great Mississippi fish kill of the early 1960s. The latter was described in detail then by reporter Frank Graham, Jr., in a fine book, *Disaster by Default*.[38] Between 1960 and 1963, roughly 10 to 15 million fishes—including several kinds of catfishes, Menhaden, mullet, Sea Trout, drumfish, shad, and Buffalo Fish—were killed in the lower Mississippi. It was a catastrophe for the local fishing industry. Water birds were also killed.

The culprit was a chemical relative of DDT called endrin, which was entering the water in runoff from sprayed and dusted agricultural fields and in waste water from the Memphis plant of Velsicol Chemical Corporation, which manufactured the compounds. Velsicol denied any responsibility, claiming that the company's tests showed the fishes had died of dropsy—a statement that had an amusing side, since dropsy is a disease that is never epidemic in fishes.

The pesticide industry, which was then and remains today one of the least socially responsible businesses, promoted the notion that trying to control the use of pesticides was a communist plot! They fought bitterly against imposition of any controls on pesticide manufacturing or usage. Controls have been steadily increased since the mid-1960s, however, without the dire consequences predicted by industry spokesmen.

Deleterious effects from pesticides and related compounds have been documented in organisms ranging from phytoplankton, the tiny plants that support aquatic food chains, to a great variety of predatory birds. The latter have attracted the most attention, since they have been the most prominent organisms clearly endangered—victims of bioconcentration of persistent poisons and subsequent reproductive failure. Fortunately, restrictions on the use of DDT and other chlorinated hydrocarbons in the United States have led to a possible reprieve for such famous victims as the Brown Pelican, Peregrine Falcon, and Osprey.[39]

The reason that certain pesticides and their chemical relatives persist in the environment is interesting in itself. These compounds are synthetic products of human culture, and as such, present a novel challenge to the natural decomposers that ordinarily break down organic chemicals and recycle their constituent parts. The decomposers lack evolutionary experience with compounds such as DDT—they have not "learned" how to break them down rapidly.

Therefore the toxins tend to hang around for a long time, poisoning organism after organism.

Pesticide problems are worldwide and are not restricted to highly persistent poisons. Azodrin is an organophosphate compound, a group generally considered nonpersistent. It is a measure of the broad-spectrum nature of pesticides that Azodrin is used primarily as an insecticide—one famous for making pest problems in cotton fields *worse* by killing the natural enemies of cotton pests.[40] An outbreak of the Levant Vole, a mouselike animal, occurred in 1975–76 in the northern Huleh Valley of Israel, which caused serious damage to alfalfa fields. Farmers sprayed Azodrin to kill the voles, not realizing that it threatened wildlife because precautions on the original label were not translated into Hebrew.

The result was a slaughter of about four hundred of the Middle East's thin supply of eagles, hawks, owls, and other predatory birds that had gathered to feast on the abundant voles. They fed on the smaller birds—pipits, wagtails, larks, thrushes, buntings, and so on—that had been killed outright by the spray. They also ate voles and small birds that had fed on contaminated food. The predatory birds were the second victims of the doses of poison that killed their prey. The Azodrin also killed Jungle Cats (a kind of wildcat) and wild pigs in the fields.[41]

Use of the pesticides in less developed countries, like the use of antibiotics there, is generally much less carefully controlled than in developed countries (where controls are still often inadequate). Unhappily, little is known about their use in, or impact on, tropical ecosystems, but from what *is* known about the use of these poisons, great care is clearly called for. One very ominous sign is the development of resistance to DDT and other pesticides by the *Anopheles* mosquito, the transmitter of malaria.[42] If past human behavior is any guide, the response to the appearance of resistance may well be the use of more and deadlier pesticides. It won't solve the *Anopheles* problem, but it will greatly increase the impacts on other populations and species.

At the moment it is simply impossible to give a sound evaluation of the indirect threat to other species posed by present global patterns of insecticide use. Few natural populations are monitored to any degree at all, and fewer still are monitored closely enough to discern with certainty the cause of an observed decline. There is some evidence, for example, that impacts from pesticides on populations of soil organisms and predatory insects in general may be profound.[43]

Some small vertebrates have managed to evolve resistance to pesticides. For example, mosquito fishes in some areas have become resistant to endrin and are able to survive with so much of the poison in their tissues that they are toxic to predators! Among larger land animals, birds seem to be especially sensitive to poisons because of their relatively long lives and their often high positions in food chains, and because of the way chlorinated hydrocarbons disrupt their reproductive processes. Even where insecticides are not the prime cause of extinction, they may sufficiently reduce the reproductive success of birds to increase their vulnerability.

Consider, for instance, the Bald Eagle. This magnificent bird is the symbol of the United States. "Eagle" was the name selected for a lunar lander in the 1960s and for a jet fighter plane in the 1980s. It is also a symbol for, among other things, a rock group, a football team, a brand of whiskey, and a twenty-dollar gold piece. It is estimated that the eagle population in the contiguous forty-eight states is less than 1 percent of the population that once lived there. The principal reason is the destruction of its preferred habitat, timbered shorelines, to make way for homes and industry. Unfortunately, the presence of people going about their business distracts Bald Eagles from theirs. Nesting birds are extremely skittish, leaving the nest readily. In a long-lived bird, such behavior was once highly adaptive: they can live to breed again another day. But the eggs cool and die quickly when left, and when disturbance is frequent, successful nesting is impossible.

Moreover, the eagles migrate during the hunting season, and many of them are shot, either accidentally or deliberately. And finally, pesticides continue to weaken their eggshells, further reducing the chances of breeding success. Here, as in other cases, the pesticide potentially plays the role of a straw that breaks the camel's back.[44]

Until 1952 the Bald Eagle actually had a price on its head because of mistaken notions of predator control (the eagle's diet is primarily dead or dying fish—hence the waterfront connection). In the 1960s its serious decline was documented, and it became a rallying point for the environmental movement. From a low of perhaps of seven hundred individuals, there has been a recovery to several thousand in the contiguous forty-eight states. But habitat destruction proceeds, even in Alaska, where timbering may threaten the larger populations surviving there.[45] There has recently been an upsurge of illegal shootings by "farmers and ranchers who view them as threats to livestock and by shotgunners who find them tempting targets."[46] And the threat from pesticides and chemically similar industrial pollutants hangs over them.

The indiscriminate killing is now being prosecuted vigorously, and if judges hand down stiff sentences, the results could be salutary. Furthermore, education programs could generate peer pressure through responsible hunters, who are as fond of eagle shooters as responsible private pilots are of their confreres who fly while drunk. Stopping the development of timbered shorelines is more difficult. Difficult also is slowing the flow of poisons being sprayed on crops and elsewhere. And retrieving long-lasting poisons once they are released into the environment is impossible. Thus the future of our national symbol, at least outside Alaska, remains in doubt.

The broadcast spreading of poisons is no longer as common as it was in the bad old days of the 1950s and 1960s when it seemed that the entire United States was being blanketed with insecticides. But abuses are still common and are sometimes committed by the most unlikely agencies. One bad actor has been the National Park Service, which continues to use a wide variety of toxins in the parks, where, of all places, they should be banned under virtually all circumstances.[47]

In 1980 the Park Service came up with an especially outrageous plan—to poison thirteen prairie-dog colonies along the inside of the boundary of the Badlands National Park. The chemical they planned to use is an extremely dangerous, persistent, broad-spectrum poison known as zinc phosphide. It is toxic to all animals and remains so indefinitely in dry areas. In the presence of moisture it slowly breaks down, giving off a poisonous gas.

Why would the National Park Service wish to poison a typical native animal —one whose "towns" used to dot much of the West, a single colony of which may have totaled 450 million individuals?[48] The answer is they were under pressure from ranchers whose land abuts the park and who saw the park's prairie-dog colonies as sources of "infestation" of their land. If the ranchers did not overgraze their land, the loss from prairie dogs probably would be negligible. But even if it were not, other solutions to the problem, even including compensation payments to the ranchers, would be far preferable to assaulting the prairie dogs on such a scale. The prairie dogs are the food source of the Black-footed Ferret, the rarest endangered mammal in North America. And their colonies provide important habitat for a wide range of other species, including Burrowing Owls, Sharptailed Grouse, snakes, and salamanders.

The attitude of the ranchers is exemplified by a statement by the Board of Commissioners of Pennington County, South Dakota, which adjoins the park in the northwest: "It is our opinion that prairie dogs are nothing more than a rodent (prairie rat might be a more appropriate name), and to continue to allow them to infest good grazing land and to re-infest private land is beyond our comprehension. We do not share the concern for the kit fox or the black-footed ferret if it has to be at the expense of the private landowners that border the national park and other federal and state lands."[49]

That the value of preserving prairie dogs or the ferret is beyond the comprehension of ranchers is one more indication of the dominant values of our society and a partial condemnation of the American education system. But that the National Park Service, presumably dedicated to preserving what is left of American natural ecosystems, should consider giving in to the pressures of such groups is shocking.

Again, there are encouraging signs that the spreading of poisons in ill-conceived and usually unsuccessful programs to control rodents, Coyotes, and the like are on the wane. The questionable effectiveness of the programs has become apparent, along with the unanticipated costs. In any cost-benefit analysis that includes ecosystem values, they have been outright disasters.[50]

On November 8, 1979, Secretary of the Interior Cecil Andrus took a step that could go a long way toward ending the indiscriminate coating of the West with poisons. He stopped all use of and research on the poison sodium fluoroacetate (Compound 1080), which has slaughtered untold millions of wild animals, including eagles. He also endorsed and encouraged the use of nonlethal predator controls—such as repellent chemicals to make livestock distasteful to Coyotes and other attackers. And he affirmed the maintenance of wildlife, including predators, as an important function of federal lands.[51]

The battle is far from won, however. The livestock industry remains unconvinced and has vowed to fight—and it is a powerful lobby. A battle there will be. As distinguished biologist Stanley A. Cain put it: ". . . goals other than those of western ranchers in the livestock industry will have to be sought in the same manner that these ranchers have attained their strength—in the political arena."[52]

Another poison spraying program promoted by grazing interests is that used in what is euphemistically called "sagebrush control." This actually is the conversion of more than 15,000 square miles of natural sagebrush communities into an artificial grassland. Programs have been undertaken in about 10 to 12 percent of the nation's principal sagebrush area—concentrated in the states of Nevada, Oregon, Idaho, Wyoming, and Colorado. About half of the conversion is to be accomplished and maintained by spraying with herbicides. Some areas, where there are few "desirable" grasses (those favored by cattle) under the sagebrush, are seeded with nonnative grasses.

There are various reasons for concern about these large-scale projects. The greatest apprehension has focused on possible impacts on prominent species of wildlife—especially Mule Deer, Sage Grouse, Pronghorn Antelope, and songbirds—but there seems little reason for serious concern there.[53] As usual, the probable extinctions of numerous populations of a great variety of native plants, insects, and other less prominent organisms are ignored, as are possible ecosystemic effects such as limiting nutrient recycling.[54] As in so many other activities, human beings are transforming nature on a gigantic scale without the slightest notion of what the long-term consequences might be.

It is important to note that many large-scale human modifications of ecosystems are necessary and (at least from a homocentric perspective) desirable. Just as one cannot reasonably deplore the conversion of some natural ecosystems to croplands, so some improvement of range for cattle or other domestic animals is obviously acceptable in principle. Large amounts of land in the western United States are suitable for grazing but not for cultivation of crops. In a food-short world, that would appear to be a prime use of such land, and since grass-fed beef is both more healthful to eat and less wasteful of food resources than grain-finished beef, it is desirable to leave cattle on the range, as opposed to fattening them in feedlots, as long as possible.

Nevertheless, the overall *scale* of the activities and the techniques used to carry out the range improvement scheme are of legitimate concern. Truly sound decisions about the extent of brush control and methods used can be made only through careful investigations that go beyond a concern for maximizing beef yield in the immediate future on one side and studies of effects only on large wild animals on the other. *Any* practice that involves spraying thousands of square miles with poisons must be viewed with suspicion, no matter how apparently worthy the cause in which the spraying is done and no matter how superficially harmless in other respects the poison may appear.

There is considerable reason to worry about herbicide use in general today: the annual sales volume of these poisons in the United States is now greater than that of insecticides. Questions about their direct impact on human health have stirred considerable controversy.[55] Equally serious, and mostly still unanswered, questions can be raised about the impacts of herbicides on ecosystems, especially soil flora and fauna, and the contributions they make to the extinction of populations and species.

INDUSTRIAL SOCIETY: SPEWING

Many of the toxic substances that assault the populations of other species are not deliberately sprayed over the landscape, but are released inadvertently or in the processes of "waste disposal." These include a vast array of chemicals, some of the most important of which are very similar to pesticides. For example, polychorinated biphenyls (PCBs) are compounds similar in chemical structure to the family of pesticides that includes DDT, and they have similar impacts on living systems, including being implicated in eggshell thinning in birds. PCBs have been used for some fifty years in a wide variety of industrial applications, such as in transformers, plasticizers, paint additives, and hydraulic fluids. Through various mishaps, they have escaped into the environment —vaporizing from plastics, leaking out of transformers—and have spread worldwide to contaminate virtually all organisms from Antarctic penguins and the denizens of two-mile-deep oceanic trenches to ourselves. Although the production of PCBs was halted in the United States in 1977, three-quarters of a million tons have ended up in dumps and landfills. They will be around for a long time, helping to loosen the rivets in Spaceship Earth.[56]

The greatest public concern over the environment comes from the substances that human activities spew into air and water. People quite properly have been alarmed by the Love Canal affair in Niagara, New York, where poorly understood toxic substances were dumped into a landfill, then an elementary school was built on the site and houses constructed adjacent to it.

The public was further horrified to learn that there are thousands of "Love Canals" scattered around the United States[57] and probably in other industrialized countries as well. People are rightly afraid of the possibility of developing emphysema or heart disease from air pollution, diarrhea from water pollution, or cancer from chemicals leaching from old dumps.

Yet the impacts of these forms of pollution on other species and thus on ecosystem services are little noted, although in the long run these impacts almost certainly are much more threatening to human health and happiness. Sulfur dioxide (SO_2) gas is produced when coal and fuel oil are burned. Oxides of sulfur have been implicated in the high rates of emphysema, acute and chronic asthma, and bronchitis seen in polluted cities. But sulfur dioxide does more than rot human lungs. In both broad-leaved and evergreen plants, sulfur oxides inhibit growth and cause crucial cells in the leaves to collapse or become distorted. Such effects on growing plants have implications not only for natural ecosystems but also for the crops and forests that humanity exploits directly.

Similarly, oxidants—substances like ozone and PAN (peroxyacetylnitrate) that readily give up an oxygen atom in chemical reactions—found in air pollution affect not only human health but the health of plants.[58] Air pollution has wiped out vegetation in the vicinity of some industrial operations, and with the plants, of course, go all of the animal populations dependent on them. Air pollution has even been killing pine trees in the Sierra Nevada of California, many miles from Los Angeles, its primary source. And oxidants are apparently responsible for a reduction of the plant species richness, and thus the animal species richness, of coastal shrubland communities in southern California.[59]

Serious as these consequences are, it appears that an indirect effect of air pollution creates much more serious and general jeopardy for other organisms. The oxides of sulfur and nitrogen that are injected into the atmosphere from factory and power-plant stacks and auto exhausts undergo chemical reactions there that convert them into sulfuric and nitric acids. As a result of the presence of these powerful acids, rains over large parts of eastern North America and Europe are ten to a thousand times as acidic as rains from unpolluted skies.[60] At Pitlochry, Scotland, there was an historic event on April 10, 1974: a downpour of rain as acid as vinegar. Even in the Colorado Rockies, the rainfall is becoming significantly more acid than precipitation from unpolluted skies.[61]

The impacts of acid rains on populations and species of other organisms are not fully understood, but the evidence available is ominous. Certain freshwater ecosystems, those occurring in areas with granite, quartz, and chemically similar kinds of rocks, are especially vulnerable. These rocks occur in the Rockies, the Appalachians, and much of Canada, New England, and northern Europe. In southern Norway, fish populations are in distress over a wide area.

Populations of bacteria, phytoplankton, zooplankton, and all the animals dependent on them in rivers and lakes are also depressed.[62]

In the Adirondacks, not only are the rains acidifying the water but the nitric acid is reacting with the soil to release large amounts of aluminum, which is washed into the lakes. The acids build up in the snowpack during the winter, and in the spring melt they pour into lakes in concentrations lethal to fishes. Then a flush of aluminum pollution follows. The result: all fish populations in three hundred Adirondack lakes are extinct, and the Brook Trout and other species may have been wiped out over the whole area.

The situation is deteriorating further north also. Canadian scientists have identified 48,000 lakes that will be incapable of supporting life in two decades if current trends continue.[63] Spotted Salamanders can no longer live in snow-melt ponds in upstate New York because the winter snows are too acid.[64] Acid rains have already destroyed a third of the Nova Scotia spawning rivers of the Atlantic Salmon—adding their weight to the pesticides, dams, other pollution, overfishing, and poaching that are already pushing this commercially valuable species toward extinction.[65]

Acid rains also may have profound effects on terrestrial ecosystems. They are damaging to microorganisms in the soil, including those involved in the crucial nitrogen cycle. Acid precipitation can change the rate at which toxins are mobilized in the soil and may worsen the effects of other pollutants. The total impact of acid rains on forest ecosystems cannot be predicted with any accuracy yet. Indeed, it may take fifty years to test whether the reports of stunting of forest growth are accurate.[66] But there is reason to believe that populations of forest species from salamanders to oak trees are being gradually snuffed out as a result of acid dropping from the sky.

It seems almost superfluous in this context to point out that water pollution of the more usual sort has already led to the extinction of countless populations of other organisms. A single example will suffice. The French explorer Père Marquette was greatly impressed by the Illinois Valley in 1673: "We have seen nothing like this river that we enter, as regards to its fertility of soil, its prairies and woods; its cattle, elk, deer, wildcats, bustards, swans, ducks, parroquets, and even beaver."

Three centuries later, in 1980, ecologist Don Moll described the river in quite different terms. On approaching closely, one is struck by

the odor from the bloated bodies of carp and gar washing against the shore. They move on thick, oily green water in rhythm to the backwash tides generated by the barge and motorboat traffic moving continually through the narrow channel. No rooted aquatic vegetation is visible and the only birds in view are grackles and ring-billed gulls, garbage eaters working the bodies along the shore."[67]

Population growth in Illinois put diverse pressures on the river. Its fertile valley was cleared, drained, and put to the plow. A killing blow came in 1871 when the flow of the Chicago River was artificially reversed to carry the city's sewage into the Illinois River system rather than into Lake Michigan, the source of its drinking water. Sewage, even if it does not contain toxins but only organic wastes, can overwhelm the decomposing capacity of natural ecosystems. Gradually a sportsman's paradise, one that used to be the destination of special trains called "fisherman's specials," was destroyed—and along with it commercial fishing for turtles, mussels, and fishes. Between 1900 and 1920, the northern hundred miles of the river became a biological desert with virtually no dissolved oxygen in the water—a prime consequence of organic pollution overload.

Some species of the river's diverse turtle fauna, those requiring clean sand for egg laying, have disappeared. A few turtle species have been able to thrive under the new conditions. Not only are they untroubled by the polluted water, but they have been relieved of the presence of bass, pike, herons, and other turtle predators. Turtle species that once ate plants as adults now eat drowned insects and other carrion. Map Turtles are feasting on an introduced, pollution-resistant animal "weed"—an Asiatic clam.

The story of the Illinois River has been repeated in freshwater systems all over the world. The Rhine is loaded with poisons and has suffered massive fish kills. In the Danube, populations of important fish species have been greatly reduced. Lake Baikal is under threat in the Soviet Union, despite efforts to stop pollution. Rivers in Central America are choked with silt as the soil washes off denuded uplands. Japan's rivers are brimful of industrial wastes. Streams in Queensland, Australia, are polluted with sugar-mill wastes. On every continent, a steady movement of feces, chlorinated hydrocarbons, mercury, cadmium, chromium, acids, alkalis, fertilizers, detergents, waste oil, carbamate insecticides, pulp wastes, and silt roll toward the sea—passing through the estuaries that are so vital to marine fisheries, threatening countless populations of aquatic organisms with extinction.[68]

Digging, Leaking, and Flooding

Mining and development of minerals and energy have wide-ranging deleterious effects on the habitats of other species both through direct attacks and through pollution. Mining has the greatest variety of effects. Many wastes, especially from mining metals, contain toxic substances. These poisons find their way into freshwater ecosystems when rain percolates through mine tailings (the piles of refuse left after ore is mined and/or milled), when mines are pumped dry, or when water is used in ore processing.

One well-advertised problem is the increase in the acidity of streams and lakes from acid mining drainage—a problem once thought to be more or less confined to coal mining but now recognized to be much more widespread.[69] As in the case of acid rains, the impact on aquatic organisms can be severe— in extreme cases exterminating all life.

Heavy-metal pollution is also common, especially with zinc, copper, lead, cadmium, chromium, and mercury, and its influence on the biota of freshwater ecosystems is often profound. For example, lead mines in Cardiganshire, Wales, were closed by 1921, but the impoverished rivers only gradually regained some of their diversity. Recent surveys comparing mine-influenced rivers with "clean" ones indicate that the species diversity in the former is more or less permanently depressed. The long-term influence of coal mines appears to be similar to that of lead and other metal mines.[70]

Mine effluents can also lead to premature aging of lakes and to other processes resulting in a dramatic reduction of dissolved oxygen in lakes and streams. Dropping oxygen levels has a lethal effect on many animals. Some of the most economically desirable fishes, such as trout, are especially sensitive to lowered oxygen levels.

Cleaning up water pollution from mines can be extremely difficult. Leakage from a small tailings pond at the Keystone mine near Crested Butte, Colorado, has killed all the fishes in Coal Creek below the pond. An expensive attempt by American Metals Climax Corporation (AMAX), which now owns the property, to clean up the mess has so far been a failure.

AMAX is interested in mining a huge, low-grade molybdenum ore body near Crested Butte in an operation that many people fear will lead to the destruction of the entire area and have an enormous effect on the flora and fauna of one of the biologically richest areas in Colorado. One problem is that huge tailings ponds, eventually several square miles in extent, will have to be built. The tailings are mixed with water to make a slurry that is then pumped into pockets behind enormous dams. Not only will the ecosystems at the pond site be destroyed, but blowing dust from the dried surface of the impoundments, possibly toxic to many organisms, is likely to be widespread. Old tailings ponds have proven to be very difficult to stabilize with plantings, especially at high altitude. (Sites for the AMAX molybdenum venture are likely to be above 8,000 feet.)

Like other mining operations, this proposed project will create air pollution from both point sources (mill vents, chimneys, engine exhausts) and nonpoint sources (tailings ponds, stockpiles) in addition to water pollution. The impact of mine-generated air pollution on other species is less well understood than that of the water pollution. But a major part of the air pollution is dust, and the seriousness of dust from mines (as well as other sources such as dirt roads) is generally underestimated. Dust can function as a quite effective insecticide, for instance. We strongly suspect that declines in butterfly populations in some areas of Colorado are in part due to increased automobile traffic on dusty roads

—and of course those populations are just indicators of what may be happening there to insects in general.

Anyone who has flown low over the state of Kentucky can hardly fail to be impressed with the area of that state ruined by that most environmentally destructive form of mining—stripping. Strip-mining is less directly hazardous to miners than underground mining, but in the long run its costs to society will almost certainly surpass those of underground operations. When an area is strip-mined, the local ecosystem is simply peeled away by giant machines to expose the desired mineral (usually coal) beneath. Except for paving over areas, few human activities are more directly and completely destructive of habitat. And since strip-mining also carries with it the other environmental burdens of mining, it may be, area for area, the single most destructive activity short of war. By 1980 it was estimated that over 4,000 square miles had been strip-mined for coal in the United States. The total affected area was probably more than 10,000 square miles.[71] With the oil shortage and increasing substitution of coal, the area will undoubtedly expand rapidly—especially as strip-mining of coal and oil shale in the western United States and the Athabasca tar sands in Alberta comes into full swing.

Reclamation of strip-mined land is both expensive and difficult. Runoff tends to make soil unsuitable for revegetation; and in much of the West, lack of rainfall is a problem. Costs of full reclamation, where such is possible, was pushing $10,000 per acre in 1980. And full reclamation in one sense is *never* possible, since the original ecosystem cannot be restored. Even the best team of biologists in the world with unlimited funds could never reestablish an ecosystem—if for no other reason than that no ecosystem on Earth is completely understood and any genetically unique populations stripped away are gone forever.

We remember well a conversation with a banker from Gillette, Wyoming, an area undergoing gross changes due to a strip-mining boom, who assured us that local ecosystems had been completely restored. It soon became apparent that grass growing on the site and the sighting of an occasional antelope or rabbit was "complete restoration" to his mind. In many cases, that seems to be about the best that can be hoped for—and much better than usually occurs. Of all the land that had been strip-mined for coal in the United States by 1980, almost none had been fully reclaimed. Until the mid-1960s, in fact, even the most basic reclamation was rare; what was done was essentially cosmetic, to serve the public relations needs of the coal companies.

The scramble to strip-mine coal in Montana and other states, to extract petroleum from oil shale in Colorado, Utah, and Wyoming, and to drill for gas and oil in the so-called overthrust belt of Montana, Idaho, Wyoming, and Utah all bode ill for conservation efforts. So does exploration for and mining of uranium and various other minerals throughout the West. Damage is already substantial in Colorado, where heavy metals from mining operations are already entering the headwaters of America's two great river systems, the Colorado and the Mississippi. But what has happened so far is nothing compared to what will happen if the helter-skelter exploitation of the West's remaining mineral wealth continues as planned.

The impacts of mining and drilling of course are not limited to the United States—they are worldwide except for Antarctica. And even that may not last. The mineral and fuel resources that have been found in recoverable quantities on all other continents and continental shelves are certain to be found in Antarctica also.[72] It may only be a matter of time before the Antarctic penguins are threatened by mining operations and oil spills.

Oil leaking out of sinking tankers or purposely spilled from tankers being cleaned has already threatened penguins outside of Antarctica. Oil pollution, in conjunction with development encroaching on breeding areas and competition from the fishing industry, is assaulting the populations of the South African Blackfooted Penguin. Westbound supertankers, which cannot get through the Suez Canal, pass by the islands on which the penguins breed, carrying loads of crude oil. Large spills have been common, coating the birds with oil. Although South African conservationists have treated the birds, a great many die at sea where they cannot be helped, and many die in spite of the care. Populations have been reduced several hundredfold.[73] In 1974 populations of another penguin species, the Magellanic, suffered heavy losses when the Shell supertanker *Metuchen* went aground in the Strait of Magellan. It was the second largest oil spill up to that time, and many thousands of the birds died. But unlike the Blackfooted Penguin, the Magellanic Penguin is not yet endangered.

The global flow of oil into the oceans from human sources is now estimated to be about ten times greater than the inputs from natural seeps. The general impact of oil spills on oceanic ecosystems remains controversial. Studies of spills off California and Massachusetts have shown very different levels of impact. At the very least, it is clear that adding more oil to the oceans can have deleterious effects on populations of a wide variety of organisms, including fish, shellfish, and other marine invertebrates.[74]

One relatively benign method of mobilizing energy for humanity is the construction of dams for hydroelectric power. But dams have their ecological costs as well as the potential for catastrophic failures. Ironically, two of the most famous endangered species in North America are endangered by dams. The tiny Snail Darter fish is threatened by the Tellico Dam in Tennessee. It has become a symbol for human oppression of obscure organisms—and its story will be told in Chapter 8 in its political context. The Furbish Lousewort —a yellow-flowered plant in the snapdragon family—has eighteen known populations in the United States. Thirteen would be flooded out by the proposed Dickey-Lincoln Dam in northern Maine, another boondoggle project of the Army Corps of Engineers.[75]

The hydroelectric project would also destroy the habitat of many other species, prominent among them being Bald Eagles, Ospreys, Lynx, Bobcats, Otters, Martens, Moose, and Blueback Trout. It would flood almost 140 square miles of valuable timber. High-voltage lines radiating from the project would scar another 400 square miles of wilderness. And for what? To replace about one-half of 1 percent of New England's oil consumption with electricity and provide 68 permanent jobs at a cost to taxpayers of one billion dollars. There are of course alternatives that are both much cheaper and much more environmentally sound.[76]

All over the United States, dams not only flood out natural populations but also divert waters from their natural courses and modify or destroy riverine habitats that are centers of biological diversity in otherwise dry and therefore less diverse areas. The degree of damming is astounding. The Platte River alone is now dammed forty-two times in the three states it crosses: Colorado, Wyoming, and Nebraska.[77] People and wildlife are competing for the river's water, and as usual the wildlife is losing. Moreover, manipulation of water in the West not only destroys habitat but also unfortunately may form traps for other organisms. For example, more than 3,000 miles of concrete-lined irrigation canals built by the Bureau of Reclamation and private interests form death traps that drown birds, snakes, Coyotes, badgers, Bighorn Sheep, antelopes, and deer, just to name a few.[78]

Dams take their toll worldwide. In India, a Hubbardia grass is presumed extinct because a dam diverted water from the waterfall that provided life-giving spray. On Mauritius, a spectacular Crinum Lily will become extinct as soon as leaks in a dam are plugged and its lake is fully flooded.[79] In the Soviet Union, so many dams have been built on the Volga that it has been described as "not so much a river as a 2,300-mile chain of reservoirs, created by hydroelectric power dams."[80] Lenin once said, "Communism is Soviet power plus electrification of the whole country." As a result, the symbol of Mother Russia was transformed, and the Caspian, fed by the Volga, began to dry up. And the three famous species of Caspian sturgeon were cut off from their spawning grounds by the dams. When a little pollution was added, the sturgeon species went into a severe decline.

The sturgeon, of course, are the source of those wonderful fish eggs called caviar. With their supply of caviar threatened, one Soviet reaction showed that technological optimists are not confined to the western side of the Iron Curtain. Russian chemists have produced several kinds of ersatz caviar. One is a charming mixture of sunflower-seed oil, casein, tea extract, and iron chloride. A standard joke in Russia has become, "Isn't it marvelous that our superb

Soviet scientists have produced an artificial caviar that is absolutely indistinguishable from the real thing—except by taste!"

Soviet environmentalists have rallied to the aid of the beleaguered sturgeon, however. Distinguished Soviet poet Andrei Voznesensky finished one of his poems with a defiant

> Technological counterrevolutionaries!
> Refuse to eat synthetic caviar!

The Russian government has responded, and a comprehensive program to save the sturgeon has begun. There has been a massive attempt to clean up the industries of the Volga and Ural basins to abate the pollution. And after a great deal of pressuring, the managers of the dams are beginning to consider the fishes in their operations—they are required to increase the flow during the breeding season.

But according to the newspaper *Sotsialisticheskaya Industria,* the effort still falls short of permitting natural spawning. The gap has been closed, however, by the operations of hatcheries where the fishes are stripped of their eggs. These are artificially fertilized, and the hatchlings are released, to return later as adults, the females full of the raw material for a commodity now worth a hundred dollars a pound wholesale.[81]

DEVELOPMENT: CUTTING DOWN

The single human activity that most threatens other species the world around is the cutting down of forests. The seriousness of deforestation as an attack on diversity varies a great deal from forest to forest and with the amount and pattern of cutting. For example, some organisms, such as herbaceous plants, many butterflies, many birds, and deer, are denizens of forest edges or areas of second-growth forest. Therefore a certain amount of forest cutting can increase the habitat available to such species. There is little question, for example, that the biological diversity of both Europe and North America was first increased by the advent of farming, and in some areas it is still higher today than it was at the time of Christ.[82]

Temperate Forests

The point is long since past in most of the temperate zone, however, at which forest cutting could still significantly enhance diversity. The trend is very much in the other direction. Virgin forest habitat is shrinking, and as it does, the ranges of a wide variety of organisms are being reduced and fragmented. The Bald Eagle is far from alone in this regard even among prominent organisms. In the United States, Montana loggers, one of whose mottoes is "The only good tree is a stump," are threatening the best remaining habitat of the Wolf and Grizzly Bear. The situation across the border in Canada is similarly grim.[83]

Thirty to 40 percent of the land in the north temperate zone is still covered by forests, most of which are coniferous (evergreen). The greatest remaining reserves are in the Soviet Union, though vast tracts also remain in North America.[84] But coniferous forests harbor relatively few species of plants and animals and cannot be considered prime reservoirs of diversity. Deciduous forests—those with broad-leaved trees that drop their leaves in winter—have a much greater variety of both flora and fauna.[85]

Because they drop their leaves annually, deciduous forests generate a soil rich in organic matter and nutrients, which in turn supports a very species-rich soil ecosystem, loaded with bacteria, fungi, roundworms, earthworms, mites, insects, and so on. When such a forest is cleared, the soil quality is jeopardized not only by increased wind and rain erosion from the loss of protective canopy, but from the loss of the trees themselves. The living trees act as giant pumps, bringing nutrients up from their roots deep in the ground and redepositing them in the upper layers of soil as they drop their leaves.

The fine soils are a key to the richness of temperate deciduous forests and to farms established when the forests are cleared (although they still do not approach the extraordinary quality of the prairie soils that are the basis of America's agricultural productivity). But great care must be paid to soil husbandry if the richness is not to be lost. In Europe, appropriate soil conservation is a strongly ingrained tradition (as it is not in the United States);[86] Bavarian farmers have planted and harvested the same ex-forest soil for centuries while maintaining its fertility.

Throughout the temperate zone there is a general commitment to reforestation at a rate that will prevent the gross area occupied by trees from decreasing further. But, beyond that, there is little agreement on how this is to be done —and especially on how the *quality* of the forests is to be maintained. This is, of course, especially important from the point of view of other species. A rich old forest in the southeastern United States, containing a great diversity of trees and organisms associated with them, may be cut down and replaced with a monoculture of pine trees to be harvested for pulp. The loss of diversity will be great, even though forest *area* is conserved. Similarly, many temperate forests are lumbered in ways that do not protect the soil; the result is that silt and nutrients are flushed out, polluting streams and reducing the regenerative capacity of the forest.

There is considerable evidence that forests can be harvested, even clear-cut, and they can regenerate quite successfully, but a certain amount of knowledge and care are necessary.[87] Unhappily, many forests are still treated as "terrestrial whales." They are harvested with too little concern for sustainable yields of timber and too much concern (from the standpoint of maintaining either those yields or species diversity) for maximizing the stream of income from the resource. Forests in the temperate zones can serve humanity directly and indirectly by providing timber at the same time that they maintain their ecosystem services and serve as invaluable reservoirs of diversity—*if* they are managed intelligently.

Tropical Forests

The fate of tropical forests will be the major factor that determines the biological wealth of Earth in the future. Those extraordinarily vulnerable ecosystems are the greatest single reservoir of biotic diversity on the planet. A reasonable assumption is that about two-thirds of the species of the tropics occur in the rainforests. If this is correct, then something on the order of two-fifths to one-half of *all* species on Earth occur in the rainforests, which occupy only 6 percent of Earth's land surface. These crucial reservoirs remain largely uncatalogued: only about 15 percent of their species have even been named, and very little is known about their biology. As Peter Raven has pointed out: "Billions of dollars have been spent on the exploration of the moon, and we now know far more about the moon than we do about the rainforests of say, western Colombia. The moon will be there far longer than these forests. . . ."[88]

The prospects for tropical rainforests (or as they are sometimes called, tropical moist forests) are much less bright than those of the temperate forests, as a report by ecologist Norman Myers recently made clear.[89] Part of the reason for Myers' dismal prognosis lies in the characteristics of the forest ecosystems themselves.

In spite of their lush appearance, tropical rainforests, the "jungles" of popular fiction, generally grow on very poor soil, not the deep, rich soil found in a temperate deciduous forest. When leaves drop from rainforest trees and decay, the nutrients released are immediately absorbed by the shallow network of roots and redeposited in the trees. The basis of the extraordinarily complex food webs of the rainforest—and thus of their amazing species diversity—is the ability of the trees to retain essential nutrients within the system.[90] This makes tropical rainforests much more vulnerable to irreversible damage than temperate-zone forests.

If a small clearing is made in a rainforest and the chopped-down trees are burned by a slash-and-burn agriculturalist, no permanent damage is done. The root network penetrates the clearing and retrieves the nutrients from the ashes. The surrounding trees protect the open area from the full force of the weather. When the clearing is abandoned by the farmer, the forest will be able to reinvade the area and heal the wound.

Suppose, however, that the clearing is made larger—as often happens when population pressures force the farmers to return before the clearing has been fully reoccupied by forest. Then the remaining ash will be less nutrient-rich, and a larger area will have to be planted to compensate for the resultant lower level of productivity. When a larger area is exposed, the root mat cannot penetrate to the center, and the soil is more exposed to being baked by the tropical sun and lashed by the torrential rains that are characteristic of rainforest areas. The downpours wash the nutrients from the thin soil, and they cannot be recovered by the trees as they would be in an intact rainforest ecosystem.[91]

In areas where the soils are rich in iron, this nutrient is one of the last to go. When exposed to sun and oxygen, these so-called *lateritic* soils undergo chemical changes that convert them to a rocklike substance called *laterite* (from the Latin word for brick). Lateritic soils underlie 5 to 10 percent of tropical rainforests.

The frequently catastrophic loss of nutrients and laterization (where it occurs) make cleared areas of rainforest very poor places to practice agriculture and also very difficult to reforest—in sharp contrast to temperate forests. To oversimplify somewhat, much tropical deforestation is not only irreversible but also a colossal waste because it does not lead to any sustainable human activity in the cleared areas.

Sadly, the rainforests are the major ecosystems now under the most determined assault by humanity, as well as being among the most priceless. They are therefore the most crucial front on which the battle to save the Earth's biota must be waged. But how rapidly tropical forests are being destroyed is not known with high accuracy, and estimates have been the subject of considerable controversy. In 1976, forester Adrian Sommer made a pioneering study of the rate of destruction of tropical rainforest clearance, and cautiously estimated that about 42,000 square miles of forest were being lost per year—about 50 acres a minute.[92]

More recent estimates and projections for the loss of the forests have been as low as a still-staggering 25 acres per minute between 1975 and the year 2000. Translated to an annual rate, this comes to some thirteen million acres of forest lost per year—or an area roughly the size of West Virginia.

Furthermore, the statistics cited deal with the "conversion" of the forests, and conversion can cover a variety of different processes with very different impacts on biological diversity. At its most benign, conversion can mean simply selectively logging out the most desirable trees—leaving the forest essentially intact except for a small change in the relative abundance of species and the disturbance caused by the operations, which, if small enough, may be quickly healed. In such operations, it appears that disturbance of the forest animals is also minimal.[93] At the other extreme, the forest is totally removed and is replaced with farms or towns, or in some cases a virtual desert. Such was the fate of 300 square miles of Espirito Santo State in Brazil when the forest trees were all cut down for logs, pasture, and banana plantations. In 1978 the area was quickly becoming desertified.[94]

No matter which figures one accepts or which assumptions one makes, however, there is no room for complacency about the world's foremost reservoir of diversity. Even the *lowest* of the estimated rates of today implies disastrous losses of populations and species in the near future. Assuming conservatively that four million square miles of relatively undisturbed tropical rainforests remain, and that they will be destroyed at the constant "low" rate of 25 acres a minute, then half the forests will be destroyed in a hundred years, and they will be totally gone in two centuries. Lest one decide that this is therefore a problem that can safely be left to our grandchildren, consider the following.

First of all, the *biotic diversity*—that is, the numbers of populations and species—will disappear much more rapidly than the forests themselves. The principal reasons are that rainforests in different parts of the world have completely different floras and faunas and the rates of conversion vary greatly from area to area. An especially rich rainforest might represent only 2 percent of Earth's rainforest area but contain 4 percent of rainforest species.

For instance, at present rates, nearly all the lowland forests of the Philippines, peninsular Malaysia, Indonesia, and much of the rest of Southeast Asia will be gone by the turn of the century. This is an area of extraordinary interest to loggers because the dominant trees of the forest, dipterocarps, produce very light, high-quality lumber especially suitable for making veneers and plywood. It is also an area of extraordinarily great biological diversity. The magnitude of the loss of species in Southeast Asia will be far out of proportion to the size of the forest area that disappears.

Another reason that rainforest diversity will decay at a disproportionate rate is that habitat fragmentation will cause extinctions of many organisms that require large areas for survival or are sensitive to the impacts of disturbance and pollution impinging on the shrinking islands of forest. As we will see in Chapter 9, such islands will be subject to a progressive loss of diversity as they become smaller and farther apart.

Second, it would be highly imprudent to operate on the conservative assumption that the rate of destruction can be held to the current rate. The countries in which the rainforests occur are mostly poor and have fast-growing populations that are encroaching on the forests as they grow. And as the rich nations struggle to maintain their affluence in a world of dwindling resources, the pressures they also put on the rainforests seem certain to increase.

For an estimate that may be more realistic, assume that the rate of conversion of the tropical moist forests over the next few decades will grow exponentially and in proportion to the exponential rate of human population growth in the poor nations.[95] With the rate of attack on diversity thus escalating, there could be a catastrophic loss of species and populations within fifteen to twenty years, and the majority of rainforest species would have disappeared within thirty to fifty years.[96] These are much shorter times than the hundred and two hundred years of the conservative "current rates continue" estimate for the disappearance of half and all of the forests themselves. But even a century or two is a very short time in the life of a species like *Homo sapiens,* which has been around for millennia and plans to be around for millennia more.

Since the more pessimistic estimates of population and species losses may also be the more realistic ones, there is not the slightest time to waste if Earth's greatest reservoir of biological diversity is not to be drained. Somehow the growing pressures on tropical rainforests must be relaxed as rapidly as possible.

The pressures primarily come from farming, lumbering, ranching, and to a lesser extent, cutting for firewood. The main assault on the tropical forests comes from an increasing number of farmers, growing either crops for their own subsistence or cash crops to serve local or foreign markets. They are quickly converting vast areas from sustainable slash-and-burn rotation patterns to a permanently destructive clearing pattern. The downward spiral we have described is well underway in tropical forest areas globally, pushing today's rainforest peoples toward the fate that long ago may have swallowed the Khmer and classic Mayan civilizations.[97] With the world food situation already marginal and tropical populations expanding, it is clear that if a laissez-faire approach to the rainforests continues, the rate of conversion to agriculture can only accelerate rapidly. The fates of the chimp habitat that once stretched continuously east from the shores of Lake Tanganyika and of the watershed home of the *Heliconius* butterflies in Trinidad will quickly overtake most of the tropical rainforests.

Almost half the wood harvested worldwide annually is used as firewood. Of the rest, about four-fifths is employed for construction of buildings, making furniture, and other "solid wood" uses—about two-thirds of it in rich countries. The other fifth of nonfuel use is to make pulp to turn into newspapers, books, and paper cartons; almost 90 percent of it is bound for markets in the rich countries. By the turn of the century, the total annual harvest of wood is projected to be much more than double today's, with use for solid wood and pulp making up two-thirds of the total.

This projection particularly bodes ill for the tropical moist forests. Although they contain as much wood as the larger temperate forests (because the trees are more massive), at present rainforests contribute only about a tenth of the timber-pulp harvest. There is little question that they will be exploited increasingly in the future to meet the growing demand. This is even more likely since recent technological advances have made it possible to convert the diverse tropical forests to pulp, hardwoods and all. The rate of attack on Southeast Asia's forests has increased as Japan, hungry for pulp, has turned in that direction to meet its needs. As forests from Tasmania to Indonesia, Malaysia, and the Philippines are being converted to wood chips, the outlook for their inhabitants from the Orangutan and the Sumatran Rhino to the tiniest mite is grim indeed.

Because of the structure of tropical forests, the impact of logging, without clear-cutting, is often considerably greater than that in temperate forests. The most commercially desirable trees frequently are the tallest—giants with spreading crowns up to fifty feet across, which (unlike those in temperate forests) are connected to other trees by long, strong vines. When such a tree falls, several of its neighbors may be broken or dragged down with it. Moreover, trees in the species-rich rainforest are surrounded by myriad enemies—especially bacteria, fungi, and insects. A seemingly trivial injury such as a broken branch or bark rubbed away can be the starting point for a lethal attack. In Southeast Asia, survey after survey has shown that between one-third and two-thirds of trees bypassed but injured by loggers have been killed as a result.

Furthermore, large areas are damaged by the dragging of logs, the building of roads to haul them away on, and the creation of depots for stockpiling them. In some cases, almost a third of the logged area may be completely denuded, and in much of that area the soil is compacted or otherwise damaged by heavy machinery.

The third major force destroying rainforests is ranching. This is concentrated almost entirely in the Western Hemisphere, where large tracts are being converted to rangeland for the grazing of cattle. In the Amazon Basin and Central America, it is the principal reason for forest clearing. In the dozen-year period prior to 1978, 30,000 square miles of Amazon land in Brazil were cleared to make room for 336 ranches running 6 million head of cattle. In Central America between 1950 and 1975, the area of grazing land more than doubled; almost all of the increase was gained by clearing virgin rainforests.

Why the sudden urge among our Latin American neighbors to emulate Texans? The answer is pure economics. The rich countries keep increasing their demand for "noninflationary" beef, and the rainforests are being sacrificed to keep up the flow of meat destined almost exclusively to be hamburgers served by fast-food chains. Beef can be produced at very low cost in poor countries because of the availability of cheap labor and cheap land ("useless" rainforest). Grass-fed beef can be raised in Latin America at a quarter the price at which it can be raised in Colorado. Although imports from Latin America only amount to 1 or 2 percent of U.S. beef consumption, they cut a nickel or so off the price of a hamburger. Ironically, though, while more than a quarter of all Central American forests have been destroyed in the past twenty years to produce beef for the United States, per-capita consumption of beef in Central American nations has dropped steadily.[98]

The prospects are that clearing of forest to make rangeland will accelerate. The demand for beef will only go up, and the temptation to cash in will be irresistible for many poor countries. Unfortunately, most rainforest areas are not very desirable for grazing once they are cleared; hence a lot of forest must be sacrificed to make way for each cow. And the whole grazing enterprise all too often cannot be sustained for long. Ranches established on the vast majority of tropical soils previously occupied by tropical lowland forest are proving to be ecologically unstable. Their fertility declines so quickly that raising cattle becomes unprofitable in a decade or so. Moreover, the structure of most of the soil is such that artificial fertilizers would not help—even if their use were economically feasible.[99]

When we visited Panama a decade ago, we were impressed by the distance that most forests seemed to be from roads. We were told that this was because all the most accessible plots had been stripped by people searching for firewood. But at the moment, clearing for firewood is only a relatively minor factor in overall conversion of rainforests because most firewood is gathered from woodlots, savannahs, and patches of shrub and brush. But, of course, as the population grows in the humid tropics and as other sources are exhausted, assaults on the rainforest for firewood will escalate. There are signs of this already in countries like Thailand, where the firewood and charcoal trade has been commercialized because the forest near the city of Bangkok is long gone.

In summary, then, unless dramatic moves are made to save them, most of the vast numbers of populations and species now living in tropical moist forests clearly will vanish in the lifetime of many readers of this book.

TRANSPORTING

One of the human activities that have often led to species extinctions is the transporting of organisms. Moving plants and animals from the ecosystems in which they evolved to other places where the native plants and animals have had no evolutionary experience with them has often had catastrophic effects on the recipient community.

As we have already seen, island floras and faunas seem especially susceptible to disruption by introduced species. They often have long evolved in the absence of most of the predators and competitors faced by their mainland relatives and therefore may have lost their abilities to defend themselves or win out in a scramble for resources. Natural selection is parsimonious. Individuals that put energy into defenses will have less energy to put into reproduction; in the absence of predators to coevolve with, they will be at a disadvantage compared to those that put less energy into defense and therefore are able to reproduce more. Over many generations, the loss of defensive capability in the population is the result.

This is why Hawaiian plants are relatively defenseless. In Hawaii there were no native reptilian or mammalian herbivores and virtually no native poisonous plants, while the spicy and smelly chemical compounds that signal plant defenses are generally at a low level. In the normally fragrant mint family, native Hawaiian species are odorless; most Hawaiian plants also lack the spines and thorns that are present on their mainland relatives. The raspberries of the Sandwich Isles, for example, are thornless.[100] As a result, native plants are giving way everywhere under the pressures of introduced herbivores and weeds, and several entire native genera of plants are now extinct.

Still, the Hawaiian Islands have been lucky compared to many other islands, which have suffered from the introduction of a single non-native species—the goat, sometimes more accurately described as the "horned locust." The island of St. Helena in the South Atlantic was once heavily forested; but, following timbering, goats ate the seedlings, preventing regrowth, and converted the island to a rocky wasteland by early in the last century. On Santa Catalina Island off California, goats helped to destroy forty-eight indigenous and eighteen exotic species of plants. And of course with the plants went the associated animals. On Santa Catalina, declines in many reptile, bird, and small mammal populations have all been associated with goat damage.

Goats—along with cattle, pigs, dogs, and cats—are at present destroying the native plants and animals of that birthplace of the theory of evolution, the Galápagos Islands. Goats had devoured much of the vegetation required by Land Iguanas on Santa Fe Island and had pushed the iguanas and the Galápagos Tomato close to extinction there by 1971. In that year, the Galápagos National Park Service exterminated the goats on Santa Fe Island, and now both iguana and tomato populations are recovering.[101] But their populations on the other islands are still threatened. Fortunately, the unique Marine Iguana appears to be in no danger—it escapes competition from the goats by feeding on seaweed under water.

The giant Galápagos Tortoise, which today is severely threatened by introduced species, particularly fascinated Darwin. Each island in the last million years or so had evolved its own distinctive form, so the origin of an individual tortoise was discernible to the practiced eye by shell-shape alone. Humanity had an early impact on these unfortunate beasts, as whalers and other seafarers slaughtered thousands to provision their ships. The tortoises were often taken on board alive because they could be stowed without care for many months, guaranteeing a supply of fresh meat. Between 1811 and 1844, the logbooks of American whaling ships alone record the removal of more than five thousand tortoises. Some ships took "from six to nine hundred of the smallest size of these tortoises on board when about to leave the islands for their cruising grounds."[102]

The actions of whalers and sealers in their day are understandable, but what is almost incomprehensible is the behavior of "scientific" expeditions in the late nineteenth and early twentieth centuries. As ecologist Ian Thornton wrote in his fine natural history of the islands: "Time and time again, scientists, often collecting on an island, declared the tortoises of that island to be extinct, only for some later expedition to discover survivors which were promptly skinned and carried away as precious specimens of a 'dying' race."[103] Four Duncan (Pinzon) Island expeditions in 1897, 1898, 1900, and 1901 each collected the "last survivors" of the local race. Then in 1905–06 Duncan was visited by an expedition from the California Academy of Sciences. The expedition found 86 tortoises on the island, which in an infamous act they slaughtered "for study." Sixty of the dead were females.

The academy, now a great center of Galápagos research, was almost directly responsible for the extermination of the Duncan race. But miraculously, the tortoises staggered back from the brink—only to be assaulted by black rats that began to thrive on the island. Apparently, no young tortoises have survived on the island for fifty years—all have been devoured by the rats.

Of the original fifteen races of Galápagos tortoises, four are now extinct. Populations of three of the remaining eleven may be self-sustaining—it is difficult to tell with animals that commonly live more than a hundred years. The other eight races are threatened with extinction. Eggs and young are being eaten by introduced rats, pigs, dogs, and cats. Feral goats and donkeys compete with them for their plant foods. And there still is occasional poaching. In spite of the massive direct assaults by people on these fascinating and harmless creatures, it is clear that the greatest impetus toward extinction has come not from human hunting but from human transporting of other animals to the Galápagos.[104]

The introduction of exotic plants and animals has had its most spectacular impacts on small islands, but it has also been responsible for countless population extinctions on large islands and continental areas. The native floras and faunas of New Zealand and Australia have been ravaged by imported plants and herbivores. The beautiful Eastern Bluebird of North America has gone into serious decline in the past half-century because of overwhelming competition for nesting holes from two introduced birds—the Starling and the House Sparrow.[105] Imported Cabbage Butterflies from Europe in some areas may have outcompeted populations of native Cabbage Butterflies in North America. In return, the importation of the American Gray Squirrel has extirpated the native red one from large areas of England. Dutch elm disease, brought in from continental Europe, has killed many populations of English and American elms. Another fungus, the Chestnut Blight, accidentally brought to the United

States from Asia on nursery plants, has pushed American Chestnut trees to virtual extinction.[106]

Introduced species have affected the flora and fauna virtually everywhere, although it is usually not obvious to an untrained eye. Huge areas of all continents are occupied by exotic weeds—taking space from native plants and often failing to support much in the way of local animal life (since one reason for their success is that they have left most or all of their enemies abroad in their homelands). And matching them are the domestic herbivores and animal "weeds" (such as feral cats, rabbits, and mongooses) that *Homo sapiens* has spread to the ends of the Earth. The number of populations and species they have been responsible for obliterating can never even be estimated, but it is clearly enormous.

RECREATION

Ironically, the very acts associated with appreciating the intricacy and beauty of natural ecosystems can sometimes contribute to their extinction. A continuing problem at our research area on Stanford's Jasper Ridge Biological Preserve has been to control the activities of researchers, students, and members of the public on nature walks so as to minimize the impact on the flora and fauna. High levels of use by people on foot, no matter how carefully regulated, inevitably seem to lead to the widening of trails, compacting of soils, trampling of plants, and so forth. In the limited areas of serpentine soil that are the habitat for the Bay Checkerspot, even small amounts of damage potentially threaten the survival of the butterfly. And as luck would have it, spring is when both the butterflies are flying and the floral display is at its peak. It is then that researcher and visitor activity also peaks, and it is then that the rain-softened soil is most vulnerable to damage.

In the Galápagos, we were very glad to see that the Ecuadorean government now requires tourists to remain strictly on carefully delineated trails to minimize further damage to already disturbed landscapes. We were also pleased to see the zeal with which the naturalist guides on our ship, the *Buccaneer*, repeatedly warned the tourists to keep to the paths, and proud that our group of Stanford alumni enthusiastically cooperated. They included businessmen, lawyers, physicians, schoolteachers, and housewives; and their behavior was a testimonial to the pleasure and interest that laypersons can find in Earth's diversity.

Especially vulnerable to damage by large numbers of walking human beings are some alpine tundra habitats where the lures of majestic mountain views and spectacular wildflower displays can lead to deterioration of the alpine flora and fauna. For example, concern has been expressed about the tourist impact on the extensive tundra around Trail Ridge Road in Rocky Mountain National

Park, Colorado.[107] There, at 12,000 feet, is a sample of the Arctic accessible in a few hours' drive over paved roads from Denver. Whether this area can absorb the impact of projected increases in visitors without being destroyed (as have been similar areas on Pikes Peak) is not yet clear. In the heavily utilized White Mountains of New Hampshire, hikers have destroyed two of the few remaining populations of the pretty yellow-flowered Robbins' Cinquefoil.[108]

The effects of human foot traffic can be subtle as well as substantial. A study in England showed that trampling had a rather dramatic effect on the diversity of the tiny animals in grassland litter (decaying leaves). Most interestingly, this occurred before any effects were noticeable on the living plants; thus a superficially undisturbed site may actually be undergoing rather serious damage.[109]

Trampling effects are not limited to terrestrial environments. On the barrier reef off Tague Bay, St. Croix, we have done intensive studies of schooling of juvenile grunts—small fishes that hunt alone for small animals in the sea-grass beds of the bay at night and retreat to rest in groups over the coral by day.[110] Our work required both marking of individuals and intensive observation of the schools. We and our colleagues watched one large school in water about four feet deep over the Tague Bay barrier reef so much that we became part of our own experiment. The tips of our fins kept hitting the staghorn coral and breaking off pieces. Sometimes when waves surged across the reef, we would grab the coral, and more would break off. Soon the intricate and ritualized patterns of the grunts' departure at dusk and return at dawn began to change, and over a period of months the school dispersed.

A large school has not occupied that site for years now, where we had gradually altered perhaps a six-by-six-yard patch of coral. In contrast, another schooling site we have studied in the Grenadines is in deeper water and was not disturbed. In 1979, eight years after our first observations, the site was still occupied by juvenile grunts.

Reefs are also damaged by the anchors of pleasure boats in some areas. One of our small study reefs at Palm Island in the Grenadines was totally destroyed by yacht anchors. About a fifth of the staghorn coral at Fort Jefferson National Monument in the Dry Tortugas off Florida has reportedly been damaged in that way.[111] Other recent work has shown that reefs near the Heron Island resort at the south end of Australia's Great Barrier Reef have been severely damaged by people walking on them at low tide.[112]

When machines become a part of recreation, the jeopardy of other species often increases greatly. More than half of the human-caused deaths among the endangered water-dwelling Florida Manatee come from slashing by the propellers of powerboats. The Manatee, or sea cow, Florida's official mammal, looks something like Grover Cleveland: "Same whiskers, thick wrinkled skin. Hefty too: up to 2,200 lbs of blubbery bulk."[113] The sluggish animals feed on aquatic weeds and are the gentlest of mammals—they will not attack even if their young are threatened. They are useful to humanity as well. They were long hunted for their meat, which tastes like veal, their ivorylike bone, oil, and leathery skin. And they have great potential as a biological control organism for the exotic weeds that tend to choke waterways throughout the warmer parts of the world.

Manatee young are very playful, "kissing" on the muzzle and hugging with their flippers. But unfortunately, the beasts are not amorous enough. Females reach sexual maturity at seven or eight years and males at nine or ten, but sea cows bear calves only about once in three years. That reproductive rate is not matching the attrition from the propellers (which have also scarred a significant proportion of the 800 to 1,000 sea cows in Florida) as well as from other sources. Curious divers harass the animals, and sometimes people have even killed them deliberately. Others drown in flood-control structures or by getting entangled in fishing gear. Still others are crushed by barges. As a result, these gentle animals, thought to be the basis of mermaid legends, soon may no longer exist.

When it comes to pure recreational destructiveness, however, off-road vehicles (ORVs) far surpass powerboats. There are now probably over 8 million dirt bikes, four-wheel-drive vehicles, all-terrain vehicles, dune buggies, and other ORVs (excluding snowmobiles) operating in the United States alone. They are the basis of a multi-million-dollar industry—and are thus backed by powerful economic and political forces. As University of California zoologist Robert Stebbins put it: "The American economy has found a new zone of expansion and with it has come a pervasive process of conditioning people to regard off-road driving on the nation's wild lands as an acceptable form of human activity."[114] Advertisements now show vehicles careening over the countryside, raising "rooster tails" of dust. Children are now supplied with ORV toys.

It is a rare environment indeed where a vehicle can be taken off-road without damage. Many years ago, when we were first trying to control access to Jasper Ridge, wardens were hired to patrol in vehicles. They persisted, however, in driving off the roads in pursuit of intruders. The patrols were quickly stopped because a single off-road sortie in a car did more damage than hundreds of hikers.

Standard ORVs with their knobbly tires are almost ideal devices for smashing plant life and destroying soil. Even driven with extreme care, a dirt bike will degrade about an acre of land in a twenty-mile drive; a four-wheel-drive vehicle will have the same impact in just six miles. As the Council on Environmental Quality commented: "First and foremost ORVs eat land. . . . It is because ORVs attack that relatively thin layer of disintegrated rock and organic material to which all earthly life clings—soil . . . that they can have such a devastating effect on natural resources." In many areas, where ORVs strip away the vegetation, an erosion process begins that continues the destruction. In the vicinity of Santa Cruz in California's coastal mountains, ORV trails that have been in use for only six years are eight-foot-deep gullies.[115]

A mind-boggling amount of destruction has already been caused in southwestern deserts by ORVs; destruction has just begun to be documented, from which (if the ORVs are stopped) it will take many years, if not centuries, to recover. Over large areas the delicate and important crust on the desert soil has been broken, increasing the chance of wind erosion and the generation of dust storms. This, coincidentally, degrades air quality for human beings and, in the Southwest, carries to them the fungal spores that cause valley fever (coccidioidomycosis).[116] And of course it may also contribute to local climate change.

In areas of heavy ORV use, the desert soil is completely denuded of vegetation. This of course also wipes out the animal populations dependent on the plants. Furthermore, disturbance tends to work against native plants and in favor of intrusive weeds like the Russian Thistle or Tumbleweed—which is spreading over more and more of the desert areas of the American Southwest. F. R. Fosberg, curator of botany at the United States National Museum and an international authority on weeds, wrote in 1974 to the Bureau of Land Management (BLM) about the problem in the desert land that agency manages:

> When walking [man] does not affect the desert vegetation more than do other large animals. But with his machines (ORVs) he creates disturbance that provides ideal conditions for the establishment of the exotic species that may then take over the areas around the disturbance. . . . Unless you are in favor of changing the character of the vegetation, and hence of the whole landscape of the desert areas under BLM jurisdiction, in my considered opinion as a botanist and ecologist, you must limit vehicular traffic to established roads and open the desert areas only to hiking and other less ecologically destructive forms of recreation.[117]

In spite of this good advice and much more like it from other biologists, in 1980 the BLM was still considering management plans for California deserts that included extensive access for ORVs.

Not only do the ORVs exterminate animals by exterminating plants, they attack them directly as well. Individual animals on the surface and in shallow burrows (where many desert animals hide to avoid the heat of the day) are crushed. Combined direct and indirect losses are heavy. Terrestrial vertebrates had suffered a 60 percent loss in one study that compared ORV areas with undisturbed plots. In another study, done at the Imperial Dunes southeast of the Salton Sea, the results were similar. In undisturbed areas, invertebrates, mostly insects, were twenty-four times more abundant than in areas of intensive ORV use, while kangaroo rats were five times, rabbits ten times, and lizards three times more abundant.

One great problem with ORVs is that they supply easy access to wilderness areas for unsupervised people who have had no previous experience with that country and have no conception of the damage they are doing. The problem is exemplified by the observations of naturalist Steve Zachary in the beautiful but vehicle-threatened San Francisco River Canyon of New Mexico and Arizona. He wrote to the Forest Service, which is supposed to protect the area:

> We stayed in the canyon seven days. It was a delightful experience with many birds (including a merganser with six ducklings and a nesting least bittern), bighorn sheep and many wildflowers. As Memorial Day weekend approached I noticed that the cattle down in the canyon were herded out. Then I began to see why: four-wheel drives. ORVs of all kinds started coming down the canyon. It turned a beautiful river canyon into a noisy motorized playground with no concern or sensitivity for the environment.
>
> It was a real challenge for the ORVs to cross the river. We saw ORVs stuck in the river dripping oil into it. Some of the people had guns and were shooting at anything that moved. Over the weekend we saw that common merganser again but with only three young and eventually with none. With motorized vehicles going up and down the river canyon, it's hard for something like a merganser to keep her family together. The least bittern nest was completely destroyed, tire tracks going right over the nest. I was really worried about the birds of prey since some people had big rifles.
>
> We started seeing litter throughout the canyon, and beautiful riparian habitat destroyed. We camped along the river at night and people would come by in their vehicles, shining their spotlights on the canyon cliffs. There were a group of bighorn sheep in the immediate area. They began to fire shots. I believe they were poaching or trying to poach the sheep. I can see now why the rancher moved his cattle. We saw a camp of ORVers drinking beer and throwing the cans in the river, then shooting them as they floated by.
>
> What seemed so ironic was that I was made to place a $500 bond to lead a group of three backpackers into the canyon. This performance bond was in case of any damages to the national forest. Yet all these vehicles were down in the same canyon destroying the canyon, river and wildlife, and do you think they paid a performance bond? When I returned to the Glenwood ranger station, the secretary wasn't going to give my $500 bond back until someone went down the canyon to see if my three backpackers had done any damage.[118]

The Forest Service answered Zachary in its usual "land of many abuses" style, saying the level of damage was not unacceptable.

The controversy over ORV use in the San Francisco Canyon highlights the desperate need for public education about wilderness and ecosystem values. The principal organization wanting to use the canyon for rallies, and unconcerned about the damage they do, is the Los Cruces Jeep Club. This group was involved in the 1966 "Gila Jeep War" in which members tried to invade the Gila Primitive Area—where, as in all primitive areas, ORVs are banned. They were stopped only by barricades and threats of arrests. Now the club is involved in a nationwide campaign to repeal the Wilderness Act!

BLOWING UP

The ultimate act of habitat destruction, the one that would have incalculable immediate consequences for other species of organisms as well as human beings, would be all-out thermonuclear war. Many studies have been done attempting to project the consequences of such a war, starting with Herman Kahn's "classic" *On Thermonuclear War*[119] and continuing through more recent studies by the National Academy of Sciences[120] and government departments of Defense and Energy.[121]

All of these studies share two characteristics. First, to any reader with the slightest discernment who penetrates all the jargon about "credible first strike forces," "ideal blast waves," "prompt effects," "whole-body doses," and "megadeaths," they paint a picture of indescribable horror. It is a sad commentary on society that George Bush became Vice President–elect in 1980 after openly stating his belief that a nuclear war can in some sense be "won." The second common characteristic of all of the studies is that with all the horrors they project, they greatly underestimate the true magnitude of the effects of such a war. In part this is due to inadequate analysis of the factors that they have explicitly considered.[122] But more important, they have uniformly failed to consider the true magnitude of the ecological effects that would or might follow an all-out exchange of hydrogen bombs.[123]

Suppose that the United States and the Soviet Union had a full-scale exchange in late September—not an unlikely time, since neither party would be anxious to start a war before their crops could be harvested. The effect of nuclear weapons varies with many factors, not the least of them being weather. If, for example, California were struck by a barrage of warheads during a clear, hot period, which is usual at that time of year, the results would go far beyond many millions of immediate human deaths.

A large portion of the state's forest and chaparral areas would be set ablaze —since thermonuclear warheads have the capacity to ignite everything flammable over thousands of square miles.[124] Where supplies of combustibles were sufficient, firestorms of immense dimensions might be generated, in some circumstances producing temperatures high enough to sterilize the soil.[125] It is not inconceivable that much of the state would be burned bare and that in many areas the ungerminated seeds in the soil would be destroyed. Countless populations and species of animals would go extinct, and those not exterminated would be very much reduced in size and irradiated. Repopulation from whatever areas remained relatively undamaged would be extremely slow.

Meanwhile, when the rains came in the fall, the soil necessary for the reestablishment of plants would begin to wash into the sea in the floods generated over the denuded watershed areas. In turn, the heavy load of silt would put additional enormous stress on surviving shallow-ocean and estuarine communities. They would already have been assaulted by the runoff from all the ruptured tanks of various industrial liquids that had not been consumed in fires and by the outpourings of offshore oil wells that had not automatically been capped.

Such effects might well be spread over much of the Northern Hemisphere, depending on the timing and scale of the war. If attacks were heavy and widespread, enormous amounts of smoke and dust would be lofted into the atmosphere, and Earth's crucial ozone shield would be thinned.[126] Local and global climatic changes would certainly be added to the stresses on surviving populations of human beings and other organisms. Even relatively rapid melting of polar ice caps and inundation of coastal areas cannot be excluded.

Under most current scenarios, immediate impacts in the Southern Hemisphere would be much less severe. But certainly both aid from and trade with the Northern Hemisphere would be dramatically curtailed, if not stopped entirely. It is not unlikely that the effects of weather change also would be felt in the poor countries of the Southern Hemisphere, threatening their already inadequate agricultural base. Local pressures to exploit Southern Hemisphere resources would be greatly increased, and endangered populations of species there would instantly be put in still greater jeopardy. It is difficult to imagine, for example, how African game parks would last six months after the tourist trade stopped—and with them would go most of the large African mammals. All hope of saving the tropical rainforests would almost certainly disappear. The overall effects on the Southern Hemisphere of a war confined to the northern superpowers thus would be less extreme than in the North, but still catastrophic.

If such a war occurred after all the nations of the world had nuclear arms, and if it spilled over significantly into the Southern Hemisphere, it is possible that technological society simply would not survive because too many of the repositories of human knowledge, technical skills, and other critical capital would be destroyed. Once technological society was lost, it is extremely unlikely that it could ever be reestablished. *Homo sapiens* started on the road to its present position in a world rich in concentrated resources. Forests and soils were largely intact, deposits of high-grade copper and iron ores were readily available once their value was recognized, and petroleum could be extracted from extremely shallow wells.

Although, after a thermonuclear war, the soils and forests would gradually regenerate, the minerals would not reconcentrate themselves, nor would new petroleum be formed on a time scale of any interest to our species. Therefore history might come full circle, and *Homo sapiens* would by necessity return to being a species of hunters, gatherers, and simple farmers. Whatever survives of the rest of Earth's biota would be left to recover from the impacts of the technological phase and perhaps after many millions of years to regenerate much of its diversity.

ENERGY AND HABITAT DESTRUCTION

Ecologists view rates of energy use as a crucial indicator of human impact on ecosystems. Almost all of the activities that lead to the indirect endangering of other species are energy-intensive. It takes a lot of energy to build buildings, pave roads and parking lots, cut down trees, and plow fields. It takes vast amounts of energy to power the cars for which the roads and lots are paved —and to produce the pollutants from their exhausts. Similarly, giant industrial complexes are not only built with energy but run on energy. Energy is used to tear the raw materials out of the Earth, and energy is used to process them and to create the pollution of air and water from industrial sources. Energy drives the gigantic machines that are involved in the tearing down of forests, and energy powers the off-road vehicles that demolish deserts.

On the stored capital of fossil-fuel energy, humanity has risen to the stature of a global ecological force. The lights of cities and the plumes of smoke from power plants are clearly visible from outer space. The amounts of many minerals, including iron, nitrogen, manganese, copper, zinc, nickel, lead, phosphorus, molybdenum, silver, mercury, tin, and antimony that human beings mobilize in the course of their activities is now on the same scale as, or much larger than, the amounts mobilized by all the rivers of the planet.[127] Humanity now co-opts something on the order of one-twentieth of all the photosynthesis —the primal driving process of life on the planet—for its own uses. And through its activities, *Homo sapiens* now threatens to alter the basic climatic patterns of the globe. And all of this has happened with astonishing suddenness.

The speed with which fossil-fuel-using humanity has achieved its overwhelming dominance of the planet can best be appreciated against a backdrop of geological time. Suppose that the 3½ billion years or so that life has existed on this planet were condensed into a single day, with life starting at a second or so after midnight.[128] The first organisms capable of carrying on photosynthesis would have appeared in the oceans around three o'clock in the morning. It would be about four o'clock in the afternoon before single-celled plants and animals appeared, and almost seven-thirty in the evening before the first organisms made up of many cells made their debut. It would be after nine at night before plants first invaded the land, and only an hour before midnight when the age of reptiles began. At 11:48 P.M., the world would be well into the age of mammals, and the first human beings would appear less than half a minute before midnight. Agriculture would have been invented about a twentieth of a second before midnight, and the industrial revolution started about a thousandth of a second before midnight. So on a time scale of 24 hours of life on Earth, humanity's great takeover occurred entirely within the last twentieth of a second, and the real achievement of dominance (along with the exploitation of fossil fuels) came in the last thousandth of a second.

LOOKING BACK AND LOOKING FORWARD

At the end of this dismal recitation of the ways in which humanity degrades the habitats of all the living things on Earth (including itself), it seems appropriate to recall the fates of past civilizations. Human societies have blown it ecologically many times. The thriving hydraulic civilization of the Tigris and Euphrates valleys—like many agricultural societies today—was unable to maintain its irrigation systems properly, and it went under. Overintensive farming appears to have done in the Khmers and the classic Mayans. And the Greeks, Romans, and others, with logging and goats, created the biological desolation that is today the Mediterranean basin, and partially as a result lost their dominant positions in the world. These peoples in the past, and many others, have paid the price of being insensitive to the long-term ecological effects of their activities. They managed to create ecocatastrophes even without access to bulldozers, strip-mining machines, ORVs, synthetic insecticides and herbicides, oil tankers, or thermonuclear weapons. What so horrifies us now in the light of past *local* collapses is the specter of a *global* civilization traveling precisely the same route, but equipped with those and other deadly tools.

Our recounting of the forces of habitat destruction is necessarily episodic. It is biased toward the temperate zones, for which there is much more information available, rather than the tropics, where the problems are orders of magnitude more serious. And it is badly marred by vast holes in understanding both of the distribution of impacts and, more importantly, of the responses of ecosystems to those impacts.

For any given activity or incident, it might be fairly easy to shrug off the effects. Can just a little more do any real damage to other species? So what if a few more square miles of rainforest are chopped down? There's lots of rainforest left. So what if one more mountain in Colorado is dug up, ground to a fine powder, and stuffed into the local valleys? There are other mountains and valleys there. So what if another stretch of stream is rendered lifeless by solvents pouring from an industrial plant? If things ever get bad, we can stop polluting and life will return.

Why worry if chlorinated hydrocarbons have become ubiquitous and are now found all over the world and in most living organisms? No one has ever proved that a species has gone extinct because of them. What did a Galápagos tortoise or a Hawaiian honeycreeper ever do for me? Maybe we'd be better off in a world dominated by people and their cattle, sheep, goats, and pigs. Surely one more freeway (or housing development or shopping center or factory) can't hurt, can it? After all, there's still lots of open space in Nevada and Antarctica. Why not drill a few more offshore wells? The oceans are huge and we desperately need the oil. Is there really a good reason not to dam that river? Surely no one will ever miss the Snail Darter or the Furbish Lousewort. Why not drive the dirt bike to that next butte? A few more miles can't make any difference. Why not increase our industrial capacity by a few percent? A tiny increase in the acidity of the rain can't make that much difference. Why not have three children? They're so cute, they make me feel so good, they may support me in my old age—and they are a drop in the bucket in a world pushing toward five billion people.

Such questions, taken in isolation, as they always are, continuously lead reasonable people to make decisions that contribute to the human assault on the ecological systems of Earth. And of course all of us to one degree or another continuously participate in that assault in the very process of living —buying food grown in the agricultural system, using energy directly, purchasing an enormous variety of goods that have diverse environmental impacts tracing all the way back to extraction of the original resources from which they were created. Just as the seemingly trivial act of throwing a gum wrapper away can, if done by enough individuals, convert a lovely scene into a rubbish-strewn mess, so seemingly minor actions on the part of all human beings can in the aggregate destroy the ecosystems of our planet—by exterminating their working parts. The essential message of this chapter is that all these individual acts together add up to a *scale* of assault that is still increasing and that may well already be unsustainable.

In the next chapter, we'll look at a sample of these acts in their political context, to set the stage for considering the societal actions required to dampen the surge of extinctions. Then the final two chapters will consider both tactics and strategies for turning the trends around. For if they are not changed, the fate of civilization is surely sealed.

Part IV

WHAT ARE WE DOING AND WHAT CAN WE DO?

Chapter VIII

The Politics
of Extinction

We have not inherited the Earth from our parents, we
have borrowed it from our children.

—IUCN,
World Conservation Strategy,
Introduction

THE quality and condition of life for *Homo sapiens* in the early twenty-first
century will depend heavily on humanity's success over the next two
decades in halting the loss of Earth's biological resources. If success is to be
achieved, it will only follow concerted political action on issues as complex and
little understood as any that modern society has ever faced. Despite the general
increase in environmental awareness and the passage of many protective laws,
the conservation movement is still fighting fundamentally rearguard actions.
Conservationists have become experts at harrying the enemy's advancing columns and mounting last-ditch defenses on battlegrounds chosen by the foe.

But political battles to conserve biotic diversity will not be won by rearguard
actions alone. Conservationists' victories indeed have been relatively few and
often only temporary. For every species saved—usually when it is already on
the verge of annihilation—crushing, unheralded losses are suffered as unknown populations and species are destroyed from Alaska to Zambia. The
tactical successes of the conservation movement, even with a strengthened
arsenal of legal weapons, still add up to a strategic disaster; and the enemies
of conservation grow daily stronger and more formidable.[1]

In this chapter, we briefly discuss some of the legal weapons that are now
available to conservationists and look at a sample of recent political battles
related to extinction. From these you may gain some flavor of how things are

going now and some background for thinking about how they *should* go in the future.

ENDANGERED SPECIES AND UNITED STATES LAW

The United States is generally acknowledged to be a world leader in conservation affairs; only England and possibly Sweden can claim to rival the United States in longevity and success in environmental protection and resource conservation. The mass movement for environmental protection, in which Americans by the millions became involved, is generally conceived to have begun in the 1960s. But that movement grew out of a long tradition of wildlife conservation, which goes back to Teddy Roosevelt, if not further. Most of the national parks and many state parks were created out of that tradition. Some conservation organizations, such as the Sierra Club, the Wilderness Society, the Audubon Society, and the National Wildlife Federation, were founded long before 1960.

The first important federal environmental legislation passed in the United States was the National Environmental Policy Act (NEPA), which became law on January 1, 1970. That law declares it a responsibility of the U.S. government to restore and maintain environmental quality. Among other things, NEPA requires that all federal agencies prepare an "environmental impact statement" (EIS) on any project or proposed legislation that would affect the environment. The EIS must include information on the likely environmental impact, especially unavoidable adverse effects, possible alternatives to the planned action, and "any irreversible and irretrievable commitments of resources" that would be involved.[2]

The law also provides that concerned citizens' groups may sue to halt projects for which there have been no adequate environmental impact statements. This has provided the legal basis for effective citizen actions against numerous projects—including the Alaska oil pipeline, the Florida Barge Canal, and the Everglades airport. Sometimes the process of examining the potential environmental impact has been sufficient to persuade developers and agencies to modify their plans. More often, citizens' suits have resulted in changes or even abandonment of some projects. NEPA made environmental protection a national policy and has had the effect of raising environmental consciousness in government agencies to a considerable degree. But biological resources—populations and species—are still undervalued by the majority of citizens, even in the relatively environmentally enlightened United States.

Other environmental legislation also has had some influence on natural ecosystems and the species that comprise them. A series of laws to control pollution of air and water was enacted during the 1960s and early 1970s. The

Clean Air Amendments of 1970 and the Water Pollution Control Act of 1972 are the latest and toughest of these. In late 1970, an administrative reorganization led to creation of the Environmental Protection Agency (EPA), whose primary responsibilities are to set standards for air and water quality and for emission controls and to regulate or ban release of chemicals that are dangerous to human health or the environment. The EPA has power to monitor and regulate emissions of polluting substances, but apart from the effects of those on natural ecological systems (which are sometimes very serious), the agency's power to protect ecosystems is very limited.

The real breakthrough in protection of species and natural ecosystems occurred in 1973 when the Endangered Species Act was passed. This was a tough, uncompromising law, which went far beyond earlier, relatively defensive attempts. This was also the first law in the United States to recognize the vital importance of preserving the *habitat* of an endangered species—the ecosystem on which the species depends. The Endangered Species Act provided that a species or subspecies "endangered throughout all or a significant portion of its range" be listed with the Secretary of the Interior or, in the case of marine species, with the Secretary of Commerce. The Fish and Wildlife Service (FWS) of Interior and the National Marine Fisheries Service (NMFS) of Commerce were to issue regulations to protect endangered or threatened (likely to become endangered) species.[3]

Under the law, all federal agencies " . . . shall, in consultation with the Secretary . . . [take] such action necessary to insure that actions authorized, funded, or carried out by them do not jeopardize the continued existence of such endangered species and threatened species or result in the destruction or modification of habitat of such species which is determined by the Secretary, after consultation as appropriate with the affected states, to be critical."

The law also prohibited killing, capturing, importing, exporting, or selling any endangered or threatened species, including plants. Federal agencies are required to consult with FWS or NMFS on all projects before starting them. If a project is initiated without modification to protect an endangered species, the law provides for citizens' suits to enjoin the project for violation of the Endangered Species Act. Since so many environmentally destructive activities (such as mining, logging, dam building, recreational development, etc.) take place on federal lands, are at least partially federally financed, or are subject to federal regulations of one sort or another, the act is potentially a powerful force in defense of species and ecosystems.

The existence of this array of environmental laws and an officially pro-environment policy in the government have not by any means brought a halt to environmental deterioration in the United States, though they may have slowed down the rates of destruction and damage somewhat. To a large degree, they have simply provided tools for legal action by concerned citizens to stop, delay, or modify actions they see as objectionable. But the forces that want to "develop" everything in sight for a profit are no less determined to do so than they ever were. And they still have plenty of friends in public office.

The Infamous Snail Darter

Pork-barrel politics in the United States continues to be among the greatest threats to the environment and to the preservation of species, and the celebrated case of the Snail Darter vs. Tellico Dam is a prime example.

The saga began in the late 1960s, when the Tennessee Valley Authority (TVA) initiated its Tellico Development Project in a valley of the Little Tennessee River in eastern Tennessee. The project included—but was by no means limited to—a dam. The valley was lovely,[4] a productive farming area; the river was the last free-flowing stretch of water in the region, described by local fishermen as the "best trout stream in eastern Tennessee." The valley is also a sacred place to the Cherokee Indians and was their homeland until they were banished to Oklahoma in the early nineteenth century. It is therefore a site of considerable archaeological significance.

TVA's plan was to turn the valley, nearly half of which would be flooded by the dam, into a booming recreational area. The dam, one of seventy TVA dams in the region, would also contribute to the regional hydroelectric power grid and to flood control. But the local people—especially the farming families who would be displaced and anglers and others who enjoyed the recreational advantages of the river as it was—were not pleased. They enlisted the aid of an environmental group, the Environmental Defense Fund (EDF), and successfully sued in 1971 to stop the project on the grounds that TVA had failed to file a proper environmental impact statement. TVA had claimed that, as a private agency, it was not obliged to do so.

An impact statement that explored the environmental effects of the project and alternatives to it was eventually produced and accepted. The possible existence in the river of several rare species of fishes, which would be threatened by the dam and reservoir, was mentioned in the report. Meanwhile the project was delayed for a year and a half. During that time, the 1973 Endangered Species Act was passed by Congress. In that same year, a University of Tennessee biologist, David Etnier, while surveying the fish fauna of the Little Tennessee River, scooped up in his hand a small, previously unknown member of the perch family, which fed on snails.[5] The Snail Darter *(Percina tanasi)* is a tan, three-inch-long fish that has been found only in the fast waters of the Little Tennessee above the dam site. It was very clear that completion of the dam and filling of the reservoir would completely destroy the habitat of the little fish.

But the TVA, once the first injunction had been lifted, was accelerating completion of the Tellico project, particularly the building of the dam, despite continued local opposition. It also fought against citizens' petitions to put the Snail Darter on the Endangered Species list and thus under protection of the act, claiming that the law could not be applied retroactively. When that round was lost, TVA grudgingly cooperated by attempting—not very successfully—to transplant the little fish to other nearby streams. TVA officials also engaged in the required consultations with the Fish and Wildlife Service, but according to an official of the FWS, they refused to consider any alternative to completing the dam. So the citizens' groups and EDF sued again.[6]

The district court found that the Snail Darter would indeed be eradicated by the Tellico dam, but declined to stop the project because it was 80 percent complete and there were "no alternatives to impounding the reservoir, short of scrapping the entire project"—a conclusion not supported by evidence presented in the trial that the roads already built would be useful and that the remaining land could remain in agricultural production. In early 1977 the district court's decision was overturned by a Circuit Court of Appeals which granted an injunction on the dam portion of the development project.

At this point, the press went to town with the story of an obscure three-inch fish that had halted a $120-million-plus dam project. Most news reports sided with TVA and overlooked all the other strong arguments against the Tellico dam besides the saving of the Snail Darter—the losses of valuable and productive farmland, recreational values, and a priceless archaeological site. Further, they depicted the dam as if it were the entire project, whereas the dam itself only accounted for about $22.5 million and was by no means an essential part of the development. The rest of the funds have been spent on buying up the valley's land and in building roads and other facilities for recreation. Newspapers presented the issue as a case of fervent environmentalism carried to unreasonable lengths. The *Washington Star* called it "the sort of thing that could give environmentalists a bad name"; Art Buchwald dubbed it "Jaws III."[7]

In June 1978, the Supreme Court upheld the injunction against the dam on the basis of the Endangered Species Act, as written. But Chief Justice Warren Burger, in writing the decision, virtually invited Congress to amend the law to allow exceptions. Proposals to amend the Endangered Species Act, many of which would have gutted the law, had already been introduced in Congress. Authorization for the act was due to expire in 1978 anyway, and this was seen by its enemies as an opportunity to finish it off.

The amendment that was eventually passed in 1978 was cosponsored by Senator John Culver of Iowa, who wanted to preserve the act as far as possible, and Senator Howard Baker of Tennessee, who was bent on ensuring completion of the Tellico dam. The amended law created an "Endangered Species Committee," composed of the secretaries of Agriculture, Interior, and the Army, the chairman of the Council of Economic Advisors, the heads of the Environmental Protection Agency and the National Oceanic and Atmospheric Administration, and a representative from the affected state. This group, which soon became known as the God Committee because its decisions would spell life or extinction for an endangered species, can be convened to resolve "irresolvable conflicts" when all other efforts have failed. It can grant exemptions to the act where no "reasonable and prudent" alternative exists, when the project is of national and regional significance, and when the benefits "clearly outweigh" those of alternative actions. The first case to be reviewed by the God Committee was the Tellico project.

There exists considerable doubt whether the God Committee amendment was really necessary even from the point of view of the most pro-development politician. In the five years between passage of the original Endangered Species Act in 1973 and the amendment in 1978, only a handful of development projects had run into serious difficulty, while thousands of consultations with the U.S. Fish and Wildlife Service on endangered species had taken place. There were 4,500 consultations in fiscal 1977 alone. Of the cases that seemed most insoluble (one being Tellico), recalcitrance on the part of the project agency was the basic cause. It is also instructive that the organizations threatening these endangered species were governmental—not giant corporations, which in this day and age are often more sensitive to public sentiment than government bureaucrats.

Moreover, it is arguable whether Tellico itself was really irresolvable. Consultations with FWS had hardly been conducted in good faith, since TVA had stubbornly refused to consider alternatives. Now, under new leadership with the appointment of S. David Freeman as chairman, TVA's stance changed. Alternatives to closing the dam were studied, and it was discovered that, even at that late stage, it would be profitable to leave the valley unflooded. Continued agricultural production alone would produce nearly twice the economic benefit that the completed project was expected to yield. These findings confirmed those of a little-known study by the General Accounting Office (GAO) in late 1977.

The possibility of alternatives to completing the dam and the other values that would be lost if it were, such as agricultural production and the Cherokee historical sites, were brought out at the hearings of the God Committee. In addition, new economic analyses showed that the dam's electricity output would be produced at a deficit. The whole thing was clearly a boondoggle. Not surprisingly, the committee therefore ruled unanimously in favor of the small Snail Darter. The chairman of the Council of Economic Advisors, Charles Schultze, observed that economic justification for completing the dam was at best dubious and said, "I can't see how it would be possible to say that there are no reasonable and prudent alternatives to the project."[8] Another member of the committee, Cecil Andrus, secretary of the interior, commented, "Frankly, I hate to see the Snail Darter get the credit for delaying a project that was so ill-conceived and uneconomic in the first place."[9]

Unfortunately, some of the dam's proponents were not graceful losers. The state of Tennessee and TVA accepted the decision without protest, but Senator Baker, who had cosponsored the bill that created the Committee on Endangered Species, now wanted to abolish it. And so did some congressmen. Several attempts were made in the Senate to pass bills specifically exempting the Tellico project from the Endangered Species Act, but they failed.

Then an amendment to the annual ten-billion-dollar water projects appropriation bill was added from the floor in the House in June 1979 by Tennessee Congressman John Duncan. The amendment was not read out, much less discussed. Only a few members were even in attendance. So without opposition a bill was passed that exempted the Tellico project from compliance not only with the Endangered Species Act but with *all other laws* that stood in its way, such as the National Environmental Policy Act, the Clean Water Act, the Historic Preservation Act, and various dam safety laws.[10]

The Senate tried to kill the Tellico amendment but was ultimately unsuccessful—Congress was tired of the whole subject and Baker was pushing hard. The final sorry chapter was President Carter's signing of the bill ("with regret," he said) in September 1979. Twelve hours later the dam builders were at work in Tennessee.

Protecting Endangered Species

The Endangered Species Act was passed with strong support by both houses of Congress in 1973. It was meant to solve a problem: to halt, or at least retard, the rate of species extinction due to human intervention and help preserve what little natural environment remains in the United States. Perhaps the legislators had no notion how serious the problem is—how many populations and species are already threatened, or the scale of attack our society is mounting. People may be aware of development projects in their own localities and sometimes realize not much undisturbed land or water is left. But they usually don't recognize that it is happening in *every* locality except a few that have been declared "wilderness" or national parks (and even many of those have suffered shocking amounts of encroachment).

The Endangered Species Act, as originally written, was a powerful weapon in the environment's behalf. Indeed, because of the very strength and rigidity of the law, environmentalists were hesitating to file suits based on it for fear that Congress, in retaliation, would refuse to extend the act. Deserving species were often not listed out of the same fear. Many environmentalists therefore saw the 1978 amendment as an acceptable compromise, though others saw the creation of the God Committee as an invitation to agencies to drag their heels in finding acceptable alternatives to their projects that would spare threatened species. In its first year of existence, however, the committee reviewed only one other case besides Tellico, and that too was decided more or less in favor of the environment.

The 1978 amendment also weakened the Endangered Species Act by making it much more difficult—and sometimes impossible—to list a species as endangered. Before a species is listed, the boundaries of its habitat must be described, an economic impact study must be made, and public hearings held, all within two years. Moreover, there is apparently no protection at all for invertebrate animal *populations* unless they have been formally named a subspecies.[11] The rate of listing of endangered species has declined precipitously since these provisions went into force; indeed, hundreds of species have even been withdrawn from the list.

Despite some erosion of the act's clout, however, it still exists as a potentially strong weapon in defense of species and environmental integrity. The changes may persuade most developers to accept decisions against them, since they now have ample opportunity to "prove" that the benefits of their projects far outweigh losses of other values. And they have leeway and mechanisms for devising less damaging compromises. But the weakening also encourages them to persist in pushing their boondoggles by allowing ways to get around the law, either by making minor modifications in their plans or by appeals through the God Committee.

Still, public surveys have shown steadily increasing public support for preservation of wilderness and wildlife. A 1980 survey indicated that a majority of Americans favored protecting wildlife (whether endangered species or not) even at the cost of forgoing additional jobs, housing, or other development projects.[12] Someday even the most dedicated dam builders and their politician friends may be convinced of its importance, too.

Meanwhile, what of the poor Snail Darter, betrayed by Congressional dirty tricks? The Tellico dam is filling as we write. A few hundred transplanted individuals still survive in nearby rivers, but whether the species can make it over the long haul remains to be seen. So far as we know, this is the first time that a deliberate, conscious action has been taken to push a species to extinction.

THE POLITICS OF WHALING

Whaling is a prime example of determined economic interests destroying a biological resource. Here not just one but *twelve* species are endangered in varying degrees. Several—the Southern Right whale, the Bowhead, the Humpback, and the gigantic Blue Whale—are "commercially extinct," which means they are so rare that it no longer pays whalers to hunt them. That whales were becoming endangered was recognized as much as fifty years ago. But if protection of species has proven exceedingly difficult at the national level, it has seemed virtually impossible on an international basis and especially on the high seas, where there is no recognized jurisdiction.

Still, attempts to regulate whaling began early with an international agreement to ban hunting of the Southern Right and Bowhead Whales in 1935. Just after World War II, the International Whaling Commission (IWC) was established, whose role was to "provide for the conservation, development, and optimum utilization of the whale resources." Commissioners were delegates from both whaling and nonwhaling nations. The IWC was advised by a Scientific Committee, which set annual quotas for hunted species of whales. There was no means of enforcement for the quotas, so the whalers proceeded to ignore them for the most part. Nevertheless, their catches often came uncannily close to the quotas, which were based upon the scientific advisors' estimates of the sizes of whale populations.

As time went on, whaling technology improved, and whalers killed ever greater numbers of whales, despite repeated warnings from IWC scientists that stocks were being overfished. As stocks of the largest whales were depleted, the whalers turned to smaller species and also pushed them toward extinction. The IWC typically responded with too little protection too late, and the whalers often ignored what efforts it did make. The Humpback and the Blue Whale were not extended protection until 1966. The California Gray Whale, which has been protected since 1947, is the only whale population known to have made a significant recovery in population size following protection.

By the 1970s considerable public opposition to whaling had arisen, especially in industrialized nations. This found expression in the 1972 United Nations Conference on the Environment, which called for a ten-year moratorium on all commercial whaling. This resolution was later adopted by the United Nations General Assembly as well. But the IWC rejected the moratorium. Instead, in 1975 it adopted a complex management scheme based on scientific estimates of sustainable yields from each population, which at least would improve the situation if it were properly implemented.[13]

Meanwhile, public opposition to whaling continued to rise, and various governments felt its pressure. Nation after nation dropped out of the whaling business. By 1980 only a dozen or so were still involved in whaling, and most of them were now members of IWC. (Nonmembers of course are not even nominally bound by the quotas.) The glacially slow process of change in the IWC has been the result of changes in position in individual nations. These nations, in turn, have responded to pressure from their own citizens—usually conservation groups. This was the case for the United States, which gave up commercial whaling only after 1970. In the late 1970s, under pressure from citizens' groups, the Australian government established a formal commission, headed by Sir Sydney Frost, to examine the whaling issue exhaustively. It did so, utilizing input from conservationists and concerned biologists, and in 1978 came out firmly in favor of the whales.[14] The Australian government then did an abrupt about-face, outlawed Australian whaling, and became a strong force for conservation in the IWC.

Only two nations, Japan and the U.S.S.R., still maintained oceangoing whaling fleets by 1980, and their activities have been increasingly constricted by the expanding protection of various species and the steady reduction of quotas. IWC quotas declined from a total of 45,000 of all species in 1972 to less than 16,000 in 1979.

The Japanese especially, in defending their whaling activities, have sometimes resorted to extraordinary statements. Usually they plead that the whaling companies' economic survival is at stake and that the scientific evidence on stock sizes is incomplete and imprecise. They have also claimed that whale meat is an essential component of the Japanese diet, although in actuality it makes up a minute—and easily substituted for—fraction. In 1979, when a total ban on commercial whaling in the Indian Ocean was proposed in an IWC meeting, the Japanese delegation insisted that the whale populations would "rise to the ceiling" unless they were controlled by whaling![15]

Whale politics illustrate how extremely convoluted the politics of extinction can be. In the late 1970s, the United States, which should have been a powerful force in behalf of the whales in IWC negotiations, was considerably hamstrung by the whaling activities of a group of its own citizens: the Eskimos of Alaska's North Slope.

The Eskimo-Bowhead Controversy

For centuries the Inupiat group of Alaskan Eskimos has hunted the Bowhead Whale as it migrated past their shores. Traditionally, the meat, skin, bones, fat, baleen, and other products from the whales were vital materials for the group's survival. The hunt accordingly was an important annual ritual with religious and ceremonial overtones. Today, the meat and other products from the whales are no longer quite so essential for the survival of the Inupiat; modern technology and transport can provide substitutes. At the same time, technology has made it far easier to catch and kill a whale than it was with traditional methods. In addition, earnings from the Alaska oil pipeline project enabled more men to buy boats and equipment for hunting. So as the whales became less essential to the Inupiat as a resource (though not as a cultural element), killing of them increased. For decades until the early 1970s, each year's hunt resulted in killing no more than 15 of the huge animals. But with new boats and equipment and more people participating in the hunt—traditionally, only a few experienced men did the hunting—the kill rate soared. In 1976, 48 whales were landed, while 43 more were struck and lost (most of which doubtless died). In 1977, only 26 were landed, while 79 were struck and lost.[16]

The Bowhead is one of the most endangered and least known biologically of the whale species. But its endangered status is actually not the result of Eskimo hunting; it is the result of nineteenth- and early-twentieth-century commercial whaling by Americans and Europeans, from which the population apparently still has not recovered. In 1977 the total Bowhead population was estimated at somewhere between 1,000 and 2,000 individuals. Annual losses in the neighborhood of 100 whales obviously were not sustainable. So in June 1977 the IWC put a moratorium on "aboriginal whaling," previously exempted from the quota system, of the Bowhead.

The Inupiat were outraged; they felt that their way of life was being threatened. In their view, managing the Bowhead population was their responsibility. It seemed to them that the U.S. government scientists and officials who estimated the population size were conspiring to defraud them of their rights. They put pressure on the U.S. government to file an objection to the IWC decision (which, if filed within ninety days, would waive responsibility to observe the quota). But the State Department, reasonably enough, felt that filing an objection would open the way for the Russians and Japanese to file objections to other quotas and go back to all-out whaling. As a leader in international pressure to end commercial whaling altogether, the United States was caught in a very embarrassing position.

The Inupiat even sued, but unsuccessfully. In December, the U.S. delegation to the IWC hammered out a compromise whereby the Inupiat were permitted a quota of twelve taken or eighteen struck, whichever came first. The Eskimos were disgruntled, but there were no serious infractions during the 1978 hunt. Further study of the Bowhead population that year put the population at over 2,200 individuals, but revealed extremely small numbers of calves and an extremely low pregnancy rate. The IWC's Scientific Committee ruled that the only safe course was to set a quota of zero.

The Inupiat, maintaining that calves were not counted because they migrate earlier, kept up the pressure and succeeded in getting their 1979 quota raised to eighteen taken or twenty-seven struck. This was less than they hoped for, however, and they stalked out of the IWC meeting, again threatening to ignore the quotas and regulations. But, because of bad weather and ice conditions, they failed even to make the quota.

At the June 1979 meeting of the IWC, there was widespread hope that a worldwide moratorium on commercial whaling might at last be instated. The United States was to be a leader in this movement; Congress had passed a resolution calling for it. But the United States was also committed to maintaining the Eskimos' quota of Bowheads—even in the face of the Scientific Committee's statements that not only was a zero quota the only safe one, but that a continued decline in Bowhead numbers seemed likely *even if none were killed* from then on.

Following some heavy behind-closed-doors negotiating, some considerably watered-down rulings emerged. The moratorium on coastal whaling, which had been separated from deep-sea whaling, failed. The moratorium on deep-sea factory whaling passed, but Japan had succeeded in getting an exemption for Minke Whales. A ten-year moratorium on whaling in the entire Indian Ocean south to the 55th parallel was passed. Quotas were reduced on several whale stocks. And the Inupiat were granted a Bowhead quota of eighteen and twenty-six.[17] Clearly, the United States had bargained away the full moratorium—which the U.S. itself had proposed to the meeting—for the Eskimos' right to continue hunting Bowheads.

In the spring 1980 hunt, the Inupiat whale hunters, despite their dissatisfaction, ended the hunt when twenty-six whales had been struck. (Unfortunately, not all the crews were notified in time, and five more were struck, resulting in cancellation of the fall hunt.) The Eskimos have also been cooperating with the scientific studies on the whale's population dynamics. Staff members of Friends of the Earth, who have been involved in negotiations with the Inupiat over their whaling and who observed the 1980 hunt, strongly feel that involving the Eskimos in the conservation studies of the Bowhead and educating them about the broader global issue of whaling in general, while permitting them to hunt a limited number of whales, are essential to gaining their cooperation in the long term. Without their cooperation, there is no hope for Bowhead Whales.[18]

Meanwhile, a new and far greater threat to the Bowhead has developed: proposals to drill for oil in the Beaufort Sea, through which the whales migrate each year. Climatic conditions for drilling are extremely hazardous in that area, and therefore oil spills, which would be disastrous for the whales, are highly likely. The Inupiat are already cooperating with scientific studies on the environmental impacts of such accidents, especially on the whales, and they could also prove to be valuable allies politically as this and other resource exploitation decisions are made in Alaska.

Pirate Whalers

If the only threat to the whales today were from the legitimate whalers who are subject to regulation by their own nations and by the IWC, there would be good reason to hope that the whales might eventually be spared. Regulations have continually been tightened, whalers have been more careful in recent years to stay within their assigned limits, and there has been a steady attrition in groups engaged in whaling as it has become less and less economic. Unfortunately, as legitimate whalers have become relatively law-abiding and responsible, there has been a rise in illegal activity.

Some conservation organizations concerned with the whaling issue—notably the international People's Trust for Endangered Species and U.S.-based Monitor—have kept track of these activities and traced ownership of the pirate whaling vessels through a series of dummy companies and changed registrations. The information was needed to help the IWC enforce its regulations. The whaling vessel *Sierra* was the first (beginning operations in 1968) and for a decade the only outlaw whaling ship. According to a report by the People's Trust,[19] the *Sierra* was capable of killing and processing 40 to 45 whales on each six-week trip, and she hunted year-round. In 1978 she was joined by another vessel, the *MV Tonna,* which acted as a factory ship while *Sierra* did the killing, more than doubling the take. But in July 1978, attempting to hoist aboard an 80-ton Fin Whale, the *Tonna* listed severely, flooded her engine room, and sank, drowning the captain and several crew members.

In 1979, a refitted Japanese trawler, renamed the *Cape Fisher,* joined the *Sierra.* The two ships together were capable of taking 1,200 whales per year. Meanwhile, two additional whalers were being refitted in South Africa. A South African company, the Sierra Fishing Agency, acted as "agents" for the pirate whalers. Exact ownership of the operation was difficult to determine, but it was finally established as the Taiyo Fishery Company, a huge Japanese-based multinational corporation.[20] There has been no mystery about where the products were sold: Japan. Captains and crew members were mostly Norwegian and South African, but four Japanese "meat inspectors" were aboard the *Sierra.* Their job was to ensure that only the choicest cuts were saved. The rest, 80 percent of the carcass, was discarded. In 1978, Japan imported over nine million dollars worth of whale products from the *Sierra*'s operation. It is estimated that some 4,000 whales have been killed each year by the pirates.

Killing of whales by the pirates is reportedly even more brutal than that of legitimate whalers. Former crewmen on *Sierra* and *Tonna* have testified that average death-times of the whales were two hours; one took three hours to die. Often the whales were inflated while still alive. (Normal procedure is inflation after death to keep the whale from sinking to the sea bottom before it can be hoisted aboard ship.) Many whales have been struck and lost, partly because grenade-tipped harpoons are not used as they "ruin too much meat." Nor have the pirates observed such niceties as leaving alone the endangered and protected species or females; any whale they find is fair game.

In 1979, militant conservationists aboard an icebreaker trawler, the *Sea Shepherd,* owned by the Fund for Animals, went after the *Sierra.* In mid-July they found her 180 miles west of Portugal and pursued her to the Portuguese port of Leixos. The conservationists were deceived into entering the harbor ahead of *Sierra* and, once in, were refused clearance to leave again. The *Sierra* remained outside, preparing to sail again. The young Canadian leader of the conservationists, Paul Watson, put his captain ashore and invited the crew of twenty volunteers to remain ashore if they had no stomach for risking their lives and a jail term. All but two—an Australian and an American—remained ashore.[21]

Manned by three men, the *Sea Shepherd* steamed out of the harbor without clearance and went straight for the *Sierra,* aiming for the harpoon gun mounted in the ship's prow. In Watson's words: "We glanced across the bow, causing minimal damage to her harpoon. But the blow did have the effect of warning them. A shudder went through the *Sierra* and her crew began pouring out of hatches like disturbed termites."

The *Sea Shepherd* executed a tight 360-degree turn and went again for the *Sierra*'s forward port side. Watson wrote:

> ... I could see the horrified amazement on the faces of the crew. I could see Captain Arvid Nordengen, the big Norwegian Captain, stand staring, cursing and helpless. I briefly glimpsed a rifle being raised and then we hit.
>   We hardly felt the impact on the *Sea Shepherd.* ... we were practically on top of the whaler, pushing her far over to starboard. With our engines pushing us and because of the angle that we had struck, we ripped her open, exposing the whale meat in her guts. We tore a six-by-eight-foot hole in the whaler, and as we pulled out we slammed full against her port side, staving in forty-five feet of her hull.

The *Sierra* fled into the harbor, leaking and listing badly. The *Sea Shepherd* headed for England, but was soon stopped by a Portuguese destroyer. Watson and his two colleagues were detained but, to their surprise, not arrested. The Portuguese media and public showed enthusiastic sympathy for the *Sea Shepherd*'s action; the authorities decided instead to consider whether Portugal should continue being a way station for the pirates to send their products to Japan.

The Sierra Fishing Agency (now illegal in South Africa) filed suit in Portugal against the *Sea Shepherd*'s operators, then demanded a $7 million settlement. The Portuguese removed Watson's and his two colleagues' passports pending resolution of the case. Unwilling to cool their heels in Portugal while the pirate whalers went back into action, the three separately slipped out of Portugal, vowing to get their ship back and "sail again in defense of the helpless and gentle giants of the deep."[22]

It soon appeared that business for the pirates had fallen on difficult times. The United States had passed a law that any nation involved in pirate whaling would lose its fishing rights in U.S. waters—a measure aimed particularly at

Japan, a heavy user of U.S. fisheries. Japan announced a ban on all imports of non-IWC sources of whale meat after July 5, 1979. South Africa, immediately after the *Sierra* was disabled, passed a law forbidding its citizens to engage in any whaling activities.[23] Lloyd's of London canceled *Sierra*'s insurance. Investigations were proceeding in Norway on Norwegian involvement —which was illegal in that country. The two ships being refitted in South Africa were sold, as was the *Sierra* and apparently the *Cape Fisher*.

Unfortunately, the whales are not yet safe from the depredations of the pirate whalers. The *Sierra* mysteriously blew up and sank in Lisbon harbor in early 1980; no one knows why or who was responsible. And South Africa has forbidden the two refitted ships to leave port. But the *Cape Fisher*, with a new name (believed to be *Astrid*) has been seen in the eastern Atlantic, and four other ships are operating out of Taiwan in the Pacific. The pirates reportedly are also expanding into coastal operations through nations that are not members of IWC. And Japan has continued to import pirated whale meat by shipping it through South Korea, an IWC member nation.[24]

Meanwhile, Japan has publicly declared that whaling is no longer profitable, and the U.S.S.R. seems to be accepting the IWC's tightening strictures relatively philosophically. The quotas continue to shrink, and although a complete moratorium was again denied in 1980, the end of legal whaling may be in sight. Assuming that pirate whaling can also be stopped—and international sanctions against Japan and cooperating non-IWC member nations may be the key to this—the questions remaining are:

How much longer will it take?

Why has it taken so long?

Will the whale populations be able to recover?

THE WILDLIFE TRADE

In a recent book on policy analysis, the authors, a policy analyst and an economist, accurately observed that, economically and politically, wildlife and endangered species have value only if society attaches value to them: ". . . neither the redwoods nor the bluebirds can speak for themselves. . . . unless human beings care about redwoods, the redwoods will disappear."[25] The economically recognized values attached to such living things have indeed mostly been limited to the commercial value that can be derived from their exploitation: the price of redwood timber, for instance. Today, the soaring "values" of many rare, threatened, and endangered species, however, are a major cause of their being pushed toward extinction.

During the 1970s the wildlife trade—the shipment between countries of living animals or plants or plant or animal products—expanded explosively.

A large portion of this trade is illegal, involving rare, threatened, and endangered species or products derived from them. And the rarer or more endangered the species, the higher the price it seems to command.

By 1979, U.S. government officials estimated that the illegal imports to the United States of live endangered animal species alone amounted to $50 to $100 million worth of business per year. When it is considered that great numbers of plants and a huge volume of products from endangered species—ivory, shells, feathers, horn, skins, and furs—also find their way to the United States, and that an enormous flow of trade (including endangered species from the United States and Canada) also goes to Europe, Japan, and other countries, some grasp of the staggering dimensions of the trade is possible.[26] Also staggering are the political complexities of dealing with a problem that in some ways is similar to the international traffic in drugs but is not taken as seriously.

Efforts to control trade of endangered species and their products began early in the United States with passage in 1900 of the Lacey Act, which prohibited interstate traffic of birds and mammals taken illegally. It was later amended to prohibit importation of wildlife killed, captured, or exported illegally from another country. Importation of wild bird plumes for women's hats was banned in 1913 by the Wilson Tariff Act. And the United States signed the first international wildlife protection convention in 1911 with Russia, Japan, and Great Britain (for Canada) to restrict hunting of Pribilof Islands fur seals— a heavily hunted population that had been reduced to 5 percent of its original size. Various other pacts between individual countries have been established in the decades since.

But something broader and more effective was clearly needed by the 1960s. In 1973 the Convention on International Trade in Endangered Species of Wild Fauna and Flora (known as CITES) was established. By 1980, fifty-one nations had joined the convention, which completely prohibits international trade in the six hundred most endangered species or their products, and requires export licenses for another two hundred species. Enforcement is necessarily left up to the exporting or importing country, and there is wide variation among countries in their ability and motivation to enforce the agreement. Moreover, only nations that have signed the convention are bound by it. International regulation is even stickier in the face of economic pressures than national regulation is. Consequently, the trade in endangered wildlife has continued to boom since CITES was established.

Most of the traded species originate in developing countries, whose governments are often poorly equipped to supervise what is exported. The great need in these countries for foreign exchange and general ignorance of the importance of species preservation may induce officials to look the other way. In Kenya, even highly placed government officials were found to be actively involved in the export trade, particularly of ivory, which has had such a devastating effect on Kenya's elephant populations.[27] Kenya has put a total ban on ivory export, but other African countries are still heavily—and largely illegally—involved in the ivory traffic.

In developed countries, which are mainly importers, the problem is largely one of manpower and training. In the United States, the few import officials who are qualified to detect endangered species or products are concentrated in only a handful of ports. Yet, from 1973 to 1978, importation to the United States of at least ostensibly legal wildlife products zoomed from 4 million to 187 million items, in spite of growing restrictions.

Some imports, including living animals, are smuggled into the country; others come with false documentation or have been passed through a third (non-CITES signatory) country.

Some of the smuggling techniques are imaginative, if cruel to the hapless animals involved. Parrots have been found by U.S. Fish and Wildlife inspectors stuffed in nylon stockings and hidden in automobile doors or concealed in shipments of poisonous snakes beneath a false bottom.[28] Wastage is extremely high; perhaps one in ten of the smuggled animals survives the trip, largely because the shippers are ignorant of how to take care of them. Products, too, have often been poorly processed and end up wasted.

Despite the high rate of wastage, the illegal wildlife trade is extremely lucrative. For example, a single rare South American macaw or Asian Cockatoo may sell in the United States for as much as $8,000, and a rare cactus may go for $5,000. The Australian Sulphur-crested Cockatoo, whose export is prohibited by Australia, became very popular in the United States when one was featured regularly in the TV series *Baretta* (despite Robert Blake's testimony on talk shows that the bird was a pain to work with). Illegally imported ones sold for $5,000 in 1979. A South American Ocelot coat may sell for $40,000 in Germany.

Yet while profits are high, as in the illegal drug trade, risks have been low. Importers have seldom been caught, but when they were, the result was usually a small fine. For instance, in the late 1970s, two Australians convicted of smuggling a shipment of rare Australian parrots worth $60,000 into the United States were each fined $1,000.

Officials in the United States discovered how far-flung and well organized illegal wildlife trading operations had become when a U.S. Customs officer, Joseph O'Kane, arrested one Henry Molt, Jr., and his associates in Philadelphia in 1975. Molt specialized in importing reptiles. Inspection of Molt's office files revealed "a dazzling array of documents from all over the world—Africa, Asia, Australia—which hinted at extensive smuggling, double invoicing, and other false documentation."[29]

Although there was plenty of evidence in the Philadelphia office to convict Molt and friends, O'Kane, another customs man, John Friedrich, and U.S.

attorney Thomas Mellon decided to trace the entire network that had supplied Molt's business, and to look into its customers as well. They took a six-week world tour, visiting Fiji, Australia, Papua New Guinea, Singapore, Thailand, and Switzerland. O'Kane told a writer for *National Wildlife*: "We were able to trace it right down to the villager who said, 'I went up that tree and got it.' "[30] The three men also questioned and obtained documents from international traders in illegal wildlife in Singapore and Bangkok.

As a result of these investigations, Molt was tried for smuggling—which is a felony, as opposed to infringing the Lacey or Endangered Species Acts, which are merely misdemeanors. But Molt ended up with a relatively light sentence—only fourteen months in jail, $20,000 in fines, three years probation, and a ban on importing wildlife and travel to the countries where he formerly collected reptiles. It was the stiffest sentence ever imposed for wildlife smuggling, but the prosecution and customs people had hoped for more. Still, compared to some bird-smuggling rings, which are alleged to have Mafia connections and ties to the drug trade, Molt's was a small operation. To his credit, unlike many smugglers, Molt also cared well for his animals.[31]

Enforcement in countries of origin, mostly developing countries, is a stickier proposition. The collector or poacher is usually a poor villager or farmer who is trying to make some extra money to feed his family. His reward is comparatively small in most cases; the big money goes to the shippers, traders, and middlemen in the trade network. The great exceptions, again, are the poachers who collect ivory and rhino horn. One kilogram of ivory can fetch a year's income to an African villager, and a single African Elephant tusk may easily weigh ten kilos. Rhino poachers often are well-organized gangs of heavily armed men whose lack of scruples extends beyond their prey. In June 1979, a crack Tanzanian game warden was shot and killed by rhino poachers armed with submachine guns in the Ngorongoro Conservation Area when he and a group of patrolmen attempted to arrest them. The incredible value of rhino horn evidently "justifies" the risks and the violence.[32]

Some African countries, alarmed by the rapid decimation of their wildlife, have become relatively serious in their efforts to crack down on poaching and collecting. But enforcement is difficult in large developing countries with poor road systems and limited resources—especially when the outlaws are wily, well-armed, and dangerous.

Much of the cost of enforcement accordingly is falling increasingly on the rich countries. There is a certain justice in this, since so much of the demand for wildlife and wildlife products originates in those nations. Conservation organizations, particularly the World Wildlife Fund, raise money to augment the protection activities of poor countries. And they are sometimes quite successful. In late 1979 over a million dollars were raised in Switzerland alone to save the African Elephants. In the same spirit, we think it would be entirely appropriate for the U.S. National Institutes of Health and similar organizations in other rich countries to help subsidize reserves for monkey and ape populations in partial compensation for the exploitation of those populations for medical research.

The enormous global traffic, legal and illegal, in wildlife obviously would not exist if there were no market for the products. Who are the buyers? Shockingly, it turns out that some of the biggest buyers are people who certainly should know better: owners and operators of public zoos. The Molt case alone implicated nine major U.S. zoos, including those of St. Louis, Washington, D.C., and Philadelphia. Their excuse has been that, although zoos can get special permits to import endangered species, using smugglers or false documents was easier and involved less red tape! Unfortunately, probably for political reasons, there were no indictments of the zoo curators implicated in the Molt case. As of this writing, though, the Department of the Interior was still considering the imposition of fines of up to $5,000 against five of them.[33]

Nor is the zoo involvement limited to the United States. A branch of the International Union for the Conservation of Nature headquartered in London in 1978 discovered illegal traffic to a zoo in Japan. And the Jakarta Zoo of Indonesia reported to the group that it had been approached by a trading company masquerading as a zoo to smuggle birds of paradise into Japan. The zoo masquerade apparently often works in persuading countries with strict export laws to permit rare species to be sent out.[34]

Other major buyers of both live animals and plants are private collectors—people who maintain private zoos, aviaries, or horticultural collections. It would be stretching credulity to assume that most of these people are unaware that they are contributing to what one observer has called "the extinction business."[35]

People who purchase endangered species products, either abroad or after importation, may sometimes do so in innocence. Much more publicity clearly should be given to products that are from actually or potentially endangered species. In the 1960s such widespread publicity surrounded the endangered wild cats of Asia and Africa—tigers, leopards, Cheetahs, and so forth—that when movie actress Gina Lollobrigida appeared in public wearing a leopard coat, she received a great deal of unfavorable publicity and public criticism. Most reputable furriers in the United States and Europe have since formally agreed not to use the pelts of these animals.

But a new set of animals, whose pelts are still taken, is now threatened with extinction: South American Ocelots and Jaguars; North American Lynxes, Bobcats, Otters, and Wolves. The Pentagon a few years ago almost singlehandedly became a major threat to the Wolves when it proposed using wolf fur to line almost 280,000 parka hoods for the armed services. Someone pointed out that carrying out this plan would require killing half of the remaining Timber Wolves in the United States![36] The Pentagon, fortunately, backed down.

The U.S. government's consciousness of endangered species appears to have been raised, as evidenced by the prosecution of the Molt gang as smugglers. President Carter urged a crackdown in his 1979 environmental message, and a new Wildlife Law Section was soon afterward established in the Justice Department. The law is now being enforced against importing *any* endangered plant, animal, or product, regardless of whether its export is banned in the exporting country or whether that country is a signatory of CITES. Furthermore, Congress is considering new amendments to the Lacey Act that would provide stiffer penalties and make it easier to prove guilt than under the present law.[37]

Tougher and more effective enforcement on importation to the United States will certainly help to discourage the illegal wildlife trade—at least in this country. But much more is needed, including serious enforcement in other CITES-pact countries, both importers and exporters, and broadening of the pact to include countries not now members. China, for instance, has recently applied; its cooperation may reduce the pressure on the fast-disappearing rhinoceri—though it may be too late to save them.

The wildlife trade, of course, is not only a grave threat to already endangered populations and species. Because of its enormous and growing volume, it threatens to endanger additional species not yet in trouble. The ultimate key to solving the problem is public education. Potential buyers around the world must be convinced that purchases of endangered species and their products are not only illegal but are not in their own best interests. The hunters, trappers, collectors, and poachers who gather the contraband—whether in remote African villages, the rainforests of South America or Papua New Guinea, or North American wilderness—must also be convinced that this livelihood is no sinecure and stripping their country of its wildlife is an utterly undesirable activity. And the makers of wildlife products must be persuaded to find substitute raw materials for their crafts. Only when it is widely recognized that *real* value attaches to an endangered species only in its natural surroundings will the wildlife trade finally be ended.

THE POLITICS OF HABITAT DESTRUCTION

In the examples discussed so far, the political struggle has focused on endangered species themselves. But the essence of preserving organic diversity is not to defend populations and species one by one but to preserve *natural ecosystems* relatively intact. Minimizing habitat destruction is the key, because virtually every area has unique populations whose disappearance contributes to Earth's biological impoverishment. The crucial battles of the conservation movement, then, will not be fought over Gorillas or whales or leopards or Snail Darters. They will be fought, like most military battles, over pieces of terrain —of *habitat. Only* by saving habitat can most species and populations be preserved in the long run.

The economic momentum of "development," propelled by the expanding human population seeking ever more resources, is not easily deflected, however. The natural environment and its living components are not yet seen by powerful economic and political interests as among humanity's most priceless resources. So the unequal struggle goes on in one arena after another between the exploiters and citizens who take a longer view of what is in the best interests of humanity.

One such struggle in progress now has far more national significance than most local environmental battles. At stake is one of the biologically richest areas of habitat remaining in the Western United States. A good many other values are also on the line in this classic example of political warfare over terrain required to support other species.

AMAX vs. Crested Butte

At 9,500 feet, in a high mountain valley on the western slope of the Colorado Rockies some fifty years ago, biologist John C. Johnson founded the Rocky Mountain Biological Laboratory (R.M.B.L.), which has been a center for ecological research and teaching ever since. We have carried on an active research program there almost every summer since 1960.

The valley and surrounding area, mostly national forest land, are not only extraordinarily beautiful and so far largely unspoiled but also harbor a remarkably rich diversity of life. Flora and fauna of the Central Plains, the Far North, the Great Basin, and the desert Southwest meet here. The area has about the highest rainfall in the state, which partly accounts for its biological productivity. Within surprisingly short distances, biologists can find and study biota of life zones ranging from above-treeline tundra through spruce and aspen forests and mountain meadows to, in lower reaches toward the southwest, semi-desert.

The mammalian wildlife includes deer, Elk, Bighorn Sheep, Mountain Goats, an occasional Black Bear, Coyotes, weasels and martens, Marmots, Snowshoe Hares, ground squirrels, Pikas, shrews, and mice. The bird life is equally diverse: from owls and hawks to Yellow-bellied Sapsuckers—one of them insists on pecking our metal stovepipe at six in the morning; he isn't the most popular bird in the valley—Redwinged Blackbirds, Spotted Sandpipers, Water Ouzels, and several species of hummingbirds. The White-crowned Sparrow can be heard singing everywhere. The sparkling streams contain four species of trout and an abundance of insect life, and over the meadows fly dozens of species of butterflies.

The flora is equally diverse, ranging from hardy lichens on mountaintop rocks to sagebrush and cactus in the lower dry areas. The wildflower display in the high meadows is simply breathtaking: sunflowers, lupines, flaxes, asters, bluebells, Scarlet *Gilia,* Indian Paintbrush, penstemon, lilies, and Larkspur. In shady places among the forest trees in July, one can find in abundance Colorado's lovely state flower, the delicate columbine. And some of the inconspicuous, hardy little plants growing high on the mountains are very rare; a few are on the endangered list.

The biologists and students who come from all over the nation—and even from foreign countries—to study in this unique outdoor laboratory are not the only people to gain from its beauty and biological wealth. Local ranchers graze their cattle in the high meadows every summer. And nearly a million people from other parts of the country visit Gunnison County every year. Hikers, backpackers, mountain climbers, horseback riders, hunters, and skiers by the thousands pass through the valley. Anglers test their skills against wily Brown, Cutthroat, Rainbow, and Brook Trout.

Nine miles from the lab in a broad valley ringed by mountains is a national historical site, Crested Butte, a former mining town that has been lovingly restored and turned into a tourist center and ski resort area. In summer, Crested Butte sponsors art festivals, air shows, and a Fourth of July celebration that has become legendary.

Many young people who left the large cities of the East and Midwest in the 1960s and 1970s, looking for a simpler, quieter life closer to nature, settled in Crested Butte. There were for a while some stresses and adjustments between the new people and the old mining and ranching families, but in the late 1970s both groups became united almost unanimously in their opposition to a new threat from the outside.

A giant multinational resource development company, AMAX (originally American Metals Climax) Corporation, had discovered the world's largest deposit of molybdenum in Mount Emmons, a 12,392-foot mountain that looms directly over the town of Crested Butte on the west. Molybdenum normally occurs in very low concentrations even in "rich" deposits. Mount Emmons evidently is such a deposit—with an average "moly" concentration of less than half of 1 percent in the main ore body. Nonetheless, that ore body is valued at over $8 billion. AMAX plans to hollow out a large portion of the mountain over a period of a few decades, grind up 165 millions of tons of ore in a gargantuan ball mill, extract the moly, which must then be further refined and transported away, and deposit the powdery residue in nearby valleys, eventually filling them.[38] The gutted mountain would gradually collapse.

The mine itself, the mill site, tailings deposits, new roads, and ore and tailings transport systems (including two 4.5-mile-long tunnels through mountains) alone would directly destroy thousands of acres of land. The mine, only two miles from town, will be a large and very noisy industrial development in a presently bucolic setting. If the mill were located near the mine, as was originally planned, the noise from it would be audible over ten miles away. In addition, the entire operation would create the kinds of air and water pollution problems inevitably associated with large-scale mining.

There are also important social and economic consequences. Operations of the mine, mill, and ore transport would require great quantities of power and fuel, necessitating the building of a new power plant in the area and the importation of huge amounts of coal and diesel fuel. Local communities would be swamped by the sudden influx of thousands of construction workers, miners, and their families, within a few years doubling the population of Gunnison County, which is now only 10,000. Many other western towns have suffered severe economic and social disruption and crime waves when such development projects have brought on a boom—which is usually followed, sooner or later, by a bust.

Furthermore, there has been a feverish rise in mining claims throughout the local mountains in recent years (1,500 in 1978 alone), mainly for uranium, but also for silver and other metals.[39] There has also been extensive exploratory drilling in various places in the nearby mountains, some of which is adjacent to and visible from the Maroon Bells and Snowmass wilderness areas. These activities indicate that numerous other firms are simply waiting for the AMAX project to be approved, which will be followed by installation of the roads, the power plant, and other infrastructure that would facilitate their own mining activities. Thus, if the AMAX project goes through, it will open the way for hundreds of other operations to begin. The collective impact would be enormous, especially since many of the smaller companies could not afford to be even as environmentally conscious as AMAX is.

When AMAX first announced its plan to open a mine on Mount Emmons, the old-timers in Crested Butte were happy with the idea. To them, in their nostalgia for the old days, a mine meant one or two men and a mule, with picks and shovels and perhaps a little dynamite, digging a man-sized hole in a mountainside. But as the enormity of AMAX's industrialized mining project dawned upon them, many of these people joined the young newcomers to Crested Butte in opposing it. In early 1979, under the leadership of their militant mayor, W. Mitchell,[40] the townspeople overwhelmingly passed a resolution against the mining project, and a majority of people in Gunnison have shown similar sentiments.

But it will take much more than resolutions, negative public opinion polls, and a courageous mayor to halt the AMAX juggernaut, and the people of

Crested Butte know it. There are legal remedies available, though. Because much of the project would be on or would impinge on federal land (national forest), environmental impact studies must be made. Approval at every step must also be gained from the National Forest Service, the County Planning Commission, and the state of Colorado, as well as the town of Crested Butte, whose air quality and water supplies are directly threatened. Each of these entities legally has power to halt the project.

AMAX claims that its right to undertake the project comes from the 1872 Federal Mining Law, which granted the right to mine and develop deposits found on federal land to anyone who properly staked a claim. But the law was passed over a century ago in the days of pick-and-shovel mining in the Wild West. Mining on a scale that destroys entire mountains and fills miles-long valleys was inconceivable to its authors. Indeed, it is clear from congressional debates of the last century that the *intent* of the 1872 law was to encourage the *small* miner.[41] Under the law, a claimant held title to the ore claimed and twenty acres of land over it; today large companies get around this limitation by staking numerous adjoining claims. AMAX acquires title to tailings sites by making "trades" with the Forest Service for other land it owns in the area.[42]

AMAX, Inc., is not a newcomer to this business; it has a track record. A Connecticut-based multinational firm, AMAX produces at least 40 percent of the world's supply of molybdenum. The company got its start in the business in 1917 in a mining project known as Climax, south of Leadville, Colorado. Over the decades, a large portion of a mountain was carved out, and an even greater area was covered by tailings, which poisoned water supplies below. Climax is nothing less than a colossal environmental blight.

A newer endeavor, the Henderson mine, not far away, has been developed with much more care. AMAX is proud of its efforts to protect the environment at Henderson and has repeatedly promised to be similarly careful at Crested Butte. It has launched a public-relations campaign to persuade the local citizenry of AMAX's great concern for the environment and to perpetuate the "mountain man with pick and shovel and mule" image of mining. To earn good will in Crested Butte, AMAX has taken on the task of cleaning up Coal Creek, which runs through the town and which is thoroughly poisoned by tailings from the old Keystone mine on property now owned by AMAX. The collapsing tailings dams have been stabilized, but cleaning up the creek is proving to be a stubborn problem.

But as examination of the Henderson operation plainly shows, whatever care is taken, molybdenum mining results by its very nature in enormous destruction of natural ecosystems and loss of amenity values. A "careful" molybdenum mine is a classic case of economist Kenneth Boulding's definition of "suboptimization": doing in the best possible way something that should never be done at all.

What about the need for molybdenum? Is it really necessary to open this new mine? What are the benefits for which Americans are asked to pay in perpetuity the cost of the ruin of the Crested Butte area? After all, AMAX already is operating Henderson, which is producing a considerable amount of moly and will continue to produce it for several decades. Climax, too, is still in production. And AMAX has mines or is negotiating for them in other places, including Alaska. Moreover, molybdenum in reasonable quantities is obtained as a by-product from copper mining. AMAX, it turns out, spends quite a chunk of its billions doing research on and promoting *new* uses for moly.

Moly is mainly used as a strengthening element alloyed with steel. The steel, thus "hardened," is used in jet and automobile engines, the Arctic oil pipeline, lightweight racing bicycles, and, reportedly, ballistic missiles and armor-piercing projectiles. The United States is by far the world's leading producer of moly, and roughly half of that production is exported.[43] (Much of it comes back in German and Japanese cars.) Furthermore, a major importer is the U.S.S.R., both directly and through European third parties. Since much of this moly is used in military hardware, we wonder why its export to the Soviet Union is permitted. Certain moly-containing alloys in fact are considered "strategic materials," and their export to Russia is not allowed.[44] Should Americans be grateful to AMAX for working so hard to tear down the western United States in order to send to Russia a metal that could be returned to us in the nose cones of Soviet ICBMs?

The Crested Butte/AMAX confrontation has taken on a David-and-Goliath flavor. Crested Butte's leaders are very conscious that this struggle, superficially just one more local battle, has implications and ramifications of national and perhaps international significance. It may prove crucial to the future development and interpretation of mining laws. It is involved in questions about the United States' international balance of payments and military security. Above all, Mount Emmons is not an isolated case. Mining interests are staking claims and accelerating operations throughout the Rocky Mountains and Great Basin, from the edges of Glacier Park on the Canadian border to the desert hills of Arizona and New Mexico. There are few unspoiled areas left now for human enjoyment or where other species can live undisturbed. In a few decades, if left to their own devices, the miners could turn the entire Rocky Mountain chain into a series of Climaxes. Mayor Mitchell expressed it this way: "It's no longer a matter of losing a little piece of a great big space. We're running out of next valleys."[45]

As of early 1981, the issue was still in doubt, hinging on such things as the legality of attempts by Crested Butte to protect its water supply and the Forest Service's interpretation of the 1872 Federal Mining Law. AMAX seems determined to proceed; its opposition is equally determined to stop the project. It is an archetypal face-off of short-term versus long-term values.

If AMAX succeeds, its stockholders and some local people will make a great deal of money. Molybdenum, and hence products requiring tough steel, may be cheaper. On the other hand, if the Crested Butte area were left undisturbed, the "renewable" economic activities of the area—grazing, recreation, tourism, biological and geological research—could go on indefinitely. And these activities, which have a relatively minuscule impact on the local nonhuman inhabitants, would in the long term provide vastly greater economic benefits to the United States than the one-shot mining operations.

AMAX and the other miners would remove their treasure and abandon the area within a few decades, leaving massive habitat destruction and social decay in their wake. Like the whalers and other large corporations that specialize in exploiting natural resources, the miners have only one goal—maximizing their income stream. They are, in essence, huge agglomerations of capital moving over the face of the planet, pausing for a decade or five to devour a resource, and then moving on to devour another, leaving devastation behind them. That many of the people involved are perfectly nice, reasonable human beings is simply one more measure of the general myopia about the long-term impacts of human activities that to one degree or another afflicts all of us.

Only public awareness and concerted action will prevent mining activities from destroying the remaining biological and esthetic resources of the United States. There are alternatives; among others, mining activities could be restricted to those substantial areas of the country that have already been despoiled. The best use of the molybdenum in Mount Emmons is to hold up Mount Emmons.

Politics and the Tropics

As battles such as AMAX vs. Crested Butte and broader ones over preservation of wilderness areas throughout the United States make clear, Americans have become concerned about preserving natural areas and biological resources. Europeans have essentially no wilderness remaining, but they, too, have taken political action to save individual endangered species and populations and to halt habitat destruction in their countries. People in rich countries, though, are mostly well fed and well housed, and have leisure time. They can afford such concerns.

Yet even if they are successful in their own countries, what will be gained if the devastation of tropical biota goes on unchecked? The loss of those resources would dwarf what has been, or is likely to be, lost in the temperate zones. But the political problems of curbing destructive activities within United States borders is child's play compared to persuading millions of peasants—to whom it is a life-and-death matter—not to clear off pieces of rainforest to start new farms, even if the farms are likely to fail within a decade. To start with, the governments of developing countries will take a very dim view of people in rich countries meddling in their affairs and telling them how to use their resources. (*Buying* them is another matter, of course.)

There are small conservation movements in many developing nations, but at the moment their ability to halt the onrush of extinctions is close to nonexistent. The pressures from population growth and economic development that are powering the destruction today are overwhelming, and they will remain so until and unless there are fundamental changes in attitudes among people on both sides of the rich-poor income gap, and particularly in the rich countries. If there is to be even a small chance to relax the pressures on tropical biota, fundamental changes in the world trade system and in the basic relationships between rich and poor nations must take place. The poor must be able to see a path to a reasonable life that does not require immediate exploitation of the handiest resources, regardless of the long-term consequences—which are now at best only dimly perceived.

Meddling though it may be, people in the developed countries have a stake in what happens to the flora and fauna of the tropics. As Peter Raven commented to us,[46] the effects of the ecological crash that will follow the final destruction of the biological diversity of the tropics will almost certainly engulf the citizens of overdeveloped nations as well.

We will return to this dilemma in the last chapter. Here we must emphasize that, if the politics of extinction continues to retain its temperate-zone bias, it will mean that the conservation movement is doing the equivalent of fiddling while Rome burns. The policies of the governments of developed countries and of multinational corporations toward tropical nations must be even more the business of conservationists than their policies toward Snail Darters or Caspian Sturgeon.

The politics of extinction is, quite literally, a life-and-death struggle, various forms of which we have only sampled here. It goes on at all levels, from the farm and hamlet to development decisions at the national level. In the future, it will increasingly be an international concern as well. The assault on nature by one of its products, *Homo sapiens,* is massive and continually growing, governed largely by short-term gain and characterized by long-term blindness. Until now conservationists, the people who care what happens to wildlife and who understand its vital importance to our lives, have been a relatively powerless minority fighting a largely losing war. But they have succeeded in having laws passed and international agreements established that will help to protect species from extinction. And in some areas they have at least slowed the lethal march of habitat destruction. If there is continued insistence that these laws and agreements be observed; if others are enlisted to support the goal of preserving natural systems with all their components; if the movement broadens to encompass concerns such as the relationships between rich and poor nations, which are vital to preserving the great reservoirs of diversity in the tropics; if we all strive to instill in others, including our children, respect for and understanding of the natural world, there might be a chance of saving some of it—and ourselves.

Chapter IX

Zoos, Reserves, and Preservation: The Tactics of Conservation

Conservation is sometimes perceived as stopping
everything cold, as holding whooping cranes in higher
esteem than people. It is up to science to spread the
understanding that the choice is not between wild places
or people. Rather, it is between a rich or an impoverished
existence for Man.

—THOMAS E. LOVEJOY,
Foreword to M. Soulé and B. Wilcox,
Conservation Biology, 1980

BEFORE a political strategy for conserving organic diversity can be fashioned, there first must be an agreed-upon set of tactical goals. In the face of multiple threats to the diversity of Earth's biota, what is a sensible course of action? What should conservationists work for politically? One answer that comes immediately to mind is to take action to preserve as much as possible of that diversity under human management in zoos, botanical gardens, arboretums, national parks, and game parks. It sounds good if you say it fast, and indeed preservation under intensive management *is* part of the overall solution. Many problems, however, guarantee that closely managed survival can be only part of the solution—and perhaps not a very significant part at that. Large-scale unmanaged reserves and changed human behavior outside of reserves seem to be the principal hope for preventing a catastrophic increase in extinctions.

CAPTIVE BREEDING

Zoos, in one form or another, have been around for a long time; both the ancient Egyptians and the ancient Chinese kept menageries. The earliest zoo is usually credited to Wen, the first emperor of the Chou dynasty. More than three thousand years ago he established a "Garden of Intelligence" in which animals from different parts of the empire were exhibited for educational purposes. Both the Greeks and Romans had zoos. And, of course, the Romans gathered exotic large animals on a scale never equaled before or since—some for exhibition, but most for destruction as part of the sadistic Roman "games." Animals were hunted in the arena, forced to fight one another, or starved and then turned loose on criminals or members of illegal religions. The animals involved included lions, tigers, leopards, rhinos, elephants, hippos, and crocodiles. Nero was able to show seals being hunted by Polar Bears! The scale of the slaughter can be judged by the sacrificing of some 11,000 wild animals to celebrate a single event, Trajan's conquest of Dacia.[1]

The menageries of kings and other nobles of the Middle Ages had metamorphosed into zoos of a more or less modern type by the early nineteenth century. For instance, the Royal Zoological Society took over the animals of the British Royal Menagerie in 1829 and established the famous London Zoo in Regent's Park. There are now hundreds of zoos worldwide.

Captive Breeding in Zoos and Game Parks

Thousands of years ago, a very unusual deer with a coat that was tawny red tending toward gray was a common denizen of swamps on the plains of northeastern China. In that ancient civilization habitat destruction was well advanced long before the time of Christ. During the Shang dynasty (1766–1122 B.C.), the swamps where the deer lived were drained as the Chihli plains were brought under cultivation, and the deer ceased to exist in nature.

In 1865 the well-known French naturalist and missionary Abbé Armand David managed to peek over the wall of the strongly guarded Imperial Hunting Park south of Peking. He saw there a herd of a unique deer unknown to science. It had unusually broad hooves, a tail longer than any other deer, and the singular feature of forked hind prongs on its antlers. It was the species formerly resident on the Chihli plains, which had survived for some three millennia as a captive in game parks.

The next year the Frenchman managed to send a couple of skins to Paris, where they were described formally and given the name *Elaphurus davidianus* —Père David's Deer. More than a full century after the great Swedish biologist Linnaeus established the formal system of giving two-part latinized names to organisms, a new species was "discovered" in captivity!

Soon thereafter, a few individuals of Père David's Deer reached the West, were successfully bred, and distributed to zoos. That was fortunate indeed. The Hun Ho River flooded in 1894, and its waters brought down part of the forty-five-mile-long brick wall that enclosed the Imperial Hunting Park. Most of the deer escaped and were promptly killed and eaten by starving peasants. The survivors were almost all killed during the 1900 Boxer Rebellion. Foreign troops bivouacked in the park killed the rest of the herd and sold their meat. The few remaining deer were preserved in Peking, but all were dead by 1921. Père David's Deer was extinct in its native land, but a handful survived elsewhere.

Faced with the bad news from China in 1900, the Duke of Bedford decided to assemble as many of the deer as he could on his estate at Woburn in southern England. He managed to collect sixteen individuals between 1900 and 1901, and by 1922 the herd had increased to sixty-four.

The herd continued to increase, and deer from it were used to establish herds in zoos in various countries. In 1964 the London Zoo sent four deer back to China, where they became residents of the Peking Zoo after almost fifty years of absence. Père David's Deer is a testimonial to the ability of captive populations to serve, in the words of William Conway of the New York Zoological Society, "as last redoubts for species which have no immediate opportunity for survival in nature."[2]

Another prominent species that now may exist only in captivity is Przewalski's Horse, the last surviving species of wild horse.[3] *Equus przewalskii* has a light coat, a large head, an erect stiff mane, a long tail, and no forelock. Once widely distributed in Central Asia, this wild horse went into decline because of competition from domestic animals and hunting pressures after the acquisition of firearms by Mongolian and Chinese hunters. Furthermore, there was a great deal of interbreeding with Mongol ponies, which diluted the basic stock.

It is possible that a few survivors of Przewalski's Horse still hold out in the inhospitable fortresses of the southwest Mongolian frontier region, where there were reasonably reliable sightings as late as 1967.[4] What is certain, however, is that the species is thriving in captivity. In 1971 there were 182 individuals in 42 zoos, all but one horse born in captivity—a 50 percent population increase over 1964, when only 24 zoos maintained the species.[5]

By 1973 the captive stock had increased to 206, and consideration was given to releasing some of them in the wild. The donor herd would be the one in the Prague Zoo, which is the oldest, largest, and most productive in captivity. That herd, however, was only in its fifth to eighth generation in captivity, but it had already undergone significant changes in its physical and reproductive characteristics in the zoo environment. An especially unsettling trend from the point of view of eventual release is that foals are now often born outside of the sharply defined foaling season in the wild. Such births would have no chance of survival in the wild.[6] This problem of evolutionary change in captivity, to which we will return, is one of the many reasons that we and others have limited enthusiasm for zoo breeding as a basic conservation tool.

One large animal that was once extinct in the wild is being returned there right now from zoo herds. That animal is the Arabian Oryx, a small, almost pure white antelope with long, nearly straight horns, known from biblical times. It is described as follows in Deuteronomy:[7] "His glory is like the firstling of his bullock, and his horns are like the horns of unicorns."

The oryx once lived throughout much of the Near East, but in the middle of the last century its populations began to die out in the northern part of its range, which by then had been constricted to the drier, unsettled areas. World War I didn't help a bit: fighting occurred in much of the oryx's homeland, and modern guns were introduced to the people there.

Like the rhinos, the Arabian Oryx suffered because of local superstitions. Some Arabian peoples thought that a bullet could be expelled from a wound by eating oryx meat; and, because of the animal's endurance and strength, killing one was considered a sign of manhood. All this made little difference when hunting was done with spears or antique long guns from camel and horseback. But automatic weapons fired from automobiles were another story. By the early 1950s as many as three hundred vehicles were used in a single hunting foray. The last recorded tracks of the oryx were seen in the Nafud Desert of northern Saudi Arabia in 1954, and the final survivors in the southern part of the range, confined to Oman, appear to have been destroyed in the early 1970s.

But the Arabian Oryx was not gone from this Earth. In 1979 there were sixty-four individuals in three American institutions—a "World Herd" established at the Phoenix Zoo, another herd in the San Diego Zoo Wild Animal Park, and another at the Los Angeles Zoo. Another thirty-five or so were in zoos and private collections in the Middle East.[8] Breeding has been sufficiently successful that in 1978 four animals had been returned to a reserve in Jordan and four more went there in 1979. In 1980 plans were underway also to return a small herd to Oman.

Whether these reintroductions to the wild will be successful remains to be seen. Educational programs for tribesmen have been undertaken in Oman, and there will be police guards against poaching. One hopeful sign is that a conservationist spirit seems to be developing. When a population was found there of about twenty Arabian Tahr, an endangered goatlike animal, local Bedouin tribesmen were persuaded to become wardens and protect them. It was potentially a case of the foxes guarding the hen house, for the same people had been responsible for the Tahr's previous decimation. But the tribesmen now are paid regular salaries to protect the animals, and they take their jobs seriously. Bedouins are now being recruited into the oryx program as well. As long as there is a stable government and relative prosperity in the area, prospects for released oryx, like those of the Tahr, may be good.[9]

Not all mammal captive propagation stories have such potentially happy endings, however. Attempts at captive breeding and maintenance of small populations in zoos have very often ended in failure. After several failures in attempts to mate the pair of Giant Pandas in Washington's National Zoo, officials finally resorted to artificial insemination in 1980. Meanwhile the small remaining wild population has been decimated by earthquakes and habitat deterioration in China.[10] Efforts in South America to maintain captive herds of Vicuña, the camel-like fine-wool producer of the high Andes, have been far from successful. This endangered species, now established in new reserves, however, is doing much better in the wild than in captivity, where deaths tend to outpace births. One expert commented that "the future of wild Vicuña looks better than the future of captive specimens. . . . The problem we now have to solve is that of saving the Vicuña in captivity!"[11] But captive breeding may never work; the Vicuña is fortunate that the reserve system is working.

The fundamental disadvantages of depending on zoos as reservoirs of organic diversity are their limited capacity and their need to display as many different animals as possible to fulfill their role in public education.[12] A random sample of ten large zoos showed that the average species of mammal was represented by only three to five individuals—some of which were overage, juveniles, or nonbreeders for other reasons. To assure the more or less permanent survival of a species in captivity, geneticists have established certain criteria that must be met: a captive population of a hundred or more, at least half of them captive-born. In 1971 it was estimated that only eight mammal species met these standards. These were the Siberian Tiger, Przewalski's Horse, the Onager (Asiatic Wild Ass), the Formosa Sika (a small deer), Père David's Deer, Wisent (European Bison), Scimitar-horned Oryx (a North African desert-fringe species with curved horns), and the Addax (another North African antelope with sinuous horns).[13] To this list might be added the Mongoose Lemur, an arboreal primate from Madagascar and other Indian Ocean islands, and, of course, the Arabian Oryx. It has been estimated that present U.S. zoos could successfully maintain a maximum of perhaps a hundred species of mammals (out of more than four thousand existing species worldwide), each with a population of about a hundred and fifty individuals, in order to avoid extinction through accident or the loss of genetic variability.

The latter is a constant problem in the maintenance of captive groups of sexually reproducing organisms. It is essential that their genetic diversity be preserved, not only so that over the long run they will be able to evolve in response to changed conditions in captivity, or in the wild if they are used for restocking, but so that the short-run deleterious effects of inbreeding can be avoided.

Long experience has shown plant and animal breeders that, in sexually reproducing organisms, vigor and fecundity tend to decline as random genetic changes occur and genetic variability is lost.[14] The decay of genetic variability occurs both because of accidental loss and because of the inbreeding—the mating of close relatives—inevitable in small populations. The weakness and reproductive problems of the animals are caused by changes in their genetic makeup associated with the changed genetic composition of the population as a whole. The details are quite technical, but the results are not. Small populations tend to get stuck with a limited assortment of genes, and often they are the "wrong" genes, to boot. This in turn can lead to their extinction.

The process of estimating the population size required to keep a population from dying out because of "inbreeding depression" likewise is highly technical.[15] Under ideal conditions, fifty animals might be enough, but the hundred and fifty mentioned above is a more realistic estimate. Interestingly, attempts to breed the endangered Black-footed Ferret were stymied because individuals captured in the wild were already showing signs of inbreeding—apparently some of their population sizes are already too small. This problem, if it is widespread, may hasten the doom of the only ferret native to the New World.[16]

The question of maintaining genetic variability for future evolution involves a decision about goals. Is the objective to keep the captive organisms as similar as possible to the wild population from which they were originally taken? Or is it simply to maintain the line in captivity? As geneticist Ian Franklin put it, "Do we wish to conserve the elephant, or ensure the survival of its elephant-like descendants?"[17] Remember that in a very few generations Przewalski's Horses had evolved into something quite different from the original wild populations. It is much easier to allow populations in captivity to evolve into strains adapted to the conditions of confinement than it is to maintain a semblance of the original wild species.

Captive Breeding Outside of Zoos

The Hawaiian Goose, or Nene, Hawaii's state bird, used to be abundant on the Big Island and Maui, where it lived on the thinly vegetated slopes high on the volcanoes Mauna Loa, Mauna Kea, Hualalai, and Haleakala. As is typical of island birds, its tameness spelled its doom when guns were introduced. The hunting season, set by temperate-zone standards in autumn and winter, happened to coincide with the tropical goose's nesting season. Many thousands were shot and salted as provisions for whaler's ships. Parts of the breeding

range were invaded by sugarcane fields and grazing herbivores. And introduced mongooses, pigs, cats, and dogs raided the higher nests.

For these reasons, a population estimated at around twenty-five thousand
in the eighteenth century had plunged to only about thirty-five wild individuals
by 1942. But long before then, concern for the Nene's survival had been voiced;
hunting stopped in 1911. A small captive flock was preserved by a Hawaiian
ranch owner, Herman Shipman, beginning in 1918 with two birds. He had
recognized that the species was in jeopardy. By 1946 he had forty-two, but in
that year a tidal wave killed all but eleven on his lowland estate. He removed
these to another estate on high ground. Around 1950 Shipman was able to
supply birds for captive breeding programs to the Hawaiian Board of Agriculture and Forestry and to the Wildfowl Trust in England, where the great
conservationist Sir Peter Scott supported the program.[18]

In spite of early problems with the breeding programs, by 1955 the captive
birds probably outnumbered those still free in Hawaii. In 1960 the first release
was made—twenty from the Hawaiian captive breeding pool were freed to
supplement the fifty or so birds still living free on the island of Hawaii. Next,
in 1962, the slopes of Haleakala on Maui were restocked with thirty Nene
raised in England. In both cases, an open-top, predator-proof release pen about
an acre in extent was used in which the birds were housed with plucked
primary feathers (to prevent their flying away) to habituate them to natural
foods and conditions. As the birds' ability to fly returned, they gradually left
the pen. But two of the Maui birds were killed by a mongoose that somehow
got in the pen.[19]

Early releases, especially on Maui, were not as successful as hoped, although
some of the birds apparently bred in the wild. Subsequently, a new and more
successful technique was tried. A large fenced "Nene Park" was set up and
stocked with birds rendered permanently flightless. The offspring of these birds
had had no extensive contact with human beings and were able to fly out of
the park. That generation apparently thrived in the wild. In the late 1970s there
were more than a thousand surviving Nene, some six hundred of them living
in nature. The Nene may prove to be an outstanding conservation success story
—a species pulled back from the brink of extinction by captive breeding.[20] But
as of this writing, the appropriate studies have not been made to evaluate
reproduction success in nature or to determine whether the free populations
can maintain themselves and even increase without a constant release of additional pen-reared birds.[21] It is possible, for instance, that expensive, difficult,
and risky programs for control of introduced predators will be required to
ensure the persistence of the Nene.

In the annals of endangered species, no bird has gotten more press than the
Whooping Crane. As far back as 1937, the Aransas National Wildlife Refuge

was established in Texas especially to protect the last survivors of this tallest of North American birds, which wintered there. The birds spend their summers on breeding grounds in subarctic Canada. When the Aransas Refuge was established, some birds also wintered in Louisiana. The total population was estimated to be twenty-nine birds. The Louisiana breeders died out, and in 1942 there were twenty-eight birds at Aransas. By 1967 they had increased slowly to thirty-eight adults and five new young. Breeding success was dishearteningly slow—two eggs are often laid, but rarely is more than one chick successfully reared.[22]

Because of the low level of Whooping Crane reproduction, a controversial management idea was put forth in the late 1950s—extraction of "spare" eggs from wild nests for captive breeding. A practice program was started in 1961 with Sandhill Cranes, and in 1967 work with the Whooping Cranes began. Single eggs were removed from two-egg nests and taken to the Patuxent Wildlife Research Center in Maryland to use in stocking a captive flock. Researchers have been successful in rearing the cranes but not in getting them to mate naturally. This problem has been circumvented by artificial insemination, and in 1977 five of eleven pairs in captivity produced twenty-three eggs. That same year, seventeen pairs laid thirty-four eggs in the wild.

As an experimental program, Whooping Crane eggs are being placed in Sandhill Crane nests in Idaho, but the success of this cross-fostering experiment is still in doubt. There have been some encouraging results from releasing pen-raised juvenile Sandhill Cranes, and the Patuxent group intends to try this release technique as well. But in spite of the very great effort, the long-term chances for survival of the Whooping Crane, especially in the wild, are still uncertain.[23]

Another famous captive-breeding program has been that of ornithologist Tom Cade of Cornell University, who attempted to arrest the decline of Peregrine Falcons. Those sleek raptors, feeding at the tops of food chains, were endangered by buildups of chlorinated hydrocarbons such as DDT in their tissues. Reproduction in the wild had been greatly reduced, and the species was endangered. The entire population in the eastern United States disappeared in the 1960s. Cade and his colleagues persisted in the face of great difficulty. Although the program started in 1970, breeding success in captivity was not achieved until three years later, when twenty eggs hatched. By 1980, hundreds of hatchlings had been successfully produced, and more than two hundred pen-raised birds had been released at fifteen sites in the East. But only about 10 percent have survived to breeding age. In 1979 the first four chicks were hatched in the wild from captive-bred parents. It is an encouraging sign, but whether the Peregrine reintroduction program will ultimately be successful remains in doubt.[24]

Attempts to save species by captive breeding can be highly controversial.

Defenders of California Condors (*Gymnogyps californianus*) disagree about the best way to save the great scavenger, which can have a wingspread of over nine feet and a weight of up to twenty-three pounds. In prehistoric times this enormous vulture ranged from Florida to Texas, northeastern Mexico, Arizona, and Baja California in the south, and northward west of the Rockies to British Columbia. By the time Europeans began to push across the continent, the condor had already retreated west of the Rockies, leaving only its bones in Florida. By the turn of the century, the bird was extinct in Mexico and Canada, and its breeding range in the United States had contracted to California alone. In 1943 the distinguished California ornithologists Joseph Grinnell and Alden Miller found the birds breeding only in a small area in the mountains behind Santa Barbara in the Coast Range, across the Central Valley in the southern Sierra Nevada, and in the transverse ranges that connect the two. Habitat loss, food scarcity, shooting, poisoning, egg collecting, environmental contamination, and general harassment had taken their toll. The total population at that time was carefully estimated by Carl B. Koford and found to be about sixty birds, including only five pairs of nesting adults.[25]

By the mid-1960s the number of condors had dropped to about fifty. It was a red-letter day indeed for San Francisco Bay Area biologists when a juvenile condor appeared, soaring magnificently over the Stanford University campus, eventually to land near the Jasper Ridge Preserve. We well remember viewing it through a spotting scope as it sat, seemingly dejected, in a tree on a rather miserable day. One could almost imagine it knew the fate of its species. By 1980 the condor's numbers had declined further to around twenty to thirty birds.

The odds are clearly against the survival of this great relic of the age in which human beings first invaded the Americas. Despairing of the chances of wild survival because of continued shooting by irresponsible hunters and the fear that, as with the Peregrines, pesticides were causing breeding failures, a distinguished panel of nine ornithologists recommended a "hands-on" program to preserve the condors. The report of the panel, jointly established by the National Audubon Society and American Ornithologists' Union, has led to a proposed plan to be implemented by the U.S. Fish and Wildlife Service. It would consist of attempting to trap most of the remaining birds, holding them long enough to put small solar-powered radio transmitters on them, take blood samples from them, and determine their sex by a minor surgical operation. This would permit critical data on the biology of the condors to be gathered.

Later nine birds would be retrapped and used to start a captive breeding program. Although the California Condor has never been bred in captivity, information on related species convinced the panel that it should prove feasible. In captivity the birds quite possibly can be induced to lay an egg a year or more, instead of one every two years as they do in the wild. And they can be fed pesticide-free food, so the rate of successful hatching should be increased. After thirty to forty years or so, the plan is to release captive-reared birds into a satisfactory condor habitat, completing the job of saving the species.

Altogether it is an impressive plan, taking advantage of the best that modern technology can offer. Why then have we and many other biologists and conservationists opposed it? One basic reason is the fear that the planned level of harassment of the surviving condors is more than just an indignity: it may hasten their plunge to extinction. There is considerable risk of injury and/or behavioral modification in the trapping, surgical, and transmitter-attaching operations.

These risks were underscored in June 1980 by the bungling of two biologists, one from the Fish and Wildlife Service and one from the Audubon Society. They were making a film, and possibly in violation of their permit, they dispatched a rock climber untrained in biology to attempt to weigh a thirteen-pound condor chick. The climber managed to kill the chick, a great tragedy for the slow-breeding condor population, which since 1965 had managed to produce less than one chick per year.[26] It was hardly an auspicious beginning for a "hands-on" program.

Beyond the dangers of handling, there is the risk that, in the course of captive rearing, the condors will lose the qualities that allow them to survive in nature. Young birds are dependent upon adults for two years. They must learn the tricks of their trade from their parents—how best to utilize air currents, find food, compete with golden eagles, and so on. How these abilities could be preserved in pens is not clear, although parallel work now started on the closely related Andean Condor may provide important clues. What experience there has been with reintroduction of birds into the wild so far does not permit us to be overly sanguine about the chances of success. The panel's scientists obviously endorsed this course not because they were certain it would work, but because they were sure the alternative was extinction.

Others disagree. The Sierra Club's California Condor Advisory Committee, chaired by David Clark, director of the Point Reyes Bird Observatory, recommended in 1979 a more conservative approach, involving experiments with common vultures over a three- to four-year period before a decision is made on the condor. Before he died, Carl Koford outlined a naturalistic condor recovery plan, in which intense, careful observations would be substituted for the technological approach favored by the panel. Instead of captive breeding, he recommended a series of measures to improve environmental conditions for the condor—assuring food supply in all seasons, decreased use of pesticides, evaluation of possible reduction of competition from Golden Eagles and Turkey Vultures, and so on.[27]

There is no guarantee, of course, that the naturalistic plan would work any better than the technological one. In our opinion, it does have one overriding advantage: *it leaves the fate of the condor tightly tied to the fate of its habitat.* This is essential, for an organism with such a low reproductive rate must have available an environment that is stable over the long term. It would have extreme difficulty in repopulating *any* environment.[28] And from the point of view of preserving overall diversity, its habitat is more important than the condor itself. As Paul wrote to the Fish and Wildlife Service in February 1980:

> Even the most wildly successful captive breeding program would be for naught if there is not sufficient habitat to support the birds after re-release. . . . The condor, therefore, should be preserved not just out of intrinsic interest, compassion and its own ecological role, but most importantly because it can serve as a rallying symbol for protecting large areas of habitat and thus many other endangered organisms.

Biologist Steven Herman of Evergreen State College commented in a similar vein:

> If the emphasis shifts to a captive propagation and plans to release progeny . . . the oil companies, the mining corporations, the developers will rejoice and begin moving with newly valid arguments the afternoon of the day the first condor egg hatches in an incubator. It is likely they will begin rolling their eyes and patting their bellies with the incarceration of the second wild condor.[29]

If condor habitat cannot be protected now, while the condors are present, it is difficult indeed to imagine how it could be preserved in the absence of condors for future releases. We think the battle should be waged for the bird *and* its habitat as a unit. Even if the birds were saved in captivity, the loss of them in nature would be a disaster. As Koford noted just before he died, in the human environment:

> Wildlife is as essential as music, and appreciation does not require active viewing any more than does personal observation of the Mt. Everest summit. . . . We are trying to preserve a natural feature for the enjoyment of many future generations. Handling, marking and caging greatly diminish the recreational value of wild condors.[30]

The condors' most critical habitat has "high oil and gas" potential, contains deposits of gypsum and gold, is a region where fire and flood control measures are desired, and contains a known geothermal area. It is unfortunately also a district where permitting ORV use is contemplated.[31] Whether, even with the condor present, destructive activities related to these values can be halted is doubtful; with the condors in cages, the whole area is almost sure to go under.

But, at least as long as they remain in the wild, the condors can contribute to attempts to stem the tide of habitat destruction in California's southern mountains for a few more decades. And a few more people will get to see them as they should be seen, in the wild, a reminder of a past world where the condors soared over much of the continent and watched over mammoths, saber-toothed cats, and giant ground sloths, waiting for them to die. One can picture them waiting around the edges of the La Brea tar pits, competing with their supervulture relatives, *Teratornis merriami,* for a chance to feed on, say, a struggling American Camel. With a wingspan of up to twelve feet and weighing as much as fifty pounds, *Teratornis* may have been the largest bird ever to fly, so the *Gymnogyps* probably lost numerous meals to them. But while *Teratornis* went extinct at the end of the Pleistocene era, the California Condor soldiered on into the twentieth century to let industrial society hear, if it is interested in listening, the musical whistling the wind makes as it flows through the bird's feathers in soaring flight.

Nature writer Kenneth Brower asked the ultimate question in his usual poetic fashion:

> And what if nothing can bring the birds back? What if *Gymnogyps,* watching Los Angeles sprawl towards its last hills, has simply decided it is time to go? Perhaps feeding on ground squirrels, for a bird that once fed on mastodons, is too steep a fall from glory. If it is time for the condor to follow *Teratornis,* it should go unburdened by radio transmitters.[32]

The Importance of Habitat

In our view, captive breeding of animals will never be more than a relatively minor stopgap measure in the fight to save the diversity of species and populations. Logistic problems permit only a relatively few more or less spectacular species to be so protected, and the sample saved is likely to be genetically impoverished and to represent the variability present in only one or a few populations. Chances of successful reintroduction are always problematical, especially when removing a prominent species from the wild removes a major argument for preservation of the habitat.

Then there is the question of the allocation of scarce resources. Captive breeding is expensive, and funds going into such programs are not available for other conservation programs that might be effective in stemming the tide of extinctions. There is the danger that too much emphasis on saving prominent individual species may distract attention from absolutely crucial but less "sexy" tasks such as saving tropical rainforests.

With respect to protecting plant diversity, preservation in botanical gardens can probably make a proportionately larger contribution than captive breeding can for animals.[33] At least two California plants once extinct in nature have been reestablished from stocks saved in botanic gardens.[34] Even so, garden breeding at best can supplement programs to save or restore undisturbed habitats. Agricultural experiment stations can also help to maintain the genetic diversity of crops. Here also, however, the game will probably be won or lost in the ecosystems themselves—in this case in agricultural ecosystems.

If reintroduction of either plants or animals is ever to be successful, maintaining habitat quality must always have high priority. For example, the Large Copper butterfly went extinct in England around 1850 as its marshy habitats were destroyed. The butterfly was reintroduced into England using Dutch stock in 1927, and a colony survived with careful management at Woodwalton Fen until 1968, when a heavy July flood forced it to extinction again. The butterfly was reintroduced again in 1970 using English laboratory-reared stock. Ecologist Eric Duffy, who has intensively studied the situation, has concluded that the Dutch stock is not well adapted to the English marsh environment and that the area of the reserve is too small for the Large Copper population to survive without constant husbandry.[35]

This is not a surprising result. We have recently been involved in attempts to transfer a rare checkerspot butterfly, which is threatened by grazing and mineral exploration over much of its range, to areas where it does not occur but where the food plant for its caterpillars and adult nectar sources are abundant. Two different artificial introductions so far have produced only one new population, which at best is barely maintaining itself after three years. The message here is that even with insects, where thousands of individuals are available, the success of introductions often depends on subtle characteristics of the available habitat.

With slow-breeding species, habitat quality is even more crucial, since the investment in time and money in producing transplantable stock can be enormous. For instance, a rather comprehensive program of breeding Galápagos tortoises is now underway, and seventy-one young individuals were used to restock Pinzon (Duncan) Island in 1971. But unless the introduced rats, pigs, and dogs that threaten their nests can be exterminated or strictly controlled, it is problematical whether most of the islands will be able to support self-sustaining tortoise populations, and "current 'nursemaiding' procedures will have to be continued indefinitely."[36]

If the goal is to save biological diversity, *the major focus must be on conserving entire ecosystems.* Trying to maintain organisms in the biotic vacuums of zoos, laboratories, and botanic gardens can help a little, but nowhere near enough. Reconstructing extinct species from frozen DNA samples, sometimes proposed by the naïve, is strictly science fiction today—and is likely to remain so for a very long time, if not forever. Sizable reserves that contain more or less intact ecosystems have by far the greatest potential for effective preservation of biological diversity.

Unfortunately, the question of reserves is fraught with uncertainties. The understanding that *all* of Earth's biological diversity is now under threat—not just the diversity of some places—has sunk into the minds of biologists only within the last decade or so. Consequently, research directed at determining what sizes, shapes, spacing, distribution, and other features of areas to be set aside would most efficiently conserve natural diversity is only beginning to yield answers.

RESERVES

In spring 1980, Paul and two of his colleagues, Bruce Wilcox and Dennis Murphy, had a very interesting lunch with representatives of the Nature Conservancy, the foremost private group in the United States dedicated to the preservation of natural habitats in North America. The meeting grew out of common interests in the problems of the proper design of nature preserves. In recent years, the Conservancy has specified its mission as setting aside land for the express purpose of maximizing the conservation of biotic diversity. And in recent years, our research group's experiences with the impact of the 1976–77 California drought have led to a total revision of basic ideas about patterns of setting aside that would more or less ensure the long-term maintenance of insect diversity in temperate regions.

The hazards of underestimating the size necessary for adequate reserves in tropical rainforest areas have long been evident. It was clear from research by our group and others that many tropical animal species essential for the reproduction of specific plants patrol rather extensive "traplines" through the forest, seeking out the plant species that are their source of sustenance and that they pollinate. Large, shiny, tropical *Euglossa* bees, *Heliconius* butterflies, and hummingbirds all show this sort of behavior. For such animals, the preservation of small patches of forest is utterly inadequate. Moreover, if the animals died out, the plants dependent upon them would soon disappear as well, leading to the loss of other herbivores that live on the plants. That in turn might well affect predator populations that are dependent on the herbivores, and so on. A cascading series of extinctions could quickly reduce the diversity in any small preserve plot.

When our research on checkerspot butterflies was started, the impression was that the 1,200-acre Jasper Ridge Biological Reserve on the Stanford University campus would be more than adequate to maintain insect species like the checkerspots permanently. It was only as the pattern of natural extinctions followed later by recolonization from other populations became clear that we began to realize our assumptions might be entirely faulty. A few checkerspot butterflies do not make an entire insect fauna, however. While the behavior of this specific group may be fascinating, the sample is too limited to permit us to generalize the results to a wide variety of species and a wide variety of situations.

Paul and his colleagues recently decided to launch a broader investigation of the design of reserves for temperate-zone butterflies and plants. Unfortunately, the ideal experiments were clearly impossible. One cannot simply set aside, say, fifty or a hundred areas of varying sizes and varying spatial relationships to one another, exterminate all of the plants and butterflies in the areas outside of the reserves, and then watch the reserve "islands" for a few hundred years to see what happens to the diversity within them. Instead it would be necessary to look at "experiments" already performed by nature and use those results to infer the proper design of reserves.

Designing Reserves

Fortunately, nature has already carried out the experiments by isolating floras and faunas in high-altitude "islands" of the mountain ranges of the Great Basin deserts of Nevada and Utah for time periods that can be rather closely established. During the last glaciation, moist habitat occupied the entire basin. The return of a drier climate created the deserts and isolated the montane "islands." Using butterflies as representatives of the faunas, all that would be required (and this is a big "all") is to determine the composition of the butterfly faunas of those islands and the plant species eaten by the caterpillars in each —those would be the results of nature's experiments—and then to interpret those results. To aid in the interpretation, there already exists in ecology a set of ideas known as the theory of *island biogeography.*

Island biogeography is simply the study of the reasons that islands have less diverse floras and faunas than mainland areas, and the reasons that some islands have more diversity than others.[37] One central idea of island biogeography is straightforward: that the diversity and character of fauna and flora of any island are related to the size of the island and to its distance from the mainland. A second central idea is a little more complicated: that the number of species on an island will reach an equilibrium determined by the rates at which new species arrive and old species go extinct.[38]

A given area can support only so many individual plants and animals. And, as more species are added, the average number of individuals of each species must shrink. But as we have already seen, small populations are more vulnerable to extinction than large ones. So as the populations of the species shrink, the rate of species extinction goes up—and it will rise until it balances the rate of successful colonizations, establishing an equilibrium of species diversity.

This logical-sounding model seems to fit nature rather well. For example, in 1883 the island volcano Krakatoa blew its top in a cataclysmic eruption.[39] The island lay between Java and Sumatra, and tidal waves from the destruction of Krakatoa killed an estimated 36,000 persons on the two larger islands. All life was extinguished on what remained of Krakatoa, providing the start of a natural experiment to test island biogeographic theory some eighty years before the theory was developed!

Recolonization of Krakatoa was surprisingly quick. Spiders ballooned in first. By 1886 there were 26 species of plants present, and by 1900 there were 115 plants and 13 species of birds. In 1920 the plant species had increased to 184 and the birds to 27. By 1934 there were 272 species of plants but still only 27 bird species. And, of the birds, 5 species previously present had disappeared while 5 new ones had taken their place. It looks very much as if Krakatoa by 1920 had reached an equilibrium number of bird species according to the theory, but had not yet reached equilibrium for plants. Thus island biogeography reveals that a single reserve cannot be set aside and then continuously stocked with species from elsewhere—even if the habitat seems perfectly suitable for the introduced species. Once the reserve's carrying capacity is reached, extinctions will balance introductions.

More interesting than this, however, from the standpoint of how reserves are usually established, is what happens when a park is carved out of a continental ecosystem and the area around it is farmed or otherwise modified so the reserve becomes, in essence, an island. First of all, there will be a short-term loss of species that were present in the ecosystem as a whole. Some simply will be excluded by accident. Some animals may just wander away: for example, herds of elephants and Wildebeest often stray out of game reserves and are sometimes slaughtered because of it.

But perhaps more important are the long-term effects of becoming an island. The rate of extinction within the reserve at best will remain the same as it was when the reserve was part of a larger system. More likely, the rate of extinction will increase because the smaller area of the reserve means smaller populations of at least some species. But the reserve is now isolated; the natural ecosystem outside its borders has been destroyed or severely damaged by being developed or turned into farmland. Even such mobile animals as songbirds and butterflies often will not cross such areas. So colonization drops precipitously because species cannot freely flow in from the rest of the ecosystem. With extinction increased or unchanged and colonization much reduced, the system will move to a new, lower equilibrium number of species determined by the new extinction and colonization rates.

This leads to a phenomenon known in its milder forms as *relaxation* of the flora and fauna—a drift toward a less diverse biota in the reserve. Such a drift is well documented at Barro Colorado Island in Panama, for example. The island was formed in 1914 by the flooding of Gatun Lake in the Panama Canal. It became a reserve shortly afterwards, and forest was allowed to reoccupy previously cleared areas. Barro Colorado Island has long been the location of an important biological research station operated by the Smithsonian Institution. Since the flooding, 48 of the original 208 species of breeding birds on the island have gone extinct, about a third of them apparently because of the relaxation effect. The other extinctions were of second-growth and forest-edge birds, which died out when their habitats were destroyed by regrowth of the forest on the island.[40]

When we visited Barro Colorado Island in 1970, we found the butterfly fauna a very poor sample of the rich diversity found elsewhere in Central America. The number of ithomiine butterfly species had dropped from about 20 to about 6[41] as the forest replaced the disturbed areas where the ithomiines' food plants had grown. Other groups of butterflies dependent on clearings were also scarce.

The results of becoming an island can be much more dramatic than the losses from Barro Colorado, however. They can include the ultimate loss of most large animal species even from very large reserves. Again, nature has performed experiments that illustrate this sort of "faunal collapse." When the last great glaciation ended, the melting of ice caused a substantial rise in sea level. This in turn created many new islands. For example, Sumatra, Java, and Borneo became islands about ten thousand years ago—around the time that the first human groups took up agriculture. Before that, they were all part of the Asian mainland and presumably had a set of mammal species similar to that found today on the Malay Peninsula. Combined, Sumatra, Java, and Borneo *do* have a more or less complete sample of the Malaysian fauna. But none of the islands individually has anything like the whole set of Malaysian mammals. For example, Java lacks the Siamang Gibbon, Malay Tapir, Sumatran Rhino, Orangutan, Malay Bear, and Indian Elephant. All of these are present in Sumatra, but it is missing the Panther and Banteng (a wild ox), both of which are present in Java. Borneo lacks the tiger, the Javan Rhino, Panther, gibbon, and Malay Tapir. With a few exceptions, it is known from fossil evidence that the missing species once *were* present.[42]

The implications for reserves of this phenomenon are profound. For example, population biologists Michael Soulé, Bruce Wilcox, and Claire Holtby took the data on the collapse of the large mammal fauna on Java, Sumatra, Borneo, and four other islands that were once united with the Malay Peninsula, along with the timing and the area involved, and extrapolated from them to develop a prognosis for the large mammal faunas of African nature reserves. Although there are some admitted uncertainties in their methodology, their general conclusions clearly are both sound and ominous.

They estimate that, once the smaller African reserves are isolated, they will lose about a quarter of their large mammal species in fifty years, two-thirds of them in five hundred years, and nine-tenths in five thousand years. The largest reserves will lose in the same periods about a twentieth, a third, and three-quarters of their species.[43] To some degree, intelligent management should at least slow the rate of extinction in the reserves. At the very least, where a species occurs in several reserves and goes extinct in one, it can be repopulated by transfers of individuals from another. But of course such movements destroy the isolation that encourages speciation. Thus the price of avoiding extinction in many cases will be the prevention of further diversification.

Both relaxation and collapses in reserves can be expected for virtually all groups of organisms.[44] Our experience with butterfly populations leads us to predict that both will be shown for them in the mountaintop "islands" of the Great Basin. What is not yet evident, however, is how different the rates of collapse of various taxonomic groups in the same reserves will be. There are reasons to expect mammal faunas, for example, to collapse more rapidly than birds or reptiles.[45] But scientists have only begun to answer these questions. Much more remains to be learned about patterns of collapse among prominent animals and the all-important plants and invertebrates as well.

Many other factors complicate problems of reserve design. For example, even within related groups, some species will be more "extinction-prone" than others. One would expect carnivorous mammals to be more vulnerable than herbivorous mammals; the carnivores have relatively small populations and need a lot of space per individual. Everything else being equal, Cheetahs will disappear faster from game preserves than Thomson's Gazelles.

Within taxonomic groups also, the presence of one or more species may exclude other species from the reserve. Biogeographer Jared Diamond has done brilliant work on the birds of Papua New Guinea and neighboring areas. He has shown that many birds are distributed in a checkerboard pattern. In each area or island, one or another of a pair of closely related bird species occurs, but almost never both. The presence of one apparently "locks out" the other. Among four species of New Guinea Cuckoo-doves, the situation is very complex. Only certain combinations drawn from the four species can coexist in nature. All four are never found together, for example, and only one of four possible combinations of three species living together is ever found; the other three combinations are "forbidden." If a small reserve were stocked with one of the forbidden combinations, one or more of the species would go extinct.[46] The problems would be similar for preserving three closely similar species of *Melidictes* honey-eater birds in Papua New Guinea, where only two of the three are ever found together. To save all three species, reserves in at least two mountain areas would be required.[47]

Patchy distributions—organisms occurring in scattered units of habitat—are more characteristic of the tropics than of temperate zones, and are especially pronounced in the highly threatened tropical rainforests. The casual impression of great uniform expanses of forest is extremely misleading. Not only are the birds patchily distributed, but the other animals and the plants, including the giant forest trees, are also. Indeed, differences among soils often produce profound differences in local tropical floras,[48] and such differences in turn are crucial to the distributions of insects and other herbivores.[49]

All of this is made even more complicated by patterns of change through time. Tropical forests are always undergoing disturbances ranging from individual trees falling over, erosion at riverbanks, and mountain landslides to mass blow-downs from hurricanes and blankets of ash from volcanic activity. After a disturbance, a rich variety of plants invades disturbed areas in a more or less well-defined sequence, as the process ecologists call *succession* returns the forest toward its relatively stable climax stage.[50]

Remember that regrowth of the forest led to the local extinction of numerous species of birds in the Barro Colorado reserve and to a reduction in the diversity of butterflies. Thus, in designing reserves, one must consider not only their carrying capacity at the moment, but how changes through time are likely to affect that capacity.

The possibility of natural disaster must also be considered in setting up a system of reserves. Wherever possible, leaving all of nature's remaining eggs in one basket should be avoided. For example, there once was a northern flightless sea bird called the Great Auk—the bird to which the term "penguin" was originally applied. It was severely persecuted by people who collected its eggs and hunted it for its feathers, fat, and meat. When the bird was reduced to a few scattered colonies, a natural catastrophe overtook it: one of the safest breeding colonies off Iceland was destroyed by a volcanic eruption.[51] Similarly, the simultaneous flowering and dying out of their bamboo food plant has greatly added to the woes of the last wild Giant Pandas.

When all of these factors are considered, the secret of successful reserve design would appear to be "as large and as many as possible." Large size is mandated to minimize the rate of relaxation. Numerous sites allow reserves to encompass different habitat patches and to minimize the problem of lock-outs. But given that there are inevitable restrictions in the amount of area that can be devoted to reserves, compromises will have to be made, using the best knowledge available. Even though that knowledge right now is inadequate for more than the crudest planning, certain principles are known that can be used as guidelines. For example, given the same total area to be preserved, a circular reserve will generally be less subject to faunal or floral collapse than a long, thin one. For some kinds of organisms, several smaller reserves will maintain more species than a large one of the same total area. And if a reserve must be fragmented, it is better to retain corridors of natural habitat connecting the fragments, even if the fragments must accordingly be smaller. This will result in lower extinction rates than if the full-sized fragments were preserved in isolation from each other.[52]

If anything resembling Earth's original bountiful living resources are to be preserved—and civilization with them—establishing a generous worldwide system of reserves will be essential. The tactics of the conservation movement therefore must include efforts to have the *maximum possible area* set aside in reserves, to establish them in the best possible sites,[53] and to do the best possible job of *configuring* those reserves to minimize the inevitable decay of diversity that will occur within them.

It is crucial to this effort, of course, that people in general and decision makers in particular be made aware of its importance. And that importance can hardly be overstated. As the distinguished biologist E. O. Wilson put it:

> The endemic plants and animals of each nation should be treated as part of their heritage, as precious as their art and history. When national leaders such as former president Daniel Oduber Quiros of Costa Rica have the courage to advance the preservation of ecosystems within their domains, they should be accorded international honors up to and including the Nobel Peace Prize, in recognition of the very great contribution they make, not just to their own generations but to generations as far into the future as it is possible to imagine.[54]

President Oduber was responsible for Costa Rica's magnificent program of setting aside large areas of rainforest in national parks. We were fortunate enough to meet him shortly after the program was initiated, in December 1977 when we were doing fieldwork on *Euptychia* butterflies in Costa Rica. Later we flew low over the unbroken rainforest of the Corcovado National Park, a most impressive testament to the breadth of Oduber's vision. Fortunately, Oduber's successor is being equally farsighted about preserving his country's heritage. President Rodrigo Carazo is planning to expand Costa Rica's parks (which already contain a greater proportion of the nation's lands than do parks in any other Latin American country) until they occupy about 10 percent of the nation's territory.[55] The question remains, however, whether Costa Rica will be able to maintain such a reserve system without substantial outside help.

Brazil also has magnificent plans for some 600,000 square miles of well-sited parks and reserves (some 18 percent of its territory), as well as more than a quarter-million square miles of national forest conservation land.[56] Every assistance and encouragement should be given to the Brazilian government to find the political will and the funding to implement their farsighted *Plano Nacional de Parques*. When we visited the Amazon Basin in December 1980, the situation was far from ideal. In one area near Manaus, forest on very poor soil was being cleared, not because it could be productive, but because large corporations based in southern Brazil could gain gigantic tax writeoffs against profits made elsewhere in Brazil. And we still heard many stories of inadequate protection of established reserves. Large portions of the Amazon forest remain intact, but this is no reason for complacency. Policy must be promptly converted into practice. If it is not, the intact forest will be destroyed in a few short decades.

Still, Brazil's official policy turnaround on Amazonia, from uncontrolled exploitation to a mix of conservation and ecologically sound development, is the single most encouraging event on the extinction front in our memory. We hope Brazil's President Baptista Figueirado will follow in the footsteps of Costa Rica's Oduber and that leaders in the rich countries will move to provide Brazil with the help it needs to make its plan a reality.

Managing Reserves

Just setting aside reserves is only the first step; after that comes the problem of managing them. Of course, nature got along quite nicely with its single biological preserve, Earth, for billions of years without "management" assistance from human beings. But the reserve was large enough and environmentally varied enough to provide stability and opportunities for evolutionary experimentation to proceed. Smaller reserves lack those advantages and the safety margin of redundancy that multiple populations confer against extinction. Some degree of management is now required for all reserves.

Perhaps the greatest irony is that human management of reserves is needed to protect them against *Homo sapiens*. At the most basic level, this means protecting them from poachers—which sometimes can be very hazardous for the guards. More commonly in many poor nations, there is continuous pressure from land-poor peasants who move into parks as squatters. In Venezuela, according to one estimate, there are some 30,000 slash-and-burn farmers living illegally within national parks and other reserves. Some of these are land-hungry illegal immigrants from Colombia. Venezuela's laws would be adequate to protect the reserves if they were enforced, but they are not, because local officials tend to be sympathetic toward the squatters.[57] We, too, can only sympathize with their plight—victims as they are of overpopulation and maldistribution of resources. They bring home a desperate fact of life to which we will return in the next chapter—the problem that all reserves are ultimately doomed to destruction in the absence of sweeping changes within human society.

Management also often includes such things as brush clearance, control of water levels, and control of undesirable animal populations. Such measures are especially necessary when reserves are very small, as they are in Great Britain, where there are 129 reserves averaging just over three square miles each in size. The 62 reserves in England (excluding Scotland and Wales) are, on the average, *just over one and a half square miles in extent.*[58] Most of them are smaller than Jasper Ridge, which contains almost two square miles and probably is too small to maintain Edith's Checkerspot without periodic restocking.

Management practices can have serious consequences beyond the borders of reserves. For example, in Michigan in 1980, a fire was set as part of a program to improve habitat on national forest land for the endangered Kirtland's Warbler. The fire got out of control, and instead of burning 100 acres, it consumed 25,000. A fireman was killed, hundreds of people had to be evacuated, and twenty-five homes and cabins were destroyed.[59] A constant worry at Jasper Ridge is that a fire will start on the reserve, go out of control, and destroy expensive homes adjacent to it. The reserve and the homes are in chaparral, a plant community that is a "fire climax"—one maintained by periodic burning. It is a risky place to build a house or to have a reserve so small that it is likely to be consumed in a single blaze. The plants would return from fire-resistant seeds, but many animal species would be lost permanently because of the lack of nearby sources of colonists.

In both rich and poor nations, ways must be found to balance the multiple uses to which many reserves are dedicated. National parks in the United States, for example, are expected not only to preserve species diversity but also to serve a wide range of educational and recreational functions as well. Most of them are remote from cities and are inaccessible to many people. In contrast, more and more local and regional parks are being sited in or near cities[60]— which is basically good—but there the pressures of human activities on other organisms are especially intense.

Local, state, and most national parks, however, are a far cry from undisturbed natural areas. Furthermore, they are not very extensive in area. U.S. national parks occupy less than 1 percent of the nation's territory. The entire National Park System, which includes national monuments, historic sites, and other areas of little biological interest, is only about 3 percent. In partial compensation, there has been pressure from conservationists in the past decade to set aside large tracts as wilderness, allowing only very limited access for recreational purposes and no exploitative activities such as mining, drilling, and lumbering. The effort has been a continuing battle between environmentalists and industrial interests. Since most of the land in question is owned by the federal government, or sometimes the state, agencies that administer them have often been caught in the middle of these conflicts. The Department of the Interior, traditionally charged with facilitating exploitation of natural resources, is now looked to as a defender of the remaining wilderness areas. It has not always gracefully adjusted to its new role, especially since appropriate management in wilderness areas consists largely of minimizing human disturbance.

Some of the hottest battles were fought over Alaska's huge tracts of wilderness. Some 111 million acres of federal lands were under dispute, an area larger than California, and about 5 percent of the United States' territory. In late 1978, President Carter declared 57 million acres to be "permanently protected," when Congress had failed to act, and the remaining 54 million acres were temporarily protected. Many Alaskans were outraged by what they saw as federal interference in their right to use their own land, and they even threatened to secede from the Union. Most of the strongest opposition naturally came from developers and oil and mining interests. Finally Congress reached agreement and passed the Alaska National Interest Lands Conservation Act, which was signed by President Carter in December 1980. The new law put some 104 million acres of Alaskan land under protection in national parks, wildlife refuges, national forest, and into the National Wilderness Preservation System.[61]

Alaska is extremely fortunate that exploitation of its relatively fragile natural areas will thus be subject to control and restriction before a great deal of damage has been done. After all, wilderness designation is something that can be reversed. But once mineral and biological resources are removed, they can never be replaced. In the more accessible lower forty-eight states, pressures are far more intense and have been making themselves felt for a longer period of time. There the general effects of a larger population, as well as of direct exploitation, have more or less severely damaged environments, and the pressures are still increasing. The Rocky Mountain and Great Basin regions of the West are particularly threatened today by the accelerating exploitation of mineral and fossil-fuel deposits there.

Even supposedly already protected areas are not immune from these pressures. The situation in U.S. national parks in the contiguous forty-eight states, for instance, is already frightening. The water supply that is crucial to the Everglades National Park is badly deteriorating under the demands of population growth in southern Florida. DDT and air pollution from nearby potash and other mineral processing plants are assaulting Big Bend National Park. Similar threats have been reported at Glacier; acid rains pour on Great Smoky Mountains National Park; air pollution from coal-burning power plants is having its impact on the plants and animals at Grand Canyon;[62] and feral burros are causing great destruction in that park and elsewhere—to the detriment of many other animals.[63]

In addition, as energy and mineral resources are perceived to be growing shorter, attempts to obtain permission for mining and drilling within and adjacent to national parks are escalating. On private land within Lassen Volcano National Park, an oil company recently bulldozed a drilling pad the size of a football field—without notifying the Park Service or testing to see what effect their 4,000-foot-deep geothermal test well would have on the geysers and hot lakes in the park.

And finally, the creation and management of national parks and wilderness areas cost money, and that money may be considered by many citizens and legislators as largely wasted in an era of proliferating economic problems.[64]

Reserves other than national parks and wilderness areas are also beset by problems. Privately owned ones, such as universities' biological study areas, are all too often viewed as potential recreational areas by everyone from vandals to local government administrators looking for land to condemn for public facilities.

The situation in the United States is further complicated by the lack of a national biological survey that would establish an inventory of the nation's biological resources, locate them, and provide a basis for monitoring and managing them. In this respect, the United States is far behind Europe, Japan, and even the Soviet Union. For example, computerized mapping of the flora of the country and major elements of the fauna would be entirely feasible with only a relatively minor commitment of funds. In the absence of an adequate national survey, it is difficult to advise decision makers on where new reserves are needed to protect valuable biological resources effectively without wasting limited funds.

Management of reserves can be exceedingly complex even in the absence of outside intervention, because of the biological problems involved, because of funding problems, and because of the human interactions involved in administration.

A common difficulty with reserves administered by the government is lack of coordination between agencies. A case in point is that of the Dusky Seaside Sparrow, a small, obscure, marsh-dwelling bird resident on Florida's east-central coast. The Dusky Seaside Sparrow has just about slipped through cracks in the federal bureaucracy. In 1979, only thirteen individuals could be found of a subspecies once numbering many thousands. Of the thirteen, twelve were males and the last a juvenile whose sex was still indeterminable. Three were captured for a possible breeding attempt, and one died. No females have been found. In 1980 only four birds were located, all males. An attempt may be made to hybridize the survivors with another subspecies of Seaside Sparrow, but the Dusky appears doomed—the first subspecies known to have disappeared since the Endangered Species Act was passed in 1973.[65] Yet this little bird lived in a government reserve. What happened?

Even though it was presumably protected, the sparrow is about to go extinct because of habitat deterioration. The first attack came when a choice marsh adjacent to Cape Canaveral was flooded as part of a mosquito-control program in 1963—with permission of the Fish and Wildlife Service, which operates the reserve.[66] Despite appeals from a concerned biologist at the time, no serious effort was made to restore even a part of that marsh.

A few years later, another biologist surveying the local sparrow populations found them in another marsh area. He persuaded the Fish and Wildlife Service (FWS) to buy up the marsh—and then the Florida Department of Transportation decided to build an expressway across it from Cape Canaveral to Disney World. Real estate developers began encroaching on the marsh edges, building drainage canals; and a ditch dug to obtain landfill effectively drained it further. The FWS inexplicably waited seven years to plug up that ditch. One marsh had been drowned; another was drained dry. The final blow was a series of fires. After the first, the FWS regional office sat on a request for fire prevention equipment until too late.

Soon one more bird will become extinct, even though it lives in a reserve, because no one in the agency that should have protected it really cared. The case of the Dusky Seaside Sparrow shows how vulnerable organisms on small reserves can be, especially in the face of administrative indifference and incompetence. It also shows that appreciation of the biology of a situation does not provide assurance against losses *even within reserves.*

Where the biology is both extremely complex and little understood, as in tropical rainforests, management problems multiply. For example, in rainforests, species that have coevolved with many other species sometimes become essential to their support. The giant *Casearia* tree in the Finca La Selva Reserve in Costa Rica fits that description. It is fed upon by at least twenty-two species of fruit-eating birds, some of which are completely dependent on it for survival during a two- to six-week annual period of fruit scarcity. If the *Casearia* population were lost, several birds would also disappear, and their loss would in turn affect the other plants that are dependent upon them for seed dispersal. The identification and preservation of critical plants should have high priority in tropical reserve management.[67]

As you might by now expect, not all such plants are prominent. Herbs and shrubs may provide vital nectar supplies that maintain populations of tiny parasitic insects at a level at which they can provide adequate control of herbivorous insects. This, in turn, prevents the defoliation and possible death of trees.[68] Understanding how such coevolved food webs work not only gives valuable insights on reserve management but also provides clues on how to use biological controls in adjacent agricultural systems.

For example, farmers in Trinidad who grow passion vines to harvest passion fruit were plagued by the caterpillar of long-winged *Heliconius* butterflies closely related to one we were working on in the nearby montane forests. Use of pesticides only made the attacks worse. The source of the difficulty was obvious to us. The pesticides were killing off the ants that play a key role in controlling herbivorous insects, including *Heliconius,* in tropical forests. Being predators, the ants lacked the long evolutionary experience with plant poisons that *Heliconius* had and were thus much less able to evolve resistance quickly to synthetic toxins.

We reiterate, the smaller the reserve, the more intensive management will have to be in order to maintain stability and to retard losses of diversity. Much can be accomplished by intelligent management, but much can also be lost even in short periods of poor management. Since *Homo sapiens* can hardly be expected to manage flawlessly for millennia, and since even with good management, things can sometimes go wrong, the necessity of maximizing both the size and numbers of reserves is urgent.

There is little doubt that maintaining the integrity of reserves in the long run will depend on public education and other measures to reduce impacts on them rather than on patrols and barbed wire, which are expensive and often ineffectual. In another of the ironies in which conservation abounds, reserves themselves can be important educational tools—but only if they are left open to the public, risking the attendant damage to their flora and fauna.

One fortunate aspect of the problem is the ability of small reserves, zoos, and even museums to play a crucial role in conservation education. In many countries, the importance of saving large mammals and birds and spectacular plants like the Sequoias is recognized by much of the public. But the importance of preserving smaller, less prominent organisms is still lost on most people. For instance, one multinational corporation recently launched a nationwide advertising campaign in the United States to promote an insecticide system, telling potential customers it would "kill every bug you've got." The ads showed only beneficial insects—including a Monarch butterfly, a ladybird beetle, and a praying mantis—which the lucky buyer presumably would be able to exterminate![69] There clearly is a strong need to increase people's awareness of the roles played by obscure organisms in ecosystems and of their potential significance to humanity and great intrinsic interest and beauty.

Huge facilities are not necessary for education about such organisms; endangered species and subspecies like the Houston Toad and Bay Checkerspot butterfly could be maintained in relatively small managed reserves and minizoos. All natural history establishments could do a much better job of public education about the values associated with obscure organisms. Plans are already underway by the Xerces Society to establish an insect interpretive center on a planned butterfly reserve in Washington state. One of these days soon, every state should have several such centers, preferably focusing on the insect-plant relationships that are central to the provision of many ecosystem services and also representative of the unrecognized roles of less prominent organisms. The Xerces Society has also started a national annual Fourth of July Butterfly Count, similar to the Christmas Bird Count of the National Audubon Society. This, along with positive efforts to preserve endangered butterfly populations, should help to raise the general level of awareness of the importance of "bugs" to people.[70]

Good management of the educational aspects of reserves, zoos, and museums obviously takes a great deal of expertise. That static displays can educate about dynamic ecology is demonstrated by successes like the magnificent new Evolution Hall at the United States National Museum—largely the result of the efforts of one evolutionist, John Burns of the Smithsonian. An enormous amount of information is conveyed in a very small area. Such public educational efforts are badly needed, not only to ease the job of establishing and protecting reserves but also to spread the word about organic diversity so that conservation efforts are not limited to zoos and reserves. Rarely has H. G. Wells' famous dictum seemed so apt: "Human history becomes more and more a race between education and catastrophe."[71]

CONSERVATION BEYOND ZOOS AND RESERVES

With all the complexities and uncertainties of reserve design and management —and the political problems of setting aside and protecting reserves—it is clear that a major task for conservationists must be to make sure that areas between reserves do not become biological deserts, but supplement the functions of reserves in preserving biotic diversity at lower levels. Simply setting aside a few areas as undisturbed reserves or places of minimum disturbance is unlikely to provide the necessary security for Earth's other inhabitants. A much more general revision of land-use practices and human lifestyles will be necessary to permit other species to live successfully in close proximity to *Homo sapiens.*

Many of these principles are already widely recognized, and in some areas they are practiced. They include the development of "greenbelt" areas in and around cities, the careful, *selective* logging of tropical moist forests as well as of many temperate forests, the acceptance of some livestock losses to predators and competitors on grazing lands, and the maintenance of numerous relatively undisturbed woodlots,[72] hedgerows, and streambanks to break up the monotony of agricultural landscapes. The latter practices have been especially successful in some areas of Europe in maintaining a relatively rich and varied landscape and preserving flora and fauna in some of the world's most overpopulated areas.[73]

But a great deal more needs to be done. Wherever possible, attempts at ecosystem rehabilitation should be made. Where there is vacant land, even tiny plots, the return of native vegetation should be encouraged, and it should be restocked with its associated animals. Planting of lawns or exotic "ground cover" plants over large areas should be discouraged. Growth of native vegetation should be promoted along the rights of way of railways, highways, and power lines to whatever degree is consistent with safety and with keeping the road surfaces, tracks, and power lines themselves clear of vegetation. The use of herbicides for indiscriminate clearing in these areas should be prohibited and more precise procedures instituted—even at immediate higher cost. The public image of an ideal ecosystem must be changed from one of closely clipped lawns or golf courses spotted with occasional "specimen" trees and bordered formal flower beds to one of the more complex and varied aspect presented by natural communities.

Similarly, further efforts to impound, channelize, and place into concrete all of the flowing water of the globe must be stopped immediately. Stream courses were once, and could become again, centers for natural communities in both urban and rural areas. Channelization and concrete impoundment result in the destruction of the natural flora and fauna of both the stream and the stream-banks. The streambank destruction has further profound effects on the organisms of the stream itself—as any experienced angler can tell you. Loss of streambank vegetation causes accelerated soil erosion and gullying on adjacent land; the self-purifying capabilities of the stream are at least impaired, even if not swamped by silt; and a valuable reservoir of potential pest-control agents (birds, predatory insects, etc.) disappears with the streambank ecosystem. The justifications (erosion control is a popular one) for most "stream improvement" programs are rarely compelling—unless one considers make-work for the Army Corps of Engineers and miscellaneous construction companies a valid excuse.[74]

It is also time people learned that living on unstable land near earthquake faults, on the slopes of active volcanoes, or on floodplains is risky. As Mount St. Helens demonstrated to a startled nation, humanity only *thinks* it can conquer nature!

We are not opposed to all development under all circumstances. But many important activities—the building of homes and factories, the construction of highways and railways, farming, grazing—require total removal or drastic modification of ecosystems. To some degree these activities are bound to continue, unfortunately even in areas where there is no fundamental need for them. But part of the damage done could and should be counterbalanced by conscientious efforts to ensure that areas not directly required for important human activities be allowed to remain in as undisturbed a state as possible—or are restored to that condition. These efforts should include strict laws and strict enforcement to minimize ORV use, to limit insecticides and herbicides to employment as ecological scalpels rather than blunderbusses, and so forth. Above all, there need to be massive public education programs on the vital importance of preserving biological diversity wherever and whenever possible.

Preserving Rainforest Diversity

Special attention will have to be paid to ways to preserve tropical rainforest diversity outside of parks and reserves—so crucial is that diversity to the future of humanity. Careful planning and considerable research will be required to determine the best patterns of utilization for areas with different soil types, different climatic regimes, and different biotas. What is suitable for those areas of Amazonia where the soils are rich because of annual floods, for instance, would be utterly inappropriate for huge expanses of unflooded forests—the so-called *terra firma,* which makes up about 90 percent of the basin.[75] The small proportion of flooded soils may produce high yields under conventional agriculture, as may a few good soil areas scattered through the upland *terra firma,* but most of the land of Amazonia is unsuitable for sustained cultivation.

Farming operations in rainforest areas therefore must be undertaken with greater care than in temperate zones if systems sustainable for the long term are to be created.[76] The tendency now is too much toward clearing forest to open new lands, rather than improving yields in areas already under cultivation. Clearing appeals to the pioneering image and often has the backing of landowners who fear land reform will divide their monster estates. "Why divide my land for those hungry farmers," they say, "when there is plenty of virgin territory that is only awaiting man's technology to be opened?" In Central America, large areas of rich, level lands are used for grazing cattle, which benefits relatively few people and underutilizes the land. By growing crops on this superior land, the pressure on marginal lands and forests could be relieved.[77]

Similarly, abundant areas of ruined croplands and degraded savannahs could be used for planting forests of pine and eucalyptus to lessen the need for further clearing of virgin forest for timber. Use of tree crops can also be encouraged—for example, breadfruit could be much more widely used for food. Since trees are the natural climax vegetation in tropical moist forest areas, they are far more likely to sustain yields over the long term without fertilizers or other "modern management" techniques.[78] Furthermore, sustained-yield cultivation of tropical hardwoods in some areas is both possible and highly desirable. As botanist Willem Meijer put it, "The world cannot live by pulp alone; high quality tropical hardwood will also be needed in the future."[79]

There is likely to be considerable future mining activity in rainforest areas; in some countries there already has been a great deal in the past. More than five hundred square miles of Malaysia, for example, are now occupied by tailings from tin mines.[80] In tropical forest areas, availability of water for reclamation of tailings sites is normally plentiful, and there is reason to believe that with care the disruption caused by mining in the moist tropics could be confined to reasonable levels.

But the most urgent need is to learn more about where reserves are needed before it is too late, and to determine, at least approximately, how the areas between reserves could best be developed both to serve humanity and to conserve organic diversity. Some of that research is underway. For example, Brazil's Project Radar Amazon is using airborne radar remote sensing systems combined with ground surveys to analyze the Amazon Basin. Maps are being produced for the first time that show the characteristics of the soil, the basic geology, forest types, and agricultural potential of Amazonia. The project has been described as providing "a rare opportunity—the chance to plan a continent's development before it happens."[81]

But cheering as this is, there is still a long way to go. The task of simply cataloguing the flora and fauna has only begun, to say nothing of unraveling the intricacies of tropical ecosystems. A virgin forest in the humid tropics contains enough mysteries and problems to intrigue a small army of biologists for centuries, to say nothing of abundant material for *Wild Kingdom* or *National Geographic* specials.

Rehabilitation of Ecosystems

Once an ecosystem has been changed by either natural or human agencies, it can never be restored to precisely its previous condition. For that matter, since individual organisms continuously age and populations continuously evolve, it is impossible to enter the same ecosystem twice (just as it is impossible to step into the same river twice). Nonetheless, there are different degrees of reversibility of ecosystem damage. A small cleared area in a tropical moist forest will undergo successional changes and automatically revert to a climax state if it is not further disturbed. But a larger cleared area may undergo a downward spiral including deterioration of the soil, which effectively prevents reestablishment of the forest locally except through enormous effort by human beings—and even then restoration might prove impossible. And chopping down a substantial portion of, say, the Amazon Basin would produce climatic changes that might completely preclude regrowth of a rainforest regardless of circumstances.

The ultimate irreversible perturbation of an ecosystem is, of course, the obliteration of its component species. While genetically appropriate populations of the species exist, there is hope. It therefore is clearly to humanity's advantage to restore as many damaged ecosystems as possible *now,* before more of the basic components are lost forever.

Naturally, the first step in the restoration of any area to a more or less natural state is to remove whatever forces damaged it to begin with. For example, a basic requirement for returning many islands more or less to their original state would be the removal of goats and other introduced herbivores. If certain temperate-zone forest areas are to be helped back toward their primitive climax condition, not only will logging have to be excluded, but acid rains will have to be controlled.

Unfortunately, the state of scientific investigation of the process of ecosystem rehabilitation is even more rudimentary than the state of knowledge of ecology as a whole. It is evident, however, that just as some systems can be damaged more readily than others, some also can be rehabilitated more readily.[82] Aquatic ecosystems seem to be quite resilient. Lake Washington in the

Seattle area, for instance, was showing signs of serious overfertilization due to human activities in the late 1950s. Local residents got together and voted in 1958 for a new sewage system that diverted the fertilizing pollutants from Lake Washington. Once this was accomplished, recovery was rapid. Between 1963 and 1974 the water became clearer, populations of noxious blue-green algae decreased, and Sockeye Salmon became much more abundant.[83]

A similar success story is that of the Thames estuary in England. More than three hundred and fifty years ago, in 1620, the Bishop of London was hoping that the Thames would be cleaned up. A hundred and fifty years later another observer wrote:

> The river Thames [is] impregnated with all the filth of London and West-minster—Human excrement is the least offensive part of the concrete, which is composed of all the drugs, minerals, and poisons, used in mechanics and manufacture, enriched with the putrifying carcasses of beasts and men; and mixed with the scouring of all the wash tubs, kennels, and common services.[84]

In spite of this, the Thames still appears to have been a good fishing river in the eighteenth century. In the nineteenth, with the installation of many water closets, the flow of sewage increased dramatically, and by 1850 all commercial fishing had ceased. Only eels could survive in the most polluted stretches of the river after that, and most water birds and other waterside animals had vanished. About a century later in the 1950s, steps were started to lower the pollution load, and water quality since then has continually increased. This was followed within a decade or so by a remarkable ecological recovery of the estuary and river. By 1977 the river contained increasing populations of some ninety species of fishes, including Plaice, Sea Trout, bass, Cod, Mackerel, Mullet, and even Salmon, as well as a variety of invertebrates and large numbers of ducks and other wild fowl.[85]

At the other end of the spectrum from these quickly rebounding aquatic systems, it has been estimated that full recovery of a cut-over lowland tropical rainforest would take about a thousand years.[86] But, interestingly, it is another aquatic system that would probably need the greatest recovery time. John Cairns of the Center for Environmental Studies at Virginia Polytechnic Institute summed it up:

> It appears highly probable that the vast oceanic ecosystems are quite fragile . . . and are protected primarily by their vastness and the resultant dilution of all potentially deleterious materials. Should an entire ocean be damaged, the time required for recovery staggers the imagination, especially since many of the organisms are highly vulnerable to change and presumably would not be able to withstand the rigors of invading new areas far from their original habitat.[87]

It would behoove those who see nothing amiss with employing the oceans as a garbage dump with infinite capacity to mark his words. The coral reefs of Kaneohe Bay in Hawaii have started to recover since raw sewage is no longer piped into it. But if pollution of the oceans becomes too widespread, recovery of such systems may no longer occur in a time span of interest to *Homo sapiens.*

Finally, let us note that it is possible to increase organic diversity through the artificial enhancement of ecosystems. For example, it is well known that increasing the structural complexity of ocean bottom can create areas of high fish-species diversity. The building of concrete-block artificial reefs for scientific research in sandy areas, both on the Australian Great Barrier Reef and in the American Virgin Islands, has documented this quite nicely.[88] Anglers have known this quite well for a long time and have often deposited piles of old automobiles or other junk on the ocean bottom to increase their catches. Basically, of course, the artificial reefs provide the protection from deep-water predators that is required for certain species to persist. Australian biologist Barry Russell was able to show that the number of saber-toothed blennies (the little dash-and-grab reef predators of Chapter 1) was limited by the number of available holes. He did it by the simple expedient of adding concrete blocks with pre-drilled holes to the reef. The number of blennies increased as soon as the artificial refuges were provided.

Similarly, artificial habitat can be provided to increase the species diversity present in terrestrial ecosystems. For instance, artificial nesting sites can be provided to increase the nesting success and population density of various bird species.[89] Millions of people have done this themselves by erecting birdhouses in their yards in cities and suburbs.

But the artificial enhancement of many such ecosystems soon fades without human renewal.[90] As more is learned about ecosystems, it may be possible to find ways both to enhance their diversity and to reduce the amount of husbandry needed to maintain the higher level of diversity.

It is important above all to remember that because people have planted trees or grass in an area does not necessarily mean that a valuable ecosystem has been created. Such a simplified system usually is either very unstable, or unable to contribute many ecosystem services, or both. Tree farms, coconut-palm plantations, golf courses, and lawns preserve relatively little diversity and normally are totally inadequate surrogates for the natural systems they replace.

Preservation of Human Diversity

People are by definition components of ecosystems, and it can be argued that preservation of their diversity is a meritorious goal, which is also important to maintaining the diversity of other organisms.[91] Indigenous peoples, for example, often have a comprehensive knowledge of medical uses of other organisms. Hawaiians knew of the anticancer activity of the body fluid of certain marine worms—an activity confirmed in the laboratory. Thus preserving different human cultures will give society access to folk knowledge that would otherwise be lost. That knowledge will increase the benefits that can be extracted from other species and in the process add impetus to efforts to preserve them, too.[92]

More importantly, the secrets of successful long-term occupancy of Earth undoubtedly are not all encompassed by Western culture. Perhaps, for example, the secret of people living peacefully together is buried in the culture of the gentle Tassaday. Or maybe a Native American culture can give us some crucial insight into how to live within the constraints imposed by natural ecosystems rather than trying to conquer them—as the following quote from a letter by Chief Sealth of the Duwamish Tribe of Washington State in 1855 suggests:

> Every part of the earth is sacred to my people. Every shining pine needle, every sandy shore, every mist in the dark woods, every clearing and humming insect is holy in the memory and experience of my people. . . . the white man . . . is a stranger who comes in the night and takes from the land whatever he needs. The earth is not his brother but his enemy, and when he has conquered it, he moves on. He leaves his father's graves, and his children's birthright is forgotten . . . all Things share the same breath—the beasts, the trees, the man. The white man does not seem to notice the air he breathes. Like a man dying for many days, he is numb to the stench. . . . What is man without the beasts? If all the beasts were gone, men would die from great loneliness of spirit, for whatever happens to the beast also happens to man. All things are connected. Whatever befalls the earth befalls the sons of earth. . . . The whites, too shall pass —perhaps sooner than other tribes. Continue to contaminate your bed, and you will one night suffocate in your own waste. When the buffalo are all slaughtered, the wild horses all tamed, the secret corners of the forest heavy with the scent of many men, and the view of the ripe hills blotted by talking wires, where is the thicket? Gone. Where is the eagle? Gone. And what is it to say good-by to the swift pony and the hunt, the end of living and the beginning of survival.[93]

There are many steps that can be taken to help preserve cultural diversity, but one is of particular interest to us here. In the process of setting aside reserves, efforts can be made to "preserve" indigenous people along with their natural habitats. For instance, the native cultures of Amazonia are rapidly being destroyed by the advance of the dominant Western culture. The process has been direct, by enslavement and killing, and indirect, by habitat destruction—a pattern familiar from the behavior of human cultures toward other species. Since 1900 the number of Indian tribes living free in the Brazilian

Amazon has dropped from 260 to 143, and fewer than ten of them contain as many as 1,000 individuals. About 26 tribes were destroyed in the past decade; some 30,000 people were displaced or killed. Most remaining tribes have only a dwindling handful of survivors—the shrinking reservoir of the tribe's culture.

But how can remaining indigenous cultures be saved? Suppose each tribal area were converted into an inviolable reserve? Such a plan has now been proposed to preserve the famous Yanomamo tribe in northwest Brazil in an area of more than 50,000 square miles. And there is some chance that it and other parks may be established, in accordance with the new attitude toward Amazonia in Brazil.[94]

Sadly, it seems certain that the few remaining "primitive" peoples will at best undergo considerable acculturation, since they all have at least some contact with the dominant culture. At worst, if dramatic steps are not taken soon, their cultures may all disappear. In spite of Brazil's recent more enlightened attitude toward the preservation of Amazonia, comments once made by their minister of the interior Mauricio Rangel Reis may sum it all up: "The ideals of preserving the Indian population within its own habitat are very beautiful ideas but unrealistic."[95]

FROM TACTICS TO STRATEGY

Beautiful ideas have little power to resist "realism" in today's world. That is why attempting to implement the conservation tactics discussed in this chapter will never in themselves suffice without an overall strategy.[96] The assaults on the Snail Darter, the Furbish Lousewort, and, above all, the species of tropical forests are symbolic of what will happen in the future. If the expansionist economic "realism" of today persists, sooner or later too many rivets will be popped, and all of humanity will suffer in the cataclysm that ensues. If the course of society is not changed, then eventually every population, species, zoo, and hedgerow will stand in the path of some industrialist's or politician's version of "progress." As we stated in Chapter 1, there will be the equivalent of a Tellico dam in the future of every wild organism and for every human group or family whose land and culture are considered unimportant by those who control the bulldozers.

Remember, the tactics of conservation are important—the battles to save individual species and set up and maintain reserves—because they buy time and help to delay the day of reckoning. But we are convinced that what is required now is a brand-new *strategy* for the preservation of organic diversity, based on the five Iron Laws of Conservation:[97]

1. In conservation there is only successful defense or retreat, never a true advance—a species or an ecosystem once destroyed cannot be restored.

2. Continued human population growth and conservation are fundamentally incompatible.

3. A growthmanic economic system and conservation are also fundamentally incompatible.

4. The notion that only the short-term goals and immediate happiness of *Homo sapiens* should be considered in making moral decisions about the use of Earth is lethal, not only to nonhuman organisms but to humanity.

5. Arguments about the right to exist of nonhuman life forms, or their esthetic value and intrinsic interest, or appeals for compassion for what may be our only living companions in the universe, now mostly fall on deaf ears. Until ethical attitudes evolve further, conservation must be promoted as an issue of human well-being and, in the long run, survival.

The new strategy must be an uncompromising one that is aimed at transforming society as a whole from one that primarily assaults ecological systems to one that automatically husbands them. It is a strategy that will be considered unrealistic or infeasible by most "practical" people. The irony, of course, is that nothing could be more fundamentally *impractical* than for society to continue down its present course.

Chapter X

The Strategy of Conservation

The first need is to develop *a blueprint for the economics of peaceful stability*. The "vigorous, growing economy" all our leaders keep exhorting us to produce is not possible on an earth of fixed size, and continuing attempts to produce it are *the* basic threat to peace.

—DAVID R. BROWER,
Not Man Apart, *May* 1980

THE basic tactics of conservation are relatively simple: establish numerous large, well-sited reserves all over the globe, and manage them (and the spaces between them) appropriately. If these tactics can be successfully employed, biological impoverishment should be arrested. But how are the political, economic, and psychological resources of humanity to be mobilized to permit this ultimate victory? Unfortunately, there is no single terrifying opponent against which conservationist forces can be rallied. The biosphere is being nickeled-and-dimed to death: some overexploitation here, a little habitat destruction there, a bit more acid rain or PCB somewhere else. The enemy is not only "us" but virtually all human activities. The crisis is terribly mundane, but terribly serious. Unless appropriate steps are taken soon to preserve Earth's plants, animals, and microorganisms, humanity faces a catastrophe fully as serious as an all-out thermonuclear war.

What is called for, then, is a strategy for transforming society so that the present rapid acceleration of extinction becomes a deceleration. The transformation must be universal because it is the everyday activities of human beings that most threaten other organisms—the ways that people treat natural systems in the process of obtaining food, clothing, shelter, and other amenities. The nature of this transformation to a *sustainable* society has been the subject of a number of recent books.[1] A sustainable society can best be described as one dedicated to living within environmental constraints rather than perpetually growing with the hopeless goal of conquering nature. Here we will give only a résumé of the strategy, to emphasize our conviction that, like so many other problems facing humanity, the problem of extinction will not be solved by minor adjustments of the sociopolitical system.

POPULATION CONTROL

Halting the growth of the human population as rapidly as is humanely possible and starting a gradual decline to a permanently sustainable level are obviously essential if the populations of most other organisms are to have a chance of persisting. The laws of biology encompass *Homo sapiens* as surely as they do the Iriomote Cat or Edith's Checkerspot Butterfly. People share the bounty of solar energy as well as the space and other resources of Earth with millions of other species. If there is to be room and sustenance for the others, then people must be willing to sacrifice space and resources for them that could otherwise be used by more people. Reserves that could support people must be left for other species. That this sacrifice ultimately is in the interests of humanity—especially of future generations—makes it no less a sacrifice for those who would enjoy large families but restrain their reproduction.

Fortunately, the notion that "every cause is a lost cause without population control" is now widely accepted, even by people who are unaware of the extinction problem. As the utter incompatibility of continued human population growth with the survival of other species becomes more generally understood, it should give further impetus to population control efforts. Yet controlling the numbers of human beings—indeed, eventually reducing them to a number permanently sustainable without ecosystem deterioration—while absolutely necessary, will not be sufficient. Even a world population of two billion people, less than half the size of today's, could easily pop enough rivets to destroy civilization if everyone tried to live like Americans. It is not just how many people there are but also how they live that determines their impacts on ecosystems.[2] The everyday activities of human beings must be altered so that they no longer cause a continual erosion of organic diversity.

ECONOMIC GROWTH

"The study of mankind in the ordinary business of life" is what the great economist Alfred Marshall called his discipline. Therefore, if human activities must change, reform of the economic system is a central challenge that must be met to permit the establishment of a sustainable society.

If anything characterizes our era, it is the dominance of economic values. The quest for equity, justice, compassion, peace, comfort, or grace all tend to take a subsidiary role to the acquisition of wealth and the drive to increase the gross national product. This imbalance, which few people question, is justified primarily by equating absolute material wealth with well-being. There is, however, massive evidence that, once a certain level of amenity is reached, the relationship between wealth and well-being is tenuous at best and reversed at worst—and that *relative* prosperity becomes paramount.[3]

The drive for wealth as a mass phenomenon is a very recent development historically—only about three centuries old.[4] It has been a fundamental reason that *Homo sapiens* has become such a threat to the existence of other species. In those few centuries, a mere blink of the eye on the time scale of biological evolution, an economic system that has material growth as its primary goal has become a central feature of human culture. It is a system that has served many people well, but heavy costs have accompanied its benefits. Enormous sacrifices were made in the eighteenth and nineteenth centuries to permit the capital investment upon which a more general prosperity could be built. These sacrifices, made almost entirely by the poorest segment of society, permitted the industrial revolution. This in turn gave humanity an unprecedented ability to exploit Earth's resources and unintentionally to incur further massive costs by assaulting natural ecosystems.[5]

The industrial revolution, based on unequal sacrifices, has led also to enormous international inequities of wealth and power, indeed to the division of the planet into rich nations and poor nations. It also produced the population explosion and thermonuclear weapons. Both the benefits and the costs of the revolution are very unequally distributed among peoples today, and they are also unequally distributed over generations. For example, the benefits of the extinction of the Passenger Pigeon (in which railroads played a central role) were reaped by people in the last century; the costs are still being paid—and will continue to be paid by future generations in perpetuity.[6]

The frightening thing about economic growth is that, regardless of its past costs and benefits, it is now causing humanity to press against the physical and biological constraints within which the economic system must operate. One clear sign of this pressure is the increasing rate of extinctions and the resultant threat to the systems that support civilization. Growing numbers of people, all demanding an ever greater level of affluence, show every sign of having already overshot the carrying capacity of Earth for *Homo sapiens.*[7]

Transforming the Economic System

It is possible, indeed likely, that the growth-oriented economic system will grind to a halt willy-nilly as it encounters natural limits. Instead of conquering nature, nature will conquer it. In that case, humanity will be stuck with a failed growth system.[8] On the other hand, there could be a rationally planned transition to a steady-state economic system. The problems of converting a system designed around perpetual growth into one appropriate for a sustainable society would be substantial but solvable.[9] A planned transition obviously would be infinitely preferable from the points of view both of humanity and of the other species that call Earth home.

What barriers are there to making such a transition? The greatest one is simply that most decision makers in all societies do not recognize the urgency of getting on with it. In part this is because the economists who advise them are utterly ignorant of the constraints: if they believe there are any limits to physical growth, they think them to be in the distant future and a problem for future generations. They equate environmental problems with "pollution" and assume people can choose to live with environmental problems or without them. In their view, environmentalism is simply "a demand for more goods and services (clean air, water, and so forth) that does not differ from other consumption demands except that it can only be achieved collectively."[10] Similarly, other species are commodities that society can value or not value, depending on its desires.[11] The dependence of human beings upon ecosystem services, which in turn depend on other species, virtually never enters into the calculations of economists.[12]

So the economists and the politicians they advise tend to behave as if the world were infinite. They assume that natural resources are infinitely substitutable for one another[13] because they are ignorant of either the environmental costs of making the substitutions or of geology, physics and chemistry. Indeed, one economist (apparently having read an old book on alchemy) recently wrote that "the future quantities of a natural resource such as copper cannot be calculated even in principle . . . because copper can be made from other metals."[14]

Similarly, most economists and decision makers do not understand that the same second law of thermodynamics that gives ecosystems their pyramidal structure also puts severe limits on the total activities of humanity.[15] Because of the thermodynamic limits, it is not possible *even in theory,* with "perfect" behavior toward the environment, to keep the economic system growing forever. Even with an ecologically ideal economic system, continual growth would lead eventually to an ecosystem collapse to which all species, including humanity, would fall victim.[16] But the real world is imperfect, of course, and so every substantial increment of global material economic growth will extract its price in extinctions.

Rich World—Poor World

If everyone were leading a relatively affluent life and the human population were no longer growing, a general prescription for the extinction problem would be relatively simple: put the brakes on material growth of the economy, and put them on hard. No more land paved over; if new housing is needed, redevelop slums. No more land plowed under or grazed: where necessary, intensify husbandry on existing agricultural land. Open no more mines: recycle everything; rework old mines to make up inevitable losses in the process of recycling. Make durability, not disposability, the goal of production. Substitute cleverness and technological innovation for brute force. Learn to extract more and more good from each unit of energy used to do work and from each pound of material mined or recycled. Let sectors of the economy grow that can do so while using less and less of Earth's physical resources (such as today's computer industry); let others shrink. Keep the aggregate impact on nature constant or shrinking. All of this is clearly feasible in a democratic, essentially capitalist system. The basic requirement is for a political device to limit the flow of resources into the economic system—for example, a system of *depletion quotas* such as has been proposed by economist Herman Daly.[17] With such a device in place, most of the other necessary steps—such as growth of economic activities that use resources efficiently and shrinkage of those that do not—would follow almost automatically.

Where it is impossible to follow the above strict rules, compensate through ecosystem restoration. Return marginal farms and overgrazed land to nature; encourage a hobby for the rich of buying country estates, tearing them down, and letting succession take place. Convert neatly trimmed parks into naturally vegetated species-rich areas. Protect and enhance at all costs that unnoticed but essential sector of the economic system—other species and their environments, the sector that provides humanity with free ecosystem services.

Obviously, although the prescription is simple, its execution would be complex. How it could be done is just starting to be explored.[18] But the situation is made much more complicated because everyone is not living a relatively affluent life, and the population is still growing. To put the brakes on material economic growth now would be to condemn the majority of the world's population to lives of misery. One could hardly expect those people, in whose nations most of Earth's biological riches reside, to accept any plan that freezes them in poverty. Nor can they be expected to take a long-term view and be greatly concerned about future losses of ecosystem services. Gandhi reputedly once said, "Some people are so poor that God can only appear to them in the form of bread." Concern for elephants and cacti—indeed, for next week or for posterity in general—is a luxury available only to those whose basic human needs are met.[19]

Difficult as is the problem of limiting material growth in an inequitable world, it is not insoluble in theory. Indeed, its solution has been outlined.[20] What is required is a great reduction in the assault rich nations are now mounting on ecosystems—a retreat from overdevelopment. This, in turn, would make room for grass-roots development of the poor countries. It basically means abandoning the notion that the way to help the poor is to make the wasteful rich even richer, so that benefits will trickle down to the poor and lift them from the muck. It means staring redistribution of wealth in the eye and then getting on with it, because to hesitate too long means there won't be any wealth to redistribute. Famine, plague, thermonuclear war, and ecosystem collapse, singly or in concert, will play hell with humanity's stock of capital.[21]

Redistribution of wealth has long been recognized as an essential ingredient for achieving a sustainable society,[22] but it remains a sticky wicket. An obvious reason is that people from whom the wealth is to be taken fear they will be reduced to living in caves and cooking over buffalo chips. But the fear is groundless. Suppose, for example, that the United States decided to make the ultimate sacrifice and cut its per capita energy consumption in half so that more energy would be available to poor nations. Would that mean the end of American civilization? Not quite—it would simply move the nation to the level of energy use it had in 1940 (or to the level of France today). And, of course, technological advances would allow that energy to be used much more efficiently today than forty years ago; the same amount of energy would produce much more in the way of goods and services than it did then. Indeed, it seems likely that with skill and determination the United States could halve its energy consumption per person and enjoy a quality of life even higher than today's.

Cars would be smaller, but the air would be cleaner, life longer and more leisurely, and American ecosystems and their components in better health. The energy *saved* by each American would, on the average, be enough to double the energy available to a hundred and seventy-five people in Bangladesh or a dozen people in Egypt.[23] Redistribution would be complicated, but it could be done—and without serious inconvenience to the donors. What is required is the political will in the rich nations to mature economically rather than to try to continue growing for growth's sake ("the creed of the cancer cell," Edward Abbey once called it[24]). Also needed is the political will in the poor nations to redefine development goals and ensure that the benefits of progress are shared by all, not confined to an elite.

How, in practical terms, can such dramatic changes on the international scene be initiated? Certainly not through imposition from above by some world government modeled on the United Nations. Humanity has tried that route and seen it fail. The movement will have to grow from below, with individual nations, both rich and poor, coming to realize that their futures depend on each other and on organizing human activities in utterly novel ways that greatly reduce the pressures toward extinction.

Education about the consequences of continuing on the present course thus will have to play a major role if catastrophe is to be avoided. There are, fortunately, some encouraging signs. In mid-1980, for instance, the U.S. Council on Environmental Quality and the Department of State presented their *Global 2000 Report to the President.*[25] The document, prepared with the cooperation of numerous government agencies, foresaw "global problems of alarming proportions by the year 2000," including "a progressive degradation and impoverishment of earth's natural resource base."

Assistant Secretary of State Thomas Pickering announced that the gloomy findings of *Global 2000* were being sent both to U.S. ambassadors abroad and to foreign governments. The idea was to "help galvanize worldwide cooperation to improve prospects for the future."[26] The need for such international educational efforts is underlined by widespread ignorance, even among those who should be informed, about the consequences of progressive erosion of the world's biological resources.

For example, a survey in 1979 showed that twelve out of nineteen journalists in fourteen countries thought that no material harm would come to the world from the ever-increasing pace of extinction.[27] A well-known English columnist, Henry Fairlie, complaining about extremism in the American conservation movement, revealed his ignorance in half-truths: "Nothing of what I wear or eat or drink, except the occasional haunch of venison sent to me by a hunter, comes to me from mother nature's larder. Even the conservationists are sustained not by nature but by man's cultivation of it. . . . "[28] We can admire Fairlie's eschewing of most fish and shellfish, but not his lack of understanding of the myriad ways that mother nature sustains him and all the rest of us. If he understood, then he too would surely be appalled at the diverse ways *Homo sapiens* has devised to destroy the systems she uses to support society.

Once enough people are awakened to the problem, then various international solutions can be developed. A more comprehensive view of economics would prevail, and rich nations would perceive that changing their own behavior, restraining their own resource consumption, and helping poor nations would be in their own direct economic interest. As a single example, helping Brazil finance its planned system of rainforest reserves could be worth many billions of dollars to the United States, if only because the climatic changes that would accompany the destruction of the Brazilian forests might well greatly reduce American agricultural productivity, which in 1980 yielded more than $40 billion in export sales. Without those exports, the United States would be even harder put to pay its oil import bill than it is now. And the world's developing countries are increasingly dependent on U.S. food exports to feed their people. What happens to the Amazon Basin is a matter of global concern —and so is the related health of America's agricultural system.

There is no question that aid to the poor is the best investment rich countries could ever make; there is considerable question whether the rich will realize it in time. But overhaul of the international economic system is long overdue, and perhaps the intertwined issues of resource shortages and extinctions will at last trigger it.[29]

A UTOPIAN VIEW

Looking specifically at the problem of preserving biological diversity on this small planet while the human population and its economy are still growing, what strategy should be adopted? To begin with, something like Aldo Leopold's land ethic[30] obviously must replace the exploitive attitudes that dominate the thinking of most people today, especially in the West. Unfortunately, these attitudes are strongest in the most powerful and influential individuals —government and corporation leaders. Farmers, who live close to the land, usually have some concept of land husbandry, but economic pressures—in poor countries, often of the starkest kind—force them into practices that on a long-term basis undermine the land's productivity. They must be given assistance and encouragement to practice soil conservation, protect local watersheds, replant woodlots and forests, refrain from overgrazing, irrigate carefully, and so forth.

All people must learn to put a high value on that spice of life: variety. They must cherish and protect wilderness and not-so-wild natural areas, understanding that their existence enhances not only the quality of their own lives but the long-term productivity of their farmlands.

The International Union for the Conservation of Nature and Natural Resources (IUCN) in early 1980 published a long-range plan for preserving Earth's living resources and putting civilization on a sustainable basis.[31] The World Conservation Strategy rests on three major goals:

1. Maintaining essential ecological processes and life-support systems
2. Preserving genetic diversity
3. Utilizing species and ecosystems sustainably

The strategy then targets several areas for particular attention in coming decades. First are agricultural systems. The present widespread deterioration of land, including the process of desertification, and the erosion of genetic resources for crops must be halted and where possible reversed. Second, the destruction of forests, especially in the tropics, must be reversed. In particular, watersheds must be protected. The strategy points out that at least half the world's people are directly affected by how watersheds are managed, even though only about 10 percent actually live in the mountain watershed regions. Third, the oceans, and especially estuaries and nearshore wetlands, must be protected and preserved.

Fourth, endangered species are given special attention, both as vital resources in themselves and as components of ecosystems. The strategy recommends giving top priority to saving species (1) that are genetically very different from any other organism; (2) that are culturally or economically important or have close relatives that are; and (3) that live in a species-rich area (such as a tropical rainforest) where many can easily be preserved together. Such considerations certainly should be included when decisions on size and location of natural reserves are made.

The World Conservation Strategy as formulated is probably not strong enough to preserve Earth's biotic diversity adequately, but it is at least a good first step. The intention was that the strategy be adopted as a major set of goals by every nation and integrated into its development policies. We would be delighted to see the U.S. government adopt it as its own official policy and try to persuade others to follow suit. But so far there seems to be little interest in Washington—or in other capitals—despite the urgent need for it. The idea that natural systems and species are among the valuable common resources of an interdependent world is an idea whose time has come.

This is not all a hopeless dream, nor is the task too great. In the 1960s everyone thought that a sudden and dramatic decline in American birthrates was a hopeless dream. Yet it happened. When the time is ripe for social change, it can occur with stunning rapidity. That's where the hope lies for our fellow passengers on Spaceship Earth, and for ourselves.

There are certainly encouraging signs. The United States and many other rich countries have reached replacement reproduction or have actually ceased population growth,[32] and there have been significant reductions of growth rates in many poor nations. In a bit more than a decade, the environment has been elevated to a major political issue almost everywhere. The Snail Darter almost beat the Tellico dam—something that would have been inconceivable before 1970. Steady-state economics and zero economic growth have become issues of debate.[33] And people are looking in the right direction for solutions—the World Conservation Strategy being one example.

Admittedly it is a small start, most people thinking and talking while the world moves inexorably toward the drain. But the first few rattling rocks could presage an avalanche. What may be needed now is a great act of leadership. Just as President Kennedy put the United States on the road to space, another president could put the nation—and, one would hope, the world—on the road to treating Earth as a spaceship. Someone had better get on with the job, for *Homo sapiens* is no more immune to the effects of habitat destruction than the Chimpanzee, Bengal Tiger, Bald Eagle, Snail Darter, or Golden Gladiolus. As Ken Brower noted, referring to the California Condor: "When the *vultures* watching your civilization begin dropping dead . . . it is time to pause and wonder."[34]

Appendix

Taxonomy of Organisms Discussed

In most cases we have referred to organisms by their American common names. While this makes the text more understandable to everyone (even biologists often do not recognize the Latin or scientific names of organisms outside of their specialties), it also creates problems. For one thing, organisms often have more than one common name: the butterfly *Nymphalis antiopa* is called the Mourning Cloak in North America and the Camberwell Beauty in Great Britain. Conversely, two different organisms may have the same common name. The "robin" of North America is *Turdus migratorius*. The "robin" of England is *Erithacus rubicula*—and the same species in Holland is "roodborst," in Germany "rotkehlchen," in Switzerland "rödhake," and in France "rougegorge."

This appendix, then, is to permit the unambiguous placement of the organisms dealt with into the taxonomic scheme. Within broad classes of organisms, we have simply listed the common names used in this book alphabetically. Where the common name is that of a species or subspecies, it is followed by the Latin specific or subspecific name. After that we give the name of the taxonomic family to which the organism belongs. In the case of animals, this name always ends *-idae;* in plants usually (not always) *-aceae.* Since there are so many families of insects, we have followed the family names of the insects with the name of the orders. Thus *Euphydryas editha* is in the family Nymphalidae (four-footed butterflies), and the Nymphalidae in turn is in the order Lepidoptera (scaly-winged insects—butterflies and moths). Miscellaneous invertebrates are mostly placed into class and phylum.

In some cases two scientific names are commonly used for the same organisms, even though great efforts are made to avoid this. For example, the lion is found (mostly in the older literature) as *Felis leo* and (more recently) as *Panthera leo.* Which is "right" depends on how similar you think lions and house cats *(Felis domesticus)* must be in order to be placed in the same genus. In cases like this, and in similar cases with common names, we have indicated a second choice with parentheses—e.g. *Panthera (Felis) leo.* Similarly, where two names are commonly in use for higher categories (e.g., plant families) a second name preceded by = is given in parenthesis. The abbreviation sp. after a generic name means an unidentified species of that genus. The abbreviation spp. means species in the plural: thus pipits are *Anthus* spp.—a lot of different species of birds of the genus *Anthus.*

MAMMALS (CLASS MAMMALIA)

Addax	*Addax nasomaculatus*—Bovidae
African lion	*Panthera (Felis) leo*—Felidae
Arabian oryx	*Oryx leucoryx*—Bovidae
Arabian Tahr	*Hemitragus jayakari*—Bovidae
Armadillo	*Dasypus novemcinctus*—Dasypodidae
Baboon	*Papio* sp.—Cercopthecidae
Badger	*Taxidea taxus*—Mustelidae
Bali tiger	*Panthera (Felis) tigris balica*—Felidae
Banteng	*Bibos javanicus*—Bovidae
Bearded seal	*Erignathous barbatus*—Phocidae
Beaver	*Castor canadensis*—Castoridae
Beechey's ground squirrel	*Citellus beecheyi*—Sciuridae
Bengal tiger	*Panthera (Felis) tigris tigris*—Felidae
Bighorn sheep	*Ovis canadensis*—Bovidae
Bison (buffalo)	*Bison bison*—Bovidae
Black bear	*Ursus americanus*—Ursidae
Black rhino	*Diceros bicornis*—Rhinocerotidae
Black-footed ferret	*Mustela nigripes*—Mustelidae
Blue whale	*Sibbaldus musculus*—Balaenopteridae
Boar	*Sus scrofa*—Suidae
Bobcat	*Lynx rufus*—Felidae
Bowhead whale	*Balaena mysticetus*—Balaenidae
Buffalo (African buffalo)	*Syncerus caffer*—Bovidae
Burro	*Equus asinus* (small strain)—Equidae
Caribou	*Rangifer caribou*—Cervidae
Caspian tiger	*Panthera (Felis) tigris virgata*—Felidae
Ceylon elephant	*Elephas maximus maximus*—Elephantidae
Chamois	*Rupicapra rupicapra*—Bovidae
Cheetah	*Acinonyx jubatus*—Felidae
Chimpanzee	*Pan troglodytes*—Pongidae
Chinese tiger	*Panthera (Felis) tigris amoyensis*—Felidae
Corbett's tiger	*Panthera (Felis) tigris corbetti*—Felidae
Cottontop marmoset	*Saguinus oedipus*—Callithricidae
Coyote	*Canis latrans*—Canidae
Deer	Cervidae (in part)
Dog	*Canis familiaris*—Canidae
Dromedary	*Camelus dromedarius*—Camelidae
Eland	*Taurotragus oryx*—Bovidae
Elk	*Cervus canadensis*—Cervidae
European brown bear	*Ursus arctos arctos*—Ursidae
Formosa sika	*Cervus nippon taioreanus*—Cervidae
Foxes	*Vulpes* sp., other genera—Canidae
Gazelles	*Gazella* spp., other genera in the subfamily Antilopinae of the family Bovidae
Giant panda	*Ailuropoda melanoleuca*—Procyonidae
Gibbons	*Hylobates* spp.—Hylobatidae
Giraffe	*Giraffa camelopardalis*—Giraffidae
Goat	*Capra hircus*—Bovidae

Gorilla	*Gorilla gorilla*—Pongidae
Gray whale	*Rhachianectes glaucus*—Rhachianectidae
Great Indian rhino	*Rhinoceros unicornis*—Rhinocerotidae
Grizzly bear	*Ursus arctos horribilis*—Ursidae
Ground sloths	Several families of the order Edentata
Ground squirrels	*Citellus* sp.—Sciuridae
Guinea pig	*Cavia cutleri*—Caviidae
Harbor seal	*Phoca vitulina*—Phocidae
Hartebeest	*Alcelaphus* sp.—Bovidae
Hippo	*Hippopotamus amphibius*—Hippopotamidae
Horse	*Equus caballus*—Equidae
Humpback whale	*Megaptera novaeangliae*—Balaenopteridae
Hyena	Hyaenidae (genera *Hyaena* and *Crocuta*)
Indian elephant	*Elephas maximus*—Elephantidae
Iriomote cat	*Prionailurus (Mayailucus) iriomotensis*—Felidae
Jaguar	*Panthera (Felis) onca*—Felidae
Javan rhino	*Rhinoceros sondaicus*—Rhinocerotidae
Jungle cat	*Felis chaus*—Felidae
Kangaroos	*Macropus* spp., *Megaleia rufa*—Macropodidae
Killer whale	*Grampus orca*—Delphinidae
Koala	*Phascolarctos cinereus*—Phalangeridae
Kudu	*Tragelaphus strepsiceros* and *T. imberbis*—Bovidae
Leopard	*Panthera (Felis) pardus*
Leopard cat	*Felis bengalensis*—Felidae
Leopard seal	*Hydrurga leptonyx*—Phocidae
Levant vole	*Microtus guentheri*—Cricetidae
Lion	*Panthera (Felis) leo*
Lynx	*Lynx canadensis*—Felidae
Malay bear (sun-bear)	*Helarctos malayanus*
Malay tapir	*Tapirus indicus*—Tapiridae
Mammoths	*Mammuthus* spp.—Elephantidae
Manatee	*Trichechus manatus*—Trichechidae
Marmot	*Marmota flaviventris*—Sciuridae
Marten	*Martes americana*—Mustelidae
Minke whale	*Balaenoptera acutorostrata*—Balaenopteridae
Mongoose	*Herpestes nyula*—Viverridae
Mongoose lemur	*Lemur mongoz*—Lemuridae
Moose	*Alces americana*—Cervidae
Mountain goat	*Oreamnos americanus*—Bovidae
Mountain gorilla	*Gorilla gorilla beringei*—Pongidae
Mouse (deer mouse)	*Peromyscus maniculatus*—Cricetidae
Mule deer	*Odocoileus hemionus*—Cervidae
Musk ox	*Ovibos moschatus*—Bovidae
Ocelot	*Felis pardalis*—Felidae
Onager (Asiatic wild ass)	*Equus hemionus*—Equidae
Orangutan	*Pongo pygmaeus*—Pongidae
Oryx	*Oryx* spp.—Bovidae

Otter (river otter)	*Lutra canadensis*—Mustelidae
Owl monkeys (night apes)	*Aotes* spp.—Cebidae
Panther (leopard)	*Panthera (Felis) pardus*—Felidae
People	*Homo sapiens*—Hominidae
Père David's deer	*Elaphurus davidianus*—Cervidae
Persian fallow deer	*Dama mesopotamica*—Cervidae
Pika	*Ochotona princeps*—Ochotonidae
Pileated gibbon	*Hylobates pileatus*—Hylobatidae
Plains bison	*Bison bison bison*—Bovidae
Platypus	*Ornithorhynchus anatinus*—Ornithorhynchidae
Prairie dog	*Cynomys ludovicianus*—Sciuridae
Pribilof Islands fur seal	*Callorhinus ursinus (C. alascanus)*—Otariidae
Pronghorn antelope	*Antilocapra americana*—Antilocapridae
Przewalski's horse	*Equus przewalskii*—Equidae
Rabbit	*Sylvilagus spp.* —Leporidae
Rat (black rat)	*Rattus rattus*—Muridae
Red kangaroo	*Megaleia rufa (Macropus rufus)*—Macropodidae (female sometimes called blue flier)
Reindeer	*Rangifer tarandus*—Cervidae
Rhesus monkey	*Macaca mulara*—Cercopithecidae
Right whales	*Eubalaena* spp.—Balaenidae
Saber-toothed cats	*Smilodon* spp., other genera—Felidae
Scimitar-horned oryx	*Oryx tao*—Bovidae
Seals (hair seals)	Phocidae
Sheep (domestic)	*Ovis aries*—Bovidae
Shrews	*Sorex* spp.—Soricidae
Siamang gibbon	*Symphalangus syndactylus*—Hylobatidae
Siberian tiger	*Panthera (Felis) tigris altaica*—Felidae
Snow leopard	*Panthera (Felis) uncia*—Felidae
Snowshoe hare	*Lepus americanus*—Leporidae
Southern right whale	*Eubalaena australis*—Balaendiae
Sperm whale	*Physeter catodon*—Physeteridae
Stoat	*Mustela* spp.—Mustelidae
Stumptail macaque	*Macaca speciosa*—Cercopithecidae
Sumatran rhino	*Didermoceros sumatrensis*—Rhinocerotidae
Sumatran tiger	*Panthera (Felis) tigris sumatrae*—Felidae
Thomson's gazelle	*Gazella thomsoni*—Bovidae
Thylacene wolf (Tasmanian wolf)	*Thylacinus cynocephalus*—Dasyuridae
Tiger	*Panthera (Felis) tigris*—Felidae
Timber wolf	*Canis lupus*—Canidae
Topi	*Damaliscus korrigum*—Bovidae
Uganda kob	*Kobus kob thomasi*—Bovidae
Vicuña	*Lama vicuyna*—Camelidae
Walrus	*Odobenus rosmarus*—Odobenidae
Warthog	*Phacochoerus aethiopicus*—Suidae
Weasel	*Mustela frenata*—Mustelidae
Weddell seal	*Leptonychotes weddelli*—Phocidae
White-crested gibbon	*Hylobates hoolock*—Hylobatidae
White-handed gibbon	*Hylobates lar*—Hylobatidae

White rhino (square-lipped rhino)   *Ceratotherium simum*—Rhinocerotidae
Wild ass   *Equus asinus*—Equidae
Wildcat (bobcat)   *Lynx rufus*—Felidae
Wildebeests (gnu)   *Connochaetes taurinus*—Bovidae
Wisent (European bison)   *Bison bonasus*—Bovidae
Wolf   *Canis lupus*—Canidae
Wombat   *Phascolomis mitchelli*—Phascolomidae
Wood bison   *Bison bison athabaskae*—Bovidae
Wooly rhinoceros   *Coelodonta antiquitatis*

BIRDS (CLASS AVES)

Adelie penguin   *Pygosceles adeliae*—Spheniscidae
Attwater's prairie chicken   *Tympanuchus cupido attwateri*—Tetraonidae
Bald eagle   *Haliaeetus leucocephalus*—Accipitridae
Birds of paradise   Paradisaeidae
Blackfooted (Jackass) penguin   *Spheniscus demersus*—Spheniscidae
Boobies   *Sula* spp.—Sulidae
Brown pelican   *Pelicanus occidentalis*—Pelicanidae
Buntings   *Emberiza* spp.—Emberizidae
Burrowing owl   *Speotyto cunicularia*—Strigidae
Bustard   In 1673 quote possibly ruffled grouse, *Bonasa umbellus*—Tetraonidae. True bustards (Otidae) are confined to the Old World.

California condor   *Gymnogyps californianus*—Cathartidae
California gull   *Larus californicus*—Laridae
Carolina parakeet   *Conuropsis carolinensis*—Psittacidae
Chinstrap penguin   *Pygoscelis antarctica*—Spheniscidae
Cock of the rock   *Rupicola rupicola*—Cotingidae
Crowned crane   *Balearica pavonina*—Gruidae
Cuckoo-doves   *Macropygia* spp.—Columbidae
Darwin's finches   Geospizidae
Dinornis (Moas)   *Dinornis*—Dinornithidae
Dodo   *Rhaphus cucullatus*—Rhaphidae
Ducks   Anatidae (subfamily Anatinae)
Dusky seaside sparrow   *Ammospiza nigrescens*—Emberizidae
Eagles   Accipitridae (various species)
Eastern bluebird   *Sialia sialis*—Turdidae
Gentoo penguin   *Pygoscelis papua*—Spheniscidae
Grackles   *Quiscalus quiscula*—Icteridae
Great auk (Garefowl)   *Alca impennis*—Alcidae
Hawaiian goose (Nene)   *Branta sandvicensis*—Anatidae
Hawaiian honeycreepers   Drepanididae
Hawk   Accipiteridae (various genera including *Accipiter, Buteo,* and *Circus*)

House sparrow (English sparrow)   *Passer domesticus*—Ploceidae
Kirtland's warbler   *Dendroica kirtlandii*—Parulidae
Larks   Alaudidae
Least bittern   *Ixobrychus exilis*—Ardeidae

Magellanic penguin	*Spheniscus magellanicus*—Spheniscidae
Marabou stork	*Leptoptilos crumeniferus*—Ciconiidae
Meadowlark	*Sternella neglecta*—Icteridae
Melidectes (honey-eaters)	*Melidectes* spp.—Meliphagidae
Merganser	*Mergus merganser*—Anatidae
Mountain bluebird	*Sialia currucoides*—Turdidae
Osprey	*Pandion haliaetus*—Accipitridae
Ostrich	*Struthio camelus*—Struthionidae
Parroquet (Carolina parakeet)	*Conuropsis carolinensis*—Psittacidae
Passenger pigeon	*Ectopistes migratorius*—Columbidae
Pelicans	*Pelecanus* spp.—Pelecanidae
Peregrine falcon	*Falco peregrinus*—Falconidae
Pine siskin	*Spinus pinus*—Fringillidae
Pipits	*Anthus* spp.—Motacillidae
Red-winged blackbird	*Agelaius phoeniceus*—Icteridae
Rhea	*Rhea americana*—Rheidae
Ring-billed gull	*Larus delawarensis*—Laridae
Sage grouse	*Centrocercus urophasianus*—Tetraonidae
Sandhill crane	*Grus canadensis*—Gruidae
Seagulls	*Larus* spp.—Laridae
Sharp-tailed grouse	*Pedioecetes phasianellus*—Tetraonidae
Skua	*Stercorarius skua*—Stercorariidae
Spotted sandpiper	*Actitis macularia*—Scolopacidae
Starling	*Sturnus vulgaris*—Sturnidae
Sulphur-crested (white) cockatoo	*Kakatoë galerita*—Psittacidae
Swans	*Cygnus* spp.—Anatidae
Thrushes	Turdidae
Tick bird (oxpecker)	*Buphagus africanus*—Sturnidae
Wagtails	*Motacilla* spp.—Motacillidae
Water ouzel (dipper)	*Cinclus mexicanus*—Cinclidae
White-crowned sparrow	*Zonotrichia leucophrys*—Emberizidae
Whooping crane	*Grus americana*—Gruidae
Yellow-bellied sapsucker	*Sphyrapicus varius*—Picidae

REPTILES (CLASS REPTILIA) AND AMPHIBIANS (CLASS AMPHIBIA)

African clawed toad	*Xenopus laevis*—Pipidae
Alligator	*Alligator mississipiensis*—Crocodylidae
Arizona ridge-nosed rattlesnake	*Crotalus willardi*—Crotalidae
Brontosaurus	*Apatosaurus (Brontosaurus)* spp.—Brachiosauridae
Ceratopsean dinosaurs	Reptilia, Ceratopidae
Chinese giant salamander	*Andrias davidianus*—Cryptobranchidae
Crocodile	*Crocodylus* spp.—Crocodylidae
Cuban crocodile	*Crocodylus rhombifer*—Crocodylidae
Frog	Amphibia, Order Salientia (especially Ranidae)
Galápagos tortoise	*Geochelone (Testudo) elephantopus*—Tesdudinidae

Geckos	Reptilia, Gekkonidae
Gila monster	*Heloderma suspectum*—Helodermatidae
Houston toad	*Bufo houstonensis*—Bufonidae
Land iguana	*Conolophus subcristatus*—Iguanidae
Malayan pit viper	*Trimeresurus wagleri*—Crotalidae
Map turtle	*Graptemys geographica*—Emydidae
Marine iguana	*Amblyrhynchus cristatus*—Iguanidae
Mediterranean spur-thighed tortoise	*Testudo graeca graeca*—Testudinidae
Pterosaur (gigantic Texas)	*Quetzalcoatlus northropi*—Pterodactylidae
Salamanders	Amphibia, Order Caudata
San Joaquin whipsnake	*Masticophis flagellum ruddocki*—Colubridae
Smooth snake	*Coronella austriaca*—Colubridae
Spotted salamander	*Ambystoma maculatum*—Ambystomidae
Toads	Amphibia, Order Salientia (in part—especially Bufonidae)
Tortoise	Testudinidae
Turtle	Reptilia, Order Chelonia (several families)
Tyrannosaurus rex	*Tyrannosaurus rex*—Tyrannosauridae
Water snake	*Nerodia (= Natrix) sipedon*—Colubridae

FISHES (CLASSES CHONDRICHTHYES AND OSTEICHTHYES)

Anchoveta	*Engraulis ringens*—Engraulidae
Anchovy	Engraulis mordax—Engraulidae
Atlantic salmon	*Salmo salar*—Salmonidae
Bass	*Micropterus salmoides* and *M. dolomieui*—Centrarchidae
Blueback trout	*Salvelinus alpinus*—Salmonidae
Brook trout	*Salvelinus fontinalis*—Salmonidae
Buffalo fish	*Ictiobus cyprinella*—Catostomidae
Butterfly fishes	*Chaetodon* spp. and related genera—Chaetodontidae
Carp	*Cyprinus carpio*—Cyprinidae
Caspian sturgeon	*Huso huso*—Acipenseridae
Catfishes	*Ictalurus* spp.—Ictaluridae
Cichlids	Osteichthyes, Cichlidae
Cleaner mimic	*Aspidontis taeniatus*—Blenniidae
Cleaner wrasses	*Labroides* spp.—Labridae
Cod	*Gadus morhua*—Gadidae
Drumfish	*Aplodinotus grunniens*—Sciaenidae
Gar	*Lepisosteus* spp.—Lepisosteidae
Goatfishes	Osteichthyes, Mullidae
Grunts	*Haemulon* spp.—Pomadasyidae
Guppy	*Poecilia (Lebistes) reticulata*—Poeciliidae
Herring	*Clupea harengus*—Clupeidae
Jacks	Osteichthyes, Carangidae
Lemon Butterfly Fish	*Chaetodon miliaris*—Chaetodontidae
Mackerel	*Scomber scombrus*—Scombridae

Menhaden	*Brevoortia tyrannus*—Clupeidae
Mosquito fish	*Gambusia affinis*—Poeciliidae
Mullet	*Mugil cephalus*—Mugilidae
Pacific sardine	*Sardinops caerulea*—Clupeidae
Parrot fishes	Osteichthyes, Scaridae
Pike	*Esox lucius*—Esocidae
Plaice	*Pleuronectes platessa*—Pleuronectidae
Saber-toothed Blennie	*Plagiotremus* spp.—Blenniidae
Salmon (Atlantic)	*Salmo salar*—Salmonidae
Sea trout	*Salmo trutta*—Salmonidae
Shad	*Alosa sapidissima*—Clupeidae
Sharks	Chondrichthyes, Order Selachii (especially Carcharhinidae)
Siamese fighting fish	*Betta splendens*—Anabantidae
Smelt	*Atherina presbyter*—Atherinidae
Snail darter	*Percina (Imostoma) tanasi*—Percidae
Stingrays	Condrichthyes, Dasyatidae
Tilapia	*Tilapia mossambica*—Cichlidae
Trout	*Salmo* spp.—Salmonidae

INSECTS (CLASS INSECTA)

Alfalfa weevil	*Hypera postica*—Curculionidae (Coleoptera)
Anopheles mosquitoes	*Anopheles* spp.—Culicidae (Diptera)
Ant lions	Myrmeleontidae (Neuroptera)
Army ants	Formicidae, subfamily Dorylinae (Hymenoptera)
Atala hairstreak	*Eumaeus atala*—Lycaenidae (Lepidoptera)
Australian scale insect (cottony cushion scale)	*Icerya purchasi*—Margarodidae (Hemiptera)
Baron's checkerspot butterfly	*Euphydryas editha baroni*—Nymphalidae (Lepidoptera)
Bay checkerspot butterfly	*Euphydryas editha bayensis*—Nymphalidae (Lepidoptera)
Birdwing butterflies	*Troides* and related genera—Papilionidae (Lepidoptera)
Blackflies	Simuliidae (Diptera)
Blowflies	Calliphoridae (Diptera)
Bollworm	*Heliothis virescens*—Noctuidae (Lepidoptera)
Bombardier beetle	*Brachynus ballistarius*—Carabidae
Brine flies	Ephydridae (Diptera)
Browntail moth	*Nygmia phaeorrhaea*—Lymantriidae (Leptidoptera)
Bush fly	*Musca vetustissima*—Muscidae (Diptera)
Cactus moth	*Cactoblastis cactorum*—Pyralidae (Lepidoptera)
Chrysomia rufifacies	Calliphoridae (Diptera)
Dance flies	Empididae (Diptera)

Dermatobia (human warble-fly)   *Dermatobia hominis*—Oestridae

Dung beetles   Scarabaeidae, subfamily Scarabaeinae (Coleoptera)

Edith's checkerspot butterfly   *Euphydryas editha*—Nymphalidae (Lepidoptera)

El Segundo blue   *Philotes (Shijimiaeoides) battoides allyni*—Lycaenidae (Lepidoptera)

Euptychia butterfly   *Euptychia* spp.—Nymphalidae (Lepidoptera)

Fig wasps   Agaonidae (Hymenoptera)

Fireflies   *Photuris versicolor* (females of this species feed on males of the genera *Photinus, Photuris, Pyractomena,* and *Robopus*)—Lampyridae (Coleoptera)

Fruit flies   *Drosophila* spp.—Drosophilidae

Gall wasps   Cynipidae (Hymenoptera)

Glaucopsyche butterfly   *Glaucopsyche lygdamus*—Lycaenidae (Lepidoptera)

Gypsy moth   *Porthetria dispar*—Lymantriidae (Lepidoptera)

Hawk louse   *Ornithomyia avicularia*—Hippoboscidae (Diptera)

Heliconius butterfly   *Heliconius ethilla*—Nymphalidae (Lepidoptera)

Honey bee   *Apis mellifera*—Apidae (Hymenoptera)

Hover flies   Syrphidae (Diptera)

Imported cabbage butterfly   *Pieris rapae*—Pieridae (Lepidoptera)

Karner Blue   *Plebejus melissa samuelis*—Lycaenidae (Lepidoptera)

Lac insect   *Laccifer lacca*—Lacciferidae (Hemiptera)

Ladybird beetles   Coccinellidae (Coleoptera)

Large blue   *Maculinea arion*—Lycaenidae (Lepidoptera)

Large copper butterfly   *Lycaena dispar*—Lycaenidae (Lepidoptera)

Leaf-cutter ants   *Atta* spp.—Formicidae (Hymenoptera)

Leaf roller   *Argyrothaenia sphaleropa*—Tortricidae (Lepidoptera)

Leaf worms   *Anomis* spp.—Noctuidae (Lepidoptera)

Luesther's checkerspot butterfly   *Euphydryas editha luestherae*—Nymphalidae (Lepidoptera)

Mazarine blue   *Cyaniris semiargus*—Lycaenidae (Lepidoptera)

Monarch butterfly   *Danaus plexippus*—Nymphalidae (Lepidoptera)

Morpho butterflies   *Morpho* spp.—Nymphalidae (Lepidoptera)

Mosquito   Culicidae (Diptera)

Naja butterflies   *Naja* spp.—Nymphalidae (Lepidoptera)

Native cabbage butterflies   *Pieris* species, especially *P. protodice* and *P. occidentalis*—Pieridae (Lepidoptera)

Oregon silverspot   *Speyeria zerene hippolyte*—Nymphalidae (Lepidoptera)

Papilio xuthus   Papilionidae (Lepidoptera)

Passion vine butterfly   *Heliconius ethilla*—Nymphalidae (Lepidoptera)

Polyphemus silkmoth   *Antheraea polyphemus*—Saturniidae (Lepidoptera)

Ptiliid beetle	Ptiliidae (Coleoptera)
Rat flea	*Xenopsylla cheopis*—Pulicidae (Siphonaptera)
Rhinoceros beetle	*Oryctes rhinoceros*—Scarabaeidae (Coleoptera)
Sandia butterfly	*Sandia macfarlandi*—Lycaenidae (Lepidoptera)
Scale insects	Superfamily Coccoidea (Hemiptera)
Soldier fly	Stratiomyidae (Diptera)
Solitary bees	Colletidae, Andrenidae, Melittidae, Megachilidae, Anthophoridae (Hymenoptera)
Sthenele brown butterfly	*Cercyonis sthenele*—Nymphalidae (Lepidoptera)
Sugar cane weevil	*Rhabdoscelus obscurus*—Curculionidae (Coleoptera)
Tarantula-hawk wasps	*Pepsis* spp.—Pompilidae (Hymenoptera)
Tiger beetles	Coccinellidae (Coleoptera)
Wasps	Superfamilies Sphecoidea, Vespoidea, Pompiloidea (Hymenoptera)
White grub	*Anomala orientalis*—Scarabaeidae (Coleoptera)
Xerces butterfly	*Glaucopsyche xerces*—Lycaenidae (Lepidoptera)

OTHER INVERTEBRATES (VARIOUS PHYLA)

Asiatic clam	*Corbicula manilensis*—Class Pelecypoda, Phylum Mollusca
Brine shrimp	*Artemia* spp.—Class Crustacea, Phylum Arthropoda
Clams	Class Pelecypoda, Phylum Mollusca
Coral	Class Anthozoa (part), Phylum Coelenterata
Crabs	*Cancer, Callinectes,* other genera—Class Crustacea, Phylum Arthropoda
Earthworms	*Lumbricus,* other genera—Class Oligochaeta, Phylum Annelida
Lobsters	*Homarus, Palinurus,* other genera—Class Crustacea, Phylum Arthropoda
Mites	Order Acarina, Class Arachnida, Phylum Arthropoda
Moss animals	Phylum Bryozoa (= Ectoprocta)
Mussels	*Mytilus,* other genera—Class Pelecypoda, Phylum Mollusca
Oysters	*Ostraea,* other genera—Class Pelecypoda, Phylum Mollusca
Proboscis worms	Phylum Nemertina
Sea anemones	Class Anthozoa (part), Phylum Coelenterata
Sea cucumbers	Class Holothuroidea, Phylum Echinodermata
Sea squirts	Class Ascidiacea, Phylum Chordata
Segmented worms	Phylum Annelida
Shrimp	*Peneus, Crago,* other genera—Class Crustacea, Phylum Arthropoda

Spider mites	Tetranychidae—Order Acarina, Class Arachnida, Phylum Arthropoda
Starfish	Class Asteroidea, Phylum Echinodermata
Ticks	Ixodidae, other families—Order Acarina, Class Arachnida, Phylum Arthropoda

PLANTS (KINGDOM PLANTAE)

Adam's mistletoe	*Trilepidea adamsii*—Loranthaceae
Alfalfa (lucerne)	*Medicago sativa*—Leguminosae (= Fabaceae)
Amaranthus	Amaranthaceae
American chestnut	*Castanea dentata*—Fagaceae
American elm	*Ulmus americana*—Ulmnaceae
Asters	*Aster, Erigeron,* other genera—Compositae (= Asteraceae)
Balsa	*Ochroma pyramidale*—Bombacaceae
Bamboos	*Bambusa, Phyllostachys,* other genera—Gramineae (= Poaceae)
Bananas	*Musa* spp.—Musaceae
Barley	*Hordeum vulgare*—Gramineae (= Poaceae)
Beans	*Phaseolus* spp.—Leguminosae (= Fabaceae)
Bluebells	*Campanula* spp.—Campanulaceae
Boxwood	*Buxus sempervirens*—Buxaceae
Breadfruit	*Artocarpus altilis*—Moraceae
Bristlecone pines	*Pinus longaeva*—Pinaceae
Bulbs	Liliaceae, Amaryllidaceae, Iridaceae
Cactus	Cactaceae
California black oak	*Quercus kelloggi*—Fagaceae
California interior live oak	*Quercus wislizenii*—Fagaceae
Cameron's *Euphorbia*	*Euphorbia cameronii*—Euphorbiaceae
Caoba tree	*Persea theobromifolia*—Lauraceae (a relative of the avocado, not to be confused with mahoganey, to which the common name "caoba" is also applied)
Cassava (Manihot)	*Manihot esculenta*—Euphorbiaceae
Cinchona trees	*Cinchona* spp.—Rubiaceae
Coconuts	*Cocos nucifera*—Palmae (= Arecaceae)
Columbine	*Aquilegia* spp.—Ranunculaceae
Crinum lily	*Crinum mauritianum*—Amaryllidaceae
Daisy (Swiss)	*Doronicum cataractum*—Compositae (= Asteraceae)
Dipterocarps	Dipterocarpaceae
Foxglove	*Digitalis purpurea*—Scrophulariaceae
Fritillaria	*Fritillaria liliacea*—Liliaceae
Furbish lousewort	*Pedicularis furbishiae*—Scrophulariaceae
Godley's buttercup	*Ranunculus godleyanus*—Ranunculaceae
Golden gladiolus	*Gladiolus aureus*—Iridaceae
Guayule	*Parthenium argentatum*—Compositae (= Asteraceae)

Heathers	Ericaceae
Hevea tree	*Hevea brasiliensis*—Euphorbiaceae
Hubbardia grass	*Hubbardia heptaneuron*—Gramineae (= Poaceae)
Indian paintbrushes	*Castilleja* spp.—Scrophulariaceae
Jasmine-flowered heather	*Erica jasminiflora*—Ericaceae
Jojoba	*Simmondsia chinensis*—Buxaceae
Kniphofia umbrina lily	Liliaceae
Lapérrine's olive	*Olea laperrinei*—Oleaceae
Larkspur	*Delphinium nelsoni*—Ranunculaceae
Leucaena	*Leucaena leucocephala*—Leguminosae (= Fabaceae)
Lichens	A fungus, usually of division Ascomycota living in symbiotic association with a green alga (Division Chlorophyta) or a blue-green alga (Cyanophyta)
Lupines	*Lupinus* spp.—Leguminosae (= Fabaceae)
Mahogonies	*Swietenia* spp.—Meliaceae
Maize	*Zea mays*—Gramineae (= Poaceae)
Milkweeds	Asclepiadaceae
Millet	*Panicum miliaceum*—Gramineae (= Poaceae)
Mints	Labiatae (= Lamiaceae)
Oaks	*Quercus* spp.—Fagaceae
Opuntia cactus	*Opuntia* spp.—Cactaceae
Orchid (Russia)	*Himantoglossum caprinum*—Orchidaceae
Orchids	Orchidaceae
Peanut (groundnut)	*Arachis hypogaea*—Leguminosae (= Fabaceae)
Peas (garden)	*Pisum sativum*—Leguminosae (= Fabaceae)
Penstemons	*Penstemon* spp.—Scrophulariaceae
Peony (Balearic Islands)	*Paeonia cambessedesii*—Paeoniaceae
Periwinkle (Madagascar)	*Catharanthus (Vinca) roseus*—Apocynaceae
Pines	*Pinus* spp.—Pinaceae
Plunkett Mallee	*Eucalyptus currisii*—Myrtaceae
Potato	*Solanum tuberosum*—Solanaceae
Proteas	*Protea* spp.—Proteaceae
Quinoa (Quinua)	*Chenopodium quinoa*—Chenopodiaceae
Rafflesia arnoldii	Rafflesiaceae
Raspberries	*Rubus strigosus, R. idaeus* and other species—Rosaceae
Rauwolfia	*Rauwolfia serpentina,* and other species—Apocynaceae
Redwood (coast redwood)	*Sequoia sempervirens*—Taxodiaceae
Rice	*Oryza sativa*—Gramineae (= Poaceae)
Robbins' cinquefoil	*Potentilla robbinsiana*—Rosaceae
Russian thistle (tumbleweed)	*Salsola pestifera*—Chenopodiaceae
Rye	*Secale cereale*—Gramineae
Sagebrushes	*Artemisia* spp.—Compositae (= Asteraceae)
Saguaro cactus	*Carnegiea gigantea*—Cactaceae
Saxifrages	*Saxifraga* spp.—Saxifragaceae

Scarlet gilia	*Ipomopsis aggregata*—Polemoniaceae
Selaginella	*Selaginella horizontalis*—Selaginellaceae
Sequoia (giant sequoia)	*Sequoiadendron giganteum*—Taxodiaceae
Sorghum	*Sorghum vulgare*—Gamineae (= Poaceae)
Soybean	*Glycene max*—Leguminoseae (= Fabaceae)
Stankevicz's pine	*Pinus stankeviczi*—Pinaceae
Sugar beet	*Beta vulgaris* (some varieties)—Chenopodiaceae
Sugar cane	*Saccharum officinarum*—Gramineae (= Poaceae)
Sunflowers	*Helianthus, Wyethia,* other genera—Compositae (= Asteraceae)
Sweet potato	*Ipomoea batatas*—Convolvulaceae
Vuleito palm	*Neoveitchia storckii*—Palmae (= Arecaceae)
Wheat	*Triticum aestivum*—Gramineae (= Poaceae)
Wild Thyme	*Thymus drucei*—Labiatae (= Lamiaceae)
Yams	*Dioscorea* spp.—Dioscoreaceae

Notes

CHAPTER I

1. Jane Goodall, "Life and death at Gombe," *National Geographic* 155:592–620, May 1979.

2. A basic source on endangered animal species are the volumes of the *Red Data Book,* published in looseleaf form and continually updated by the International Union for the Conservation of Nature and Natural Resources (IUCN): Volume 1, *Mammalia*; Volume 2, *Aves*; Volume 3, *Amphibia and Reptilia*; and Volume 4, *Pisces.* For plants, there is the bound 1978 volume, *The IUCN Plant Red Data Book,* compiled by Gren Lucas and Hugh Synge. All are published at Morges, Switzerland.

3. *Origin of Species,* 1st ed., p. 109. A good brief introduction to evolution for laypeople is Frank H. T. Rhodes, *Evolution,* Golden Press, New York, 1974. A great deal of information is presented simply and accurately.

4. Estimates are from P. R. Ehrlich, A. H. Ehrlich and J. P. Holdren, *Ecoscience: Population, Resources, Environment,* W. H. Freeman, San Francisco, 1977, p. 142. The estimates are of course *very* rough.

5. Crocodilians are representatives of archosaurian (ruling) reptiles, of which the dinosaurs comprised two different groups (Saurischia and Ornithischia) that may be no more closely related to one another than they are to the crocodilians. Pterosaurs, often popularly considered dinosaurs, are a third group of archosaurs. Some scientists think, however, that the two dinosaur groups are closely related and were actually warm-blooded, and that the birds, which descended from the Saurischia, should be included with the Saurischia and Ornithischia in the class Dinosauria (R. T. Bakker and P. M.

Galton, "Dinosaur monophyly and a new class of vertebrates," *Nature* 248:165–172, 1974).

6. J. Harte and R. H. Socolow, "The Everglades: Wilderness versus rampant land development in South Florida," in Harte and Socolow, *Patient Earth,* Holt, Rinehart and Winston, New York, 1971.

7. Of course, it is impossible to state exactly the consequences of an entire group not evolving. In the absence of birds, for example, bats might have evolved into day fliers and taken over the ecological roles of insectivorous birds. What *is* certain is that without birds evolutionary history would have been very different, and one could not even state with assurance that humanity would even have appeared on the scene.

8. For the results of work we did on the significance of the colors of these fishes, see P. R. Ehrlich, T. H. Talbot, B. C. Russell, and G. R. V. Anderson, "The behaviour of chaetodontid fishes with special reference to Lorenz's 'poster colouration' hypothesis," *Journal of Zoology, London* 183:213–228, 1977.

9. G. Anderson, A. Ehrlich, P. Ehrlich, J. Roughgarden, B. Russel and F. Talbot, "The community structure of coral reef fishes," in press, *American Naturalist.*

10. J. E. Lovelock, *Gaia: A New Look at Life on Earth,* Oxford University Press, New York, 1978, pp. 97–98.

CHAPTER II

1. For a conventional view of species definition, see the standard text, *Populations, Species and Evolution: An Abridgement of Animal Species and Evolution,* by an outstanding evolutionist, Ernst Mayr (Harvard University Press, Cambridge, 1970). A more heterodox view can be found in P. R. Ehrlich, "Has the biological species concept outlived its usefulness?" *Systematic Zoology* 10:167–176, 1961; and P. R. Ehrlich and R. H. Raven, "Differentiation of populations," *Science* 165:1228–1232, 1969. Note that the differences in viewpoint expressed in these sources are about details of the evolutionary process and the classification of evolutionary products. Such differences are now frequently misrepresented in anti-evolutionist tracts as indications of the weakness of evolutionary theory. They are no such thing. The explanations of evolution in this and later chapters are necessarily simplified.

2. When a species is formally described and named by a taxonomist in a publication, the name has a required form and is latinized. For instance, the dog was named *Canis familiaris* in 1758 by the great Swedish taxonomist Linnaeus. Linnaeus was the founder of the *binomial* (two name) system of naming species. Each species name has two parts: a generic name *(Canis)* and a trivial name *(familiaris).* Our species name is *Homo sapiens*—we belong to the genus *Homo.*

Each genus belongs to a family, the name of which in animals is formed by adding -idae to the root of the name of one genus in the family. The dog belongs to the family Canidae (dogs, wolves, foxes, etc.). In turn the family Canidae belongs to the order Carnivora, which also includes the cat family, Felidae, the bear family, Ursidae, and others. The order Carnivora belongs to the class Mammalia, along with the order Primates *(our* order), the order Rodentia (rodents), and fifteen others. The order Mammalia belongs to the phylum Chordata (vertebrates and their relatives), which in turn is one of the many phyla of the Kingdom Animalia (animals). All organisms can be similarly placed in the hierarchical scientific classification scheme.

3. P. R. Ehrlich and H. K. Clench, "A new subgenus and species of *Callophrys* (s.l.)

from the southwestern United States (Lepidoptera: Lycaenidae)," *Entomological News* 71:137–141, 1960.

4. This description of human descent is oversimplified, but few biologists would contest the existence of a direct line from an australopithecine through *H. erectus* to *H. sapiens.*

5. Identical twins technically can have slight genetic differences if mutations occur after the fertilized egg divides to form two individuals.

6. J. H. Camin and P. R. Ehrlich, "Natural selection in water snakes *(Natrix sipedon)* on islands in Lake Erie," *Evolution* 12:504–511, 1958.

7. This is a very oversimplified description of geographic, or *allopatric,* speciation, the mechanism that probably is the most common in nature. For a textbook overview of speciation, see Ehrlich, Holm, and Parnell, *The Process of Evolution,* 2nd ed., McGraw-Hill, New York, 1974. For the standard scholarly treatment, see Ernst Mayr, *Population, Species and Evaluation,* op. cit. Some insight into the complexities of speciation theory can be found in G. L. Bush, "Modes of animal speciation," *Annual Review of Ecology and Systematics* 6:338–364, 1975.

8. *Origin,* p. 340. A facsimile paperback edition of the first edition of *Origin,* with a fine introduction by Ernst Mayr, was published by Harvard University Press in 1964.

9. Everyman's Library Edition, Dutton, New York, 1959, p. 365.

10. *Origin,* p. 400. Our explanation of speciation in the Galápagos is necessarily simplified.

11. Norman Myers, *The Sinking Ark,* Pergamon Press, New York, 1979; Council on Environmental Quality, *Global 2000: Entering the 21st Century,* Government Printing Office, Washington, D.C., 1980.

12. The half-billion estimate is from G. G. Simpson, "How many species?" *Evolution* 6:432, 1952. This, of course, is just a ballpark guess, even wilder than the guess that about ten million species are alive today. What scientists do agree on is that those alive today are a very small percentage of those that have ever lived.

13. L. W. Alvarez et al., "Extraterrestrial cause for Cretaceous-Tertiary extinction," *Science* 208:1095–1108, 1980. For a popular account of the asteroid hypothesis, see Stephan Jay Gould, "The belt of an asteroid," *Natural History,* June 1980, pp. 26–33. Gould is one of the best writers among scientists, and his monthly column in *Natural History* is always worth reading. There may be some validity to the hypothesis—that is, a reduction in photosynthesis could have been a factor in the extinctions. But the notion that photosynthesis was completely turned off for a decade is absurd. Most, if not all, groups of insects, for example, would have gone extinct.

14. The *Glaucopsyche* work is described in P. R. Ehrlich, D. E. Breedlove, P. F. Brussard, and M. A. Sharpe, "Weather and the 'regulation' of subalpine populations," *Ecology* 53:243–247, 1972.

15. See, for example, J. Terborgh, "Preservation of natural diversity: The problem of extinction prone species," *Bio Science* 24:715–722, 1974. In the technical literature, the slow reproducers are now usually referred to as K-selected and the fast reproducers as r-selected.

16. Information on rhinos is from G. E. Hutchinson and S. D. Ripley, "Gene dispersal and the ethology of the Rhinocerotidae," *Evolution* 8:178–179, 1954; and J. Fisher, N. Simon, and J. Vincent, *Wildlife in Danger,* Viking, New York, 1969.

17. Information on the uses of rhino horn is from wildlife ecologist Lee Talbot, reported by Janet Raloff in "Stealing a horn of plenty," *Science News* 116:346–348, 1979.

18. Rhino horn brought as much as $5,000 per kilo in Hong Kong, slightly over $150 per troy ounce (*Focus,* World Wildlife Fund Special Report, Summer 1979).

CHAPTER III

1. Details on the death of Digit are from "His name was Digit," by Dian Fossey, published by the International Primate Protection League, P.O. Drawer X, Summerville, SC 29483, undated.

2. Information on methods of whale killing from Sir Sydney Frost, chairman, *Whales and Whaling: Report of the Independent Inquiry,* Australian Government Publishing Service, Canberra, 1978.

3. John Larkin, the Melbourne *Age,* June 18, 1977, quoted in ibid.

4. Discussion of the Special Commission on Internal Pollution, London, October 1975.

5. The nation of Papua New Guinea is the first in the world to establish insect conservation as a national objective in its constitution. That country, through a system of reserves and programs of commercial breeding, is not only protecting the gorgeous birdwings but turning them into an important economic resource by rearing them for sale to butterfly collectors around the world (R. M. Pyle, "Butterflies: Now you see them," *Defenders*, January–February 1981, pp. 4–10).

6. For more on nonfading colors and a broad overview of insects, see H. V. Daly, J. T. Doyen, and P. R. Ehrlich, *Introduction to Insect Biology and Diversity,* McGraw-Hill, New York, 1978.

7. The original research on this topic was by D. Aneshansley, T. Eisner, J. Widom and B. Widom, "Biochemistry at 100° C: The explosive discharge of bombardier beetles (*Brachinus*)," *Science* 165:161–163, 1969. More recent results were reported by Eisner in a seminar to the Department of Biological Sciences, Stanford University, February 1980.

8. J. C. Lloyd, "Aggressive mimicry in *Photuris* fireflies: Signal repertoires by femmes fatales," *Science* 187:452–453, 1975.

9. The amazing sequence was worked out by E. L. Kessel in a classic paper, "The mating activities of balloon flies," *Systematic Zoology* 4:97–104, 1955. The first form of the behavior discovered (in 1875) was the last evolutionary stage. It made evolutionary sense only when the intermediate stages were uncovered. The whole behavioral sequence is put in perspective in E. O. Wilson's landmark *Sociobiology: The New Synthesis,* Harvard University Press, Cambridge, 1975.

10. See E. O. Wilson, *The Insect Societies,* Harvard University Press, Cambridge, 1971; and C. D. Michener, *The Social Behavior of Bugs: A Comparative Study,* Harvard University Press, Cambridge, 1974, for an overview.

11. Unpublished hypothesis of R. W. Holm, Stanford University.

12. *Moby Dick,* Bobbs-Merrill Educational Publishing, Indianapolis, 1964 (first published 1851), p. 189.

13. W. E. Schevill, "Underwater sounds of cetaceans," in W. N. Tavolga, ed., *Marine Bio-Acoustics,* Pergamon Press, New York, 1964; R. S. Payne and S. McVay, "Songs of Humpback Whales," *Science* 173:585–597, 1971.

14. The songs of the whales are available on a superb stereo disc by CRM Records (album SWR–11). A good hi-fi is required to get the full benefit of the whales' virtuosity; the highest and lowest notes may be lost if the speakers are inadequate. The album is accompanied by a very informative booklet.

15. Much of the information on blue whales is based on George L. Small, *The Blue Whale*, Columbia University Press, New York, 1971.

16. Reported in F. T. Bachmura, "The economics of vanishing species," *Natural Resources Journal* 11:687, 1971.

17. H. H. Iltis, P. Andrews, and O. Loucks, "Criteria for an optimum human environment," *Bulletin of Atomic Scientists* 26 (1):2–6, 1970.

18. Oxford University Press, New York, 1978. It is too bad that this controversial volume is not required reading for every person in industrial societies.

19. Bernard Dixon, "Smallpox—Imminent extinction and an unresolved dilemma," *New Scientist* 69:430–32, 1976.

20. Lynn White, Jr., "The historical roots of our ecologic crisis," *Science* 155:-1203–1207, 1967.

21. The discussion of nonhomocentrism in the Judeo-Christian tradition is largely based on environmentalist Roderick Nash's excellent article "Do rocks have rights?" *The Center Magazine*, November–December 1977. See also his *Wilderness and the American Mind*, Yale University Press, New Haven, 1967.

22. *Sand County Almanac*, Oxford University Press, New York, 1949 (reprinted 1970), p. 204.

23. C. D. Stone, *Should Trees Have Standing? Toward Legal Rights for Natural Objects*, Kaufmann, Los Altos, California, 1974.

24. "Do rocks have rights?" op. cit.

CHAPTER IV

1. P. R. Ehrlich and P. H. Raven, "Butterflies and plants: a study of coevolution," *Evolution* 18:586–608, 1964.

2. Information on origins of vincristine is from G. E. Trease and W. C. Evans, *Pharmacognosy*, 10th ed., Williams and Wilkins, Baltimore, 1972. The figure on the value of sales is from Norman Myers, "What is a species worth?" manuscript for *Science Digest*, 1980.

3. S. M. Kupscher, I. Uchida, A. R. Bronfman, R. G. Dailey, Jr., and B. Yu Fei, "Antileukemia principles isolated from Euphorbiaceae plants," *Science* 191:571, 1976.

4. W. H. Lewis and M. P. F. Elvin-Lewis, *Medical Botany: Affecting Man's Health*, Wiley, New York, 1977; an excellent comprehensive source. See also R. S. Solecki, "Shanider IV, a Neanderthal flower burial in northern Iraq," *Science* 190:880–881, 1975.

5. For example, Lewis and Elvin-Lewis, *Medical Botany*, op. cit. This is the source of some of the material in this section.

6. *The Sinking Ark*, Pergamon Press, New York, 1979, p. 70.

7. Ibid., p. 72.

8. For examples of the problems drug companies have experienced, see N. Farnsworth and R. Morris, "Higher plants—the sleeping giant of drug development," *American Journal of Pharmacology* 148:46–52, March–April 1976.

9. G. D. Ruggieri, "Drugs from the sea," *Science* 194:491–497, 1976.

10. See the symposium on marine biomedicals in the December 1969 issue of *Lloydia*.

11. Ruggieri, "Drugs from the sea," op. cit.; cytarabine is also known as cytosine arabinoside.

12. S. K. Sikes, "Observations on the ecology of arterial disease in the African elephant *(Loxodonta africana)* in Kenya and Uganda," *Symposium of the Zoological Society of London*, no. 21, pp. 251–273, 1968.

13. S. S. Cohen, "Comparative biochemistry and drug design for infectious disease," *Science* 205:964–971, 1979.

14. There is also hepatitis A (infectious hepatitis), caused by a different virus from the one that causes hepatitis B, called serum hepatitis.

15. "New vaccine may bring man and chimpanzee into tragic conflict," *Science* 200:1027–1030, 1978. Our summary of the vaccine controversy is based on this article.

16. Information on virus susceptibility from Dr. David C. Regnery, Department of Biological Sciences, Stanford University, personal communication.

17. B. Bauerle et al., "The use of snakes as a pollution indicator species," *Copeia,* no. 2, pp. 366–368, 1975.

18. National Academy of Sciences, *Underexploited Tropical Plants of Promising Economic Value,* National Academy of Sciences, Washington, D.C., 1975.

19. P. R. Ehrlich, A. H. Ehrlich, and J. P. Holdren, *Ecoscience: Population, Resources, Environment,* W. H. Freeman, San Francisco, 1977, p. 286.

20. National Academy of Sciences, *Underexploited Tropical Plants . . .,* op. cit., p. 1.

21. UN Food and Agriculture Organization, *State of Food and Agriculture, 1978,* Food and Agriculture Organization, Rome, 1979.

22. National Academy of Sciences, *Underexploited Tropical Plants . . .,* op. cit.

23. Eel grasses are not true grasses but members of the family Zosteraceae, which is sometimes included in the family Potamogetonaceae. For their food value, see R. Felger and M. Maser, "Eelgrass (*Zostera marina* L.) in the Gulf of California: Discovery of its nutritional value by the Seri Indians," *Science* 181:355–356, 1973.

24. The material on plant potential is largely from the National Academy of Sciences, *Underexploited Tropical Plants,* op. cit.

25. Ehrlich and Raven, "Butterflies and plants," op. cit.

26. *Ecoscience,* op. cit., p. 345.

27. National Academy of Sciences, *Genetic Vulnerability of Major Crops,* National Academy of Sciences, Washington, D.C., 1972.

28. The complexities of the Green Revolution are much simplified here; for details, see *Ecoscience,* op. cit., pp. 329ff.

29. *Newsweek,* "What comes naturally," September 1, 1975.

30. Bunt figures are from Norman Myers, *The Sinking Ark,* op. cit., p. 68; the story on perennial maize is from R. W. Holm, Department of Biological Sciences, Stanford University, personal communication.

31. U.S. Department of Commerce, Bureau of the Census, *Statistical Abstract of the United States, 1978.*

32. Sources: UN Food and Agriculture Organization, *State of Food and Agriculture, 1977;* and the Council on Environmental Quality, *Global 2000: Entering the Twenty-first Century,* U.S. Government Printing Office, Washington, D.C., 1980.

33. G. Fryer and T. D. Iles, *The Cichlid Fishes of the Great Lakes of Africa: Their Biology and Evolution,* T.F.H. Publications, Hong Kong, 1972.

34. M. H. Crawford, "The case for new domestic animals," *Oryx* 12:351–360, 1974.

35. Mitchell Prize paper, 1979, mimeo.

36. Crawford, "The case for new domestic animals," op. cit.

37. Paul DeBach, *Biological Control by Natural Enemies,* Cambridge University Press, London, 1974, p. 118.

38. F. J. Simmonds et al., "History of biological control," in C. B. Huffaker and

P. S. Messenger, *Theory and Practice of Biological Control,* Academic Press, New York, 1976.

39. Council on Environmental Quality, op. cit.; Erik Eckholm, "Planting for the future: Forestry for human needs," *Worldwatch Paper 26,* Worldwatch Institute, Washington, D.C., February 1979.

40. R. H. Raven, "Ethics and attitudes," in J. B. Simmons et al., *Conservation of Threatened Plants,* Plenum, New York, 1976, p. 174.

41. Thomas K. Maugh II, "Guayule and Jojoba: Agriculture in semiarid regions," *Science* 196:1189–1190, 1977, and National Academy of Sciences, "Underexploited tropical plants . . .," op. cit.

42. Ibid.

43. National Academy of Sciences, *Leucaena: Promising Forage and Tree Crops for the Tropics,* National Academy of Sciences, 1977.

44. J. D. Johnson and C. W. Hinman, "Oils and rubber from arid land plants," *Science* 208:460–463, 1980.

CHAPTER V

1. Many of the influences are so small as to be lost in the "noise" of the system.

2. The story of DDT and the cats was told by Gordon Harrison, *Natural History,* December 1968.

3. Since all natural and technological processes conserve energy—the first law of thermodynamics says that the total amount of energy remains constant even as its distribution among different energy forms changes—one might think that energy could be recycled. But the attribute of energy that makes it useful is its capacity to do work, and this attribute is not conserved. The second law of thermodynamics says that all energy transformations lead to a nonrestorable reduction in the energy's capacity to do work. The amount of this reduction in capacity is equal to or greater than the work actually done, and in real-world processes the "greater than" always prevails. The useful attribute of energy—its ability to do work—is not recyclable but can only be used once. It is for this reason that all systems, natural and technological, require for their continued operation a more or less continual inflow of "fresh" energy.

The fraction of a given quantity of energy that represents its theoretical capacity to do work often is called the *availability* of the energy. This fraction depends only on the properties of the energy itself—that is, its form (chemical, electrical, thermal, etc.) and sometimes other characteristics such as (in the case of thermal energy) the temperature difference between the energy and its surroundings. What fraction of the theoretical potential actually gets harnessed by a given energy-using process, however, depends on the details of that process. Thus, although the chemical energy in gasoline has an availability of essentially 100 percent—in theory almost all of it could be made to perform useful work—today's internal-combustion gasoline engines manage to extract as useful work only about 30 percent of this potential. (Most of the availability is lost in this case in the first step of the process, the combustion of the gasoline to form the hot gases that run the engine. The availability in these combustion products is only about half that in the gasoline.)

The difference between the quantity of energy processed and the quantity delivered as work, then, has two distinct components: the part of the energy that was unavailable to start with, and the part that is *made* unavailable by the shortcomings of the specific process used. Both kinds of losses tend to show up in the same form: the emission of waste heat (thermal energy) whose availability is small because its temperature is barely above that of the surroundings. It is important to note that even the energy delivered as work also ends up sooner or later in the environment as waste heat as the result of additional transformations that tend inevitably (according to the second law) in that direction.

4. Herbivores can outweigh the plants they feed on if the plants grow very fast so there is a rapid turnover at the producer level—just as you could exist in a system with only five pounds of food as long as that food was replaced daily. In some marine systems, the plants are single-celled algae that reproduce very rapidly, and the mass of plants is outweighed by the animals that graze on them—which do not reproduce as fast. This situation is termed an *inverted biomass pyramid.* The flow of energy, however, is always much greater through the producer trophic level than through the herbivores; energy-flow pyramids are never inverted. For more details, see E. P. Odum, *Fundamentals of Ecology,* 3rd ed., W. B. Saunders, Philadelphia, 1971.

5. Letter to the London *Times,* February 4, 1971.

6. See G. M. Woodwell, "Toxic substances and ecological cycles," *Scientific American,* March 1967.

7. P. R. Ehrlich and Peter Raven, "Butterflies and plants: a study in coevolution," *Evolution* 18:586–608, 1964.

8. The term "greenhouse effect" is actually a misnomer, since most of the warming effect in a greenhouse does not come from a process similar to that which occurs in the atmosphere—see P. R. Ehrlich, A. H. Ehrlich, and J. P. Holdren, *Ecoscience: Population, Resources, Environment,* W. H. Freeman, San Francisco, 1977, Chapter 2.

9. R. A. Bryson and W. M. Wendland, "Climatic effects of atmospheric pollution," in S. F. Singer, *Global Effects of Environmental Pollution,* Springer-Verlag, New York, 1970, p. 130.

10. W. E. Omerod, "Ecological effect of control of African trypanosomiasis," *Science* 191:815–821, 1976. See also C. Sagan, O. B. Toom, and J. B. Pollack, "Anthropogenic albedo changes and the Earth's climate," *Science* 206:1263–1368, 1979.

11. George Woodwell, "The carbon dioxide question," *Scientific American* 238(1):-34–43, January 1978.

12. "Observations for the California Energy Futures Conference," Sacramento, May 20, 1978. For a superb overview on climate and agriculture, see S. H. Schneider and L. E. Mesirow, *The Genesis Strategy: Climate and Global Survival,* Plenum, New York, 1976.

13. F. H. Bormann, "An inseparable linkage: Conservation of natural ecosystems and the conservation of fossil energy," *BioScience* 26:754–760, 1976.

14. Norman Myers, "Development rather than destruction for tropical moist forests," manuscript dated February 14, 1980, to be submitted to *New Scientist.*

15. A. H. Gentry and J. Lopez-Parodi, "Deforestation and increased flooding of the upper Amazon," *Science* 210:1354–1356, 1980.

16. For more details on soils, see *Ecoscience,* op. cit., pp. 252ff.

17. For example, the requirements of and relationships among the blowflies that devour sheep carcasses in Australia are astonishingly complex. Some species do best in freshly killed sheep; others are specialists in well-decomposed bodies. One of the latecomers, *Chrysomia rufifacies,* can prevent one of the early species from successfully maturing by competing with it for food, by driving it from the carcass, and by killing it directly. The action in what might be described as the "dead-sheep" ecosystem has

been described in detail by Australian biologists whose dedication to science overcame the messages of their olfactory organs. For a summary of the blowfly work, see the classic text by H. G. Andrewartha and L. C. Birch, *The Distribution and Abundance of Animals,* University of Chicago Press, Chicago, 1954, pp. 449ff.

18. *Ecoscience,* op. cit.

19. U.S. Department of Agriculture, *Agricultural Statistics,* U.S. Government Printing Office, 1977.

20. William Ramirez B., "Host specificity of fly wasps (Agaonidae)," *Evolution* 24:680–691, 1970.

21. C. C. Delwiche, "The nitrogen cycle," *Scientific American* 223(3):137–158, 1970.

22. For a theoretical discussion of keystone species and other topics in population biology, see Jonathan Roughgarden's excellent *Theory of Population Genetics and Evolutionary Ecology: An Introduction,* Macmillan, New York, 1979.

23. R. T. Paine, "Food web complexity and species diversity," *American Naturalist* 100:65–75, 1966.

24. S. J. McNaughton, "Diversity and stability of ecological communities: A comment on the role of empiricism in ecology," *American Naturalist* 111:515–525, 1977.

25. F. H. Bormann, op. cit., p. 759.

26. S. J. McNaughton, "Serengeti migratory wildebeest: Facilitation of energy flow by grazing," *Science* 191:92–94, 1976.

27. David Hopcraft, "Nature's technology," Mitchell Prize Paper, 1979 (mimeo).

28. D. E. Breedlove and P. R. Ehrlich, "Coevolution: Patterns of legume predation by a lycaenid butterfly," *Oecologia* 10:99–104, 1972.

29. For the details of this study and an explanation of how variability in the plant populations is maintained, see P. M. Dolinger, P. R. Ehrlich, W. L. Fitch, and D. E. Breedlove, "Alkaloid and predation patterns in Colorado lupine populations," *Oecologia* 13:191–204, 1973.

30. G. F. Edmunds, Jr., and D. N. Alstad, "Coevolution in insect herbivores and conifers," *Science* 199:941–945, 1978; K. B. Sturgeon, "Monoterpene variation in ponderosa pine xylem resin related to western pine beetle predation," *Evolution* 33:803–814, 1979.

31. J. Artie Browning, "Relevance of knowledge about natural ecosystems to development of pest management programs for agro-ecosystems," *Proceedings of the American Phytopathological Society* 1:191–199, 1975; Graham Harvey, "The Cambridge strategy," *New Scientist,* February 16, 1978, pp. 428–429.

CHAPTER VI

1. One of our pictures of the attack is reproduced in P. R. Ehrlich, A. H. Ehrlich, and J. P. Holdren, *Ecoscience: Population, Resources, Environment,* W. H. Freeman, San Francisco, 1977, p. 168.

2. The Murphy quotes are from *Logbook for Grace,* Time-Life Books, New York, 1965, p. 188.

3. Economic men view the future through dollar-colored glasses. Suppose for example, one were able to state categorically that each and every Snail Darter would be worth a thousand dollars a century from now. Would this lead to a stampede of entrepreneurs to protect the Snail Darter or even to domesticate and breed it? The answer unfortunately is no. Entrepreneurs and economists ask themselves the question "What is the value today of a thousand dollars a hundred years from now?" That

question can be rephrased as "How much money would I have to invest today to have a thousand dollars in a hundred years?" At an annual interest rate of 10 percent, the answer is about seven cents—seven cents put in a long-term deposit at 10 percent interest will be worth a cool grand in a hundred years. Ten percent is a standard rate used currently for calculations of present value. For 1980 it is a conservative rate. The formula for calculating present value (PV) is $\frac{FV}{(1+i)^t}$ where FV is the future value, i the interest rate, and t the time in years (in such calculations interest is often compounded annually, as it is here). Thus the *present value* of a Snail Darter would be negligible in spite of its great value to our grandchildren. Its future value is "discounted."

All this is not, of course, just economic double-talk. Everyone discounts the future. As time stretches out, uncertainty increases, opportunities are lost forever, life ends. Some of us might be willing to make a substantial investment with a prospective payoff in fifty years, which might benefit children or grandchildren we know personally, if not ourselves, but we would be much less willing to make an investment that would benefit more remote descendants after a century or more.

Since we live in an era dominated by economics, the value that society places on resources, including species, tends to be the present economic value. As you can see, if the discount rate remains the same, the economic value of a resource declines as the time when it will be used gets further away. Everything else being equal, there is more value to you in your child's being able to see a California Condor, or to know that a wild Snow Leopard exists somewhere, than in your great-grandchild's.

It is the economic concept of present value that makes it "sensible" purposely to drive the great whales to extinction. (See Colin W. Clark, "The economics of overexploitation," *Science* 181:630–634, 1973.) Overexploitation to the point of extinction can be "justified" economically from the standpoint of an exploiting firm or nation, especially in the case of a "common property" resource—one such as the whales with no defined owner or owners.

4. *Science* 162:1243–1248, 1968. For a more technical and optimistic view, see S. Ciriacy-Wantrup and R. Bishop, " 'Common property' as a concept in natural resource policy," *Natural Resources Journal* 15:713–727, 1975.

5. Most of the material on the herring is from *Ecoscience,* op. cit., pp. 363–364.

6. UN Food and Agriculture Organization, *State of Food and Agriculture, 1978,* Food and Agriculture Organization, Rome, 1979, pp. 1–28.

7. Colin W. Clark, *Mathematical Bioeconomics: The Optional Management of Renewable Resources,* Wiley, New York, 1976.

8. G. I. Murphy, "Population biology of the Pacific sardine *(Sardinops caerulea),* " *Proceedings of the California Academy of Sciences* 34:1–84, 1966.

9. J. H. Ryther, "Photosynthesis and fish production in the sea," *Science* 166:72–76, 1969.

10. J. L. McHugh, "Management of estuarine fisheries," in *A Symposium on Estuarine Fisheries,* American Fisheries Society, Washington, D.C., 1966.

11. *The Origin of Species,* p. 72.

12. *The Geographical Distribution of Animals,* vol. 1, Macmillan, London, 1876, p. 150.

13. *The World of Life,* Moffat, Yard and Company, New York, 1911.

14. See the fascinating volume edited by Paul S. Martin and H. E. Wright, Jr., *Pleistocene Extinction: The Search for a Cause,* Yale University Press, New Haven, 1967; and P. S. Martin, "The discovery of America," *Science* 179:968–974, 1973. Whether hunting or climate was primarily responsible for the megafaunal extinctions remains controversial. For a summary of the arguments on both sides, see evolutionist

Leigh Van Valen's excellent "Late Pleistocene extinctions," *Proceedings of the North American Paleontological Convention:* 469–495, 1969.

15. The Předmost story is based largely on material in Geoffrey Bibby, *The Testimony of the Spade,* New York, Knopf, 1956.

16. N. K. Vereschchagin, "Primitive hunters and Pleistocene extinction in the Soviet Union," in Martin and Wright, *Pleistocene Extinction,* op. cit., pp. 388–392.

17. For example, John E. Guilday, "Differential extinction during late-Pleistocene and recent time," in Martin and Wright, *Pleistocene Extinction,* op. cit., pp. 121–140, but see Van Valen, op. cit.

18. P. D. Gingerich, "Patterns of evolution in the mammalian fossil record," in A. Hallam, ed., *Patterns of Evolution as Illustrated by the Fossil Record,* Elsevier, Amsterdam, 1977, pp. 476–478.

19. J. J. Hester, "The agency of man in animal extinctions," in Martin and Wright, *Pleistocene Extinction,* op. cit., pp. 178–179.

20. Conversations of P.R.E. with Tommy Bruce, Southampton Island, Northwest Territories, 1952; and Asen Balikci, *The Netsilik Eskimo,* Natural History Press, Garden City, N.Y., 1970.

21. Material on Aivilingmiut from P.R.E., unpublished.

22. R. B. Lee and I. DeVore, "Problems in the study of hunters and gatherers," in Lee and Devore, *Man the Hunter,* Aldine, Chicago, 1968, p. 3.

23. Ibid.

24. W. Craig, "The expression of emotion in the pigeons. III. The Passenger Pigeon (*Ectopistes migratorius* Linn)," *Auk* 28:408, 1911.

25. I. L. Brisbin, "The Passenger Pigeon: A study in extinction," *Modern Game Breeding* 4:3–20, 1968.

26. The account of the Passenger Pigeon is based largely on Brisbin, op. cit., and Tim Halliday, *Vanishing Birds: Their Natural History and Conservation,* Holt, Rinehart, and Winston, New York, 1978. Halliday's book is beautifully written and illustrated and well documented.

27. T. Halliday, "The Extinction of the Passenger Pigeon *Ectopistes migratorius* and its relevance to contemporary conservation," *Biological Conservation* 17:157–162, 1980.

28. F. G. Roe, *The North American Buffalo: A Critical Study of the Species in the Wild State,* University of Toronto Press, Toronto, 1951.

29. *San Francisco Examiner and Chronicle,* October 5, 1975.

30. "I witnessed a massacre," *International Wildlife,* January–February 1980, p. 29.

31. Michael Weisskopf, "Iran's Wild Casualties," *Defenders,* April 1980.

32. "Creatures," *Audubon,* May 1980. Based on the book by an anonymous Russian bureaucrat (writing under the pseudonym Boris Komarov), *The Destruction of Nature in the Soviet Union* (N. F. Sharpe, White Plains, N.Y.).

33. Cited in Marshall, ed., *The Great Extermination: A Guide to Anglo-Australian Cupidity, Wickedness and Waste,* Heinemann, London, 1966, p. 19.

34. A. A. Burbidge, *The Status of Kangaroos and Wallabies in Australia,* Australian Government Publishing Service, Canberra, 1977; "New count method could determine kangaroos' future," *The Bulletin,* March 25, 1980; "Will U.S. encourage kangaroo slaughter?" *The Australian,* May 12, 1980. The ban was lifted in late 1980.

35. D. Cousins, "Man's exploitation of the Gorilla," *Biological Conservation* 13:-287–296, 1978.

36. Gibbon statistics from Population Reference Bureau, *World Population Data Sheet, 1979.*

37. Letter to us, February 6, 1980.

38. Letter to Russell Train, January 21, 1980.

39. *IUCN Bulletin,* September 1978, p. 52.

40. 1970 statistics are from *Biological Conservation,* vol. 4, no. 1, October 1971. The estimate for 1979 was obtained by extrapolation from the figure of over 300 million total wildlife imports in that year; *Defenders,* February 1980. The figures in the following paragraphs are from the same sources.

41. A. S. Johnson, "The snaker's game," *Defenders,* February 1980; *IUCN Red Data Book,* 1975.

42. I. F. Spellerberg, "The amphibian and reptile trade with particular reference to collecting in Europe," *Biological Conservation* 10:221–232, 1976; *IUCN Red Data Book,* 1975.

43. Halliday, *Vanishing Birds,* op. cit., p. 44.

44. F. Campbell and J. Tarr, "The international trade in plants is still unregulated," *National Parks and Conservation Magazine,* April 1980.

45. Ibid.

46. *New Scientist,* April 3, 1980.

47. Saxifrage and *Rafflesia* stories from *IUCN Bulletin,* February 1979.

48. "On the disadvantages of wearing fur," in A. J. Marshall, ed., *The Great Extermination,* op. cit.

49. The information on the Snow Leopard and tigers is from Simon and Géroudet, *Last Survivors: The Natural History of Animals in Danger of Extinction,* World Publishing Company, New York, 1970, pp. 114–131; and *IUCN Bulletin,* May 1979, 136–137.

50. For example, see Kai Curry-Lindahl, *Let Them Live: A Worldwide Survey of Animals Threatened with Extinction,* William Morrow, New York, 1972—a well-done survey of conditions about a decade ago, with good historical material.

51. Norman Myers, "The Cheetah in Africa under threat," *Environmental Affairs* 5:617–647, 1976.

52. *IUCN Red Data Book,* 1975.

53. *IUCN Bulletin,* April 1980.

54. *Sunday Nation,* Nairobi, Kenya, April 16, 1980; *IUCN Bulletin,* January/February 1980.

55. Available as a Dell paperback, New York, 1965.

56. Some of the information on Coyotes is from a seminar by R. Cassin, Department of Biological Sciences, Stanford University, June 5, 1980.

57. Information on the Ceylon Elephant is from Simon and Géroudet, *Last Survivors,* op. cit., pp. 132–139.

CHAPTER VII

1. The story of the destruction of Kaneohe Bay is told dramatically in the film *Cloud Over the Coral Reef* by Lee Tepley and R. E. Johannes.

2. A summary of the situation with the Hawaiian avifauna may be found in Tim Halliday's superb *Vanishing Birds: Their Natural History and Conservation,* Holt, Rinehart and Winston, New York, 1978.

3. For example, P. Ehrlich, R. White, M. Singer, S. McKechnie and L. Gilbert, "Checkerspot butterflies: An historical perspective," *Science* 188:221–228, 1975; P. Ehrlich, I. Brown, D. Murphy, C. Sherwood, M. Singer and R. White, "Increase, stability,

and extinction: The response of checkerspot butterfly *(Euphydryas)* populations to the California drought," *Oecologia* 46:101–105, 1980.

4. See, for example, L. Gilbert and M. Singer, "Butterfly ecology," *Annual Review of Ecology and Systematics* 6:365–397, 1975.

5. *The Naturalist on the River Amazon,* J. M. Dent, London, 1864.

6. R. M. Pyle, "Conservation of Lepidoptera in the United States," *Biological Conservation* 1:55–75, 1976. Some of the information on endangered butterflies came from this article.

7. L. Itow, "San Bruno and the butterfly bloc," *San Francisco Examiner,* May 14, 1980.

8. Lyudmila Beloussova, "Endangered Plants of the U.S.S.R.," *Biological Conservation* 12:1–11, 1977.

9. G. Lucas and H. Synge, *The IUCN Plant Red Data Book,* International Union for Conservation of Nature and Natural Resources, Morges, Switzerland, 1978, p. 345.

10. United Nations, *Concise Report on the World Population Situation, 1970–1975, and Its Long-Range Implications,* New York, 1974.

11. The story of the Golden Gladiolus is adapted from an account written and kindly sent to us by Dr. Anthony V. Hall, Project Leader, Threatened Plants Research Group, Bolus Herbarium, University of Cape Town.

12. P. R. Ehrlich, A. H. Ehrlich and J. P. Holdren, *Ecoscience: Population, Resources, Environment,* W. H. Freeman, San Francisco, 1977, p. 252.

13. D. J. Oxley et al., "The effects of roads on populations of small mammals," *Journal of Applied Ecology,* 11:51–59, 1974.

14. René Honegger, "Unknown . . . unloved . . . threatened," *Naturopa,* no. 27, pp. 13–18, 1977.

15. Stories of plants in this paragraph from *IUCN Plant Red Data Book,* pp. 317, 309, 253, and 413.

16. P. Ehrlich and L. Gilbert, "The population structure and dynamics of a tropical butterfly *Heliconius ethilla,*" *Biotropica* 5:69–82, 1973.

17. J. Muggleton and B. Benham, "Isolation and the decline of the Large Blue butterfly *(Maculinea arion)* in Great Britain," *Biological Conservation* 7:119–128, 1975.

18. For details of the stages of caterpillar development, see Daly, Doyen and Ehrlich, *Introduction to Insect Biology and Diversity,* McGraw-Hill, New York, 1978.

19. Large Blue life cycle from E. B. Ford, *Butterflies,* Collins, London, 1945; and T. C. Emmel, *Butterflies,* Knopf, New York, 1975.

20. D. Ratcliffe, "The end of the Large Blue butterfly," *New Scientist,* November 8, 1979.

21. *Provisional Atlas of British Butterflies,* cited in ibid.

22. "Ethics and attitudes," in J. B. Simmons et al., *Conservation of Threatened Plants,* Plenum, New York, 1978.

23. We are indebted to Dr. Anthony V. Hall, Project Leader, Threatened Plants Research Group, Bolus Herbarium, University of Cape Town, for sending us the account of the Cape flora on which this account is based.

24. *IUCN Bulletin,* February 1979.

25. Beloussova, op. cit.

26. *IUCN Plant Red Data Book,* p. 147.

27. *IUCN Bulletin,* op. cit.

28. Information on the euphorbia and the buttercup is from *IUCN Plant Red Data Book,* pp. 211 and 475.

29. J. S. Turner, "The decline of plants," in A. J. Marshall, ed., *The Great Extermination: A Guide to Anglo-Australian Cupidity, Wickedness and Waste,* Heinemann, London, 1966, pp. 134–155.

30. Information on the Gila Monster and Iriomote Cat is from *IUCN Red Data Book* for 1975 and 1978.

31. Information on Kenya is from Norman Myers, "Kenya's population: 4 percent growth rate," manuscript, 1980.

32. *IUCN Bulletin,* August/September 1977; *United Nations Conference on Desertification,* various publications, United Nations, New York, 1977 and 1978.

33. *IUCN Plant Red Data Book,* pp. 355–356.

34. See P. L. Fradken, "The eating of the West," *Audubon,* January 1979. For a rebuttal showing the responsible other side of the story, see the letter by Richard B. Scudder in the March 1979 issue, pp. 120–122.

35. W. L. Minckley and J. E. Deacon, "Southwestern fishes and the enigma of 'endangered species,'" *Science* 159:1424–1432, 1968.

36. G. Brechen and D. Phillips, "The ebbing tide at Mono Lake," *Sierra,* September–October 1979.

37. For a retrospective review, see P. R. Ehrlich, "Silent Spring," *Bulletin of the Atomic Scientists,* October 1979.

38. M. Evans, New York, 1966.

39. For example, D. W. Anderson et al., "Brown pelicans: Improved reproduction of the Southern California coast," *Science* 190:806–808, 1975; P. R. Spitzer et al., "Productivity of Ospreys in Connecticut–Long Island increases as DDE residues decline," *Science* 202:333–335, 1978.

40. *Ecoscience,* op. cit., pp. 644ff.

41. H. Mendelssohn and V. Paz, "Mass mortality of birds of prey caused by Azodrin, an organophosphorus insecticide," *Biological Conservation* 11:163–169, 1977.

42. G. Harrison, *Mosquitoes, Malaria and Man: A History of Hostilities Since 1880,* Dutton, New York, 1978, pp. 232ff. This is an excellent book, flawed only slightly by the lack of discussion of the development of resistance by the malarial organisms themselves to the drugs used to treat them.

43. See the discussion of pesticide impacts on ecosystems in *Ecoscience,* op. cit., Chapter 11.

44. Some of the current Bald Eagle material is from the *Today Show,* NBC, March 28, 1980.

45. Frank Graham, Jr., "Will the Bald Eagles survive to 2076?" *Audubon,* March 1976.

46. K. M. Schreiner and C. J. Senecal, "The American government's programs for endangered birds," in S. A. Temple, ed., *Endangered Birds: Management Techniques for Rescuing Threatened Species,* University of Wisconsin Press, Madison, 1978, p. 22.

47. M. Frome, "Crusade for wildlife," *Defenders,* April 1980.

48. T. L. Kimball and R. E. Johnson, "The richness of American wildlife," in Council on Environmental Quality, *Wildlife and America,* U.S. Government Printing Office, Washington, D.C., 1978, pp. 3–17.

49. Quoted in Frome, op. cit.

50. See, for example, S. A. Cain, "Predator and pest control," in Council on Environmental Quality, *Wildlife and America,* op. cit., pp. 379–395.

51. "Defenders view," *Defenders,* April 1980.

52. Cain, op. cit., p. 394.

53. T. R. Vale, "Sagebrush conversion projects: An element of contemporary environmental change in the Western United States," *Biological Conservation* 6:274–284, 1974.

54. R. Daubenmire, "Steppe vegetation of Washington," *Technical Bulletin of the Agricultural Station of Washington State University,* 62, 1970.

55. See "Dialogue," *BioScience,* vol. 29, no. 2, February 1979.

56. For information on PCBs, see *Ecoscience;* and R. W. Peterson, "Ecology: Accumulating threats to life," *Environment* 22:3–5, April 1980.

57. Michael Brown, *Laying Waste: The Poisoning of America by Toxic Chemicals,* Pantheon, New York, 1979.

58. T. T. Kozlowski, "Impacts of air pollution on forest ecosystems," *BioScience* 30:88–93, 1980. See also *Ecoscience,* op. cit., p. 661.

59. W. E. Westman, "Oxidant effects of Californian coastal sage scrub," *Science* 205:1001–1003, 1979.

60. *Ecoscience,* op. cit., Chapter 11.

61. The pH of the Pitlochry rain was 2.4—Bryan Sage, "Acid drops from fossil fuels," *New Scientist,* March 6, 1980. Acid rain measurements in the Rockies were made at the Rocky Mountain Biological Laboratory by Drs. Ron Hall and John Harte.

62. G. E. Likens et al., "Acid rains," *Scientific American,* October 1979.

63. R. W. Peterson, op. cit.; "Aluminum pollution caused by acid rain killing fish in Adirondack lakes," *BioScience,* July 1978.

64. F. Pough, "Acid precipitation and embryonic mortality of Spotted Salamanders, *Ambystoma maculatum,*" *Science* 192:68–70, 1976.

65. S. Kimber, "Empty rivers: Dashed hopes," *International Wildlife.* May–June 1980.

66. For example, Peterson, op. cit.

67. "Dirty river turtles," *Natural History,* May 1980; the Marquette quote is from the same source.

68. See, for example, Anton Lelek, "Perish in silence," *Naturopa,* vol. 28, 1977.

69. For acid mine drainage and technical discussion of other problems, see C. G. Down and J. Stocks, *Environmental Impact of Mining,* Applied Science Publishers, London, 1978.

70. Ibid., p. 115; M. Abdullah and L. Royle, "Heavy metal content of some rivers and lakes in Wales," *Nature* 238:329–330, 1972.

71. F. Stearns and J. Ross, "The pressures of urbanization and technology," in Council on Environmental Quality, *Wildlife and America,* op. cit., pp. 209–210; D. Pimentel et al., "Land degradation: Effects on food and energy resources," *Science* 1941:149–155, 1974.

72. J. H. Zumberge, "Mineral resources and geopolitics in Antarctica," *American Scientist* 67:68–77, 1979.

73. P. G. H. Frost, "Conservation of the Jackass Penguin (*Spheniscus demersus* L.), *Biological Conservation* 9:79–91, 1976; see also G. G. Simpson's *Penguins: Past and Present, Here and There,* Yale University Press, New Haven, 1976—a fine little book written with Simpson's customary breadth and brilliance.

74. D. F. Boesch et al., *Oil Spills and the Marine Environment,* Ballinger, Cambridge, 1974.

75. *IUCN Plant Red Data Book,* p. 505; R. Peterson, "A blow against boondoggles," *Audubon,* May 1980, p. 129; for one view of the corps in historical perspective, see A. Morgan, *Dams and Other Disasters,* Porter Sargeant, Boston, 1971.

76. R. W. Peterson, op. cit.

77. Bill Vogt, "Now the river is dying," *National Wildlife,* June-July, 1978.

78. Guy Bonnivier, "Drowning wildlife," *Defenders,* February 1980.

79. *IUCN Plant Red Data Book,* op. cit., pp. 227 and 63.

80. C. R. Whitney, "Where caviar comes by the ton," *International Wildlife,* November/December 1979.

81. Ibid.

82. Technically, biologists call the brute number of species in a community the *species richness,* while *species diversity* includes both richness and the equality of importance of the members of the community. For example, a butterfly community made up of four species, each with 1,000 individuals, is considered more diverse than one made up of the same four species, one with 3,700 individuals and the other three with 100 each. For a brief technical discussion, see E. R. Pianka, *Evolutionary Ecology,* 2nd ed., Harper & Row, 1978. A more extended treatment can be found in R. E. Ricklefs, *Ecology,* 2nd ed., Chiron Press, Portland, 1973.

83. Hank Fischer, "Mountain timber sales threaten Wolf, Grizzly," *Defenders,* February 1980.

84. *Ecoscience,* op. cit., p. 273.

85. *Ecoscience,* op. cit., pp. 145 ff.

86. See A. Ehrlich and P. Ehrlich, "A resource down the river," *Mother Earth News,* August/September 1980.

87. See, for example, F. H. Bormann and G. E. Likens, *Pattern and Process in a Forested Ecosystem,* Springer-Verlag, New York, 1979.

88. "The destruction of the tropics," *Frontiers* 40:22–23, July 1976.

89. The discussion of tropical moist forests (TMF) that follows is based heavily on the report prepared by Norman Myers for the Committee on Research Priorities in Tropical Biology of the U.S. National Research Council, *Conversion of Tropical Moist Forests,* National Academy of Sciences, 1980, and on *Ecoscience,* op. cit. The definition given by Myers of TMF is "evergreen or partly evergreen forests, in areas receiving not less than 100 mm of precipitation in any month for 2 out of 3 years, with mean annual temperature of $24+$ °C and essentially frost-free; in these forests some trees may be deciduous; the forests usually occur at altitudes below 1,300 m (though often in Amazonia up to 1,800 m and generally in Southeast Asia up to only 750 m); and in mature examples of these forests, there are several more or less distinctive strata [pp. 11–12]." References to the basic sources of estimates of tropical moist forest species diversity can be found on p. 14.

90. N. Stark and C. Jordan, "Nutrient retention by the root mat of an Amazonian rain forest," *Ecology* 59:434–437, 1978.

91. *Ecoscience,* op. cit., pp. 624–625.

92. "Attempt at an assessment of the world's tropical moist forests," *Unasylva* 28:112–113, 1976.

93. C. Wilson and W. Wilson, "The influence of selective logging on primates and some other animals in East Kalimantan," *Folia Primatologia* 23:245–274, 1975.

94. *Time,* May 22, 1978.

95. Assume further that 1 percent should be added to the population growth rate for impacts generated in overdeveloped countries (such as cutting of forest for wood pulp to be shipped to Japan). Then another 1 percent should be added because the diversity of species will be lost more rapidly than the forest itself. Taking a rather optimistic guess at the population growth rate of 1.5 percent annually (roughly 1 percent lower than it

is now), this would give a rate of attack on the biological diversity of the forest growing exponentially at 3.5 percent per annum. For population growth rates, see Population Reference Bureau (PRB), *1980 World Population Data Sheet,* available from Population Reference Bureau, P.O. Box 35012I, Washington, D.C. 20013.

96. The basic equation relating the proportion of the remaining forest diversity to be depleted, *D,* the current annual rate of depletion, $\dot{Q}_0$ (as a fraction of the then currently remaining diversity), the exponential rate of growth of the depletion rate, *r,* and time in years, *t,* is

$$D = \frac{\dot{Q}_0}{r} (e^{rt} - 1) \qquad (1)$$

Time in years to the complete depletion of diversity ($D = 1$) would be, by algebraic manipulation of (1) and substituting .01 for $\dot{Q}_0$:

$$t = \frac{1}{r} \ln (100\, r + 1)$$

The value $\dot{Q}_0 = .01$ assumes that diversity at the present time is being depleted at just 1 percent a year—a number close to some current estimates of rates of elimination of the forests themselves (Myers, op. cit., p. 25). Under that assumption, with $r = .035$, half of the populations and species in tropical moist forests would be extinct early in the next century, and none would be left by 2025.

Suppose a less conservative figure for $\dot{Q}_0$, say 2 percent a year, is used—a number considered respectable as near the upper bound on the conversion rate (Myers, loc. cit.). Then using $r = .035$, the diversity would be half gone by the year 2000 and exhausted by 2010. All of these projections, of course, consider no slow-up in the rate of assault because of the successful maintenance of reserves, or modification of forestry or agricultural practices to help preserve diversity.

97. *Ecoscience,* op. cit., p. 626.

98. Statement by Peter H. Raven before the Subcommittee on International Organizations, Committee on Foreign Affairs, U.S. House of Representatives, May 7, 1980, p. 8 (mimeo).

99. Ibid., pp. 9–10.

100. The information on Hawaiian plants is largely from S. Carlquist, *Hawaii: A Natural History,* Natural History Press, Garden City, N.Y., 1970; and personal observations made in collaboration with botanist Richard W. Holm.

101. K. A. Christian, "Endangered Iguanas," *BioScience* 30:76, 1980. For details on the iguanas and other aspects of Galápagos natural history, see Ian Thornton's excellent *Darwin's Islands,* Natural History Press, Garden City, N.Y., 1971.

102. Captain Benjamin Morrell, quoted in *Darwin's Islands,* op. cit., p. 135.

103. Ibid., p. 137.

104. C. G. MacFarland et al., "The Galápagos giant tortoises *(Geochelone elephantopus),* Part I: Status of surviving populations," *Biological Conservation* 6:118–133, 1974.

105. L. Zeleny, "Nesting box programs for bluebirds and other passerines," in S. A. Temple, ed., *Endangered Birds: Management Techniques for Preserving Threatened Species,* University of Wisconsin Press, Madison, 1977, p. 55.

106. For details on invasions, see biologist C. S. Elton's classic *The Ecology of Invasions by Animals and Plants,* London: Methuen, 1958.

107. J. Marr and B. Willard, "Persisting vegetation in an Alpine recreation area in the southern Rocky Mountains, Colorado," *Biological Conservation* 2:97–104, 1970.

108. *Endangered Species Technical Bulletin,* Department of the Interior, April 1980.

109. E. Duffey, "The effects of human trampling on the fauna of grassland litter," *Biological Conservation* 7:255–274, 1975.

110. P. Ehrlich and A. Ehrlich, "Coevolution: Heterotypic schooling in Caribbean reef fishes." *American Naturalist* 107:157–160, 1973; J. Ogden and P. Ehrlich, "The behavior of heterotypic resting schools of juvenile grunts (Pomadasyidae)," *Marine Biology,* pp. 273–280, 1977.

111. G. Davis, "Anchor damage to a coral reef on the coast of Florida," *Biological Conservation* 11:29, 1977.

112. D. Woodland and J. Hooper, "The effect of human trampling on coral reefs," *Biological Conservation* 11:1–3, 1977.

113. *Time,* March 24, 1980.

114. Stebbins, "Off-road vehicle impacts on desert plants and animals and BLM management prescriptions," manuscript prepared for *The California Desert: An Introduction to Natural Resources and Man's Impact,* special publication no. 5, California Native Plant Society, 2380 Ellsworth, Berkeley, CA 94704, 1980.

115. Quote and Santa Cruz information from editorial in *Wild America,* July 1979, published by the American Wilderness Alliance.

116. Documentation of desert impacts can be found in the bibliography of Stebbins, op. cit.

117. Quoted in Stebbins, op. cit.

118. Quoted in Dave Foreman, "ORVs threaten a wild canyon," *Living Wilderness,* September 1979.

119. First edition, Princeton University Press, Princeton, N.J., 1960; second edition available in paperback, Free Press, New York, 1969.

120. "Long-term worldwide effects of multiple nuclear weapons detonations," National Academy of Sciences, Washington, D.C., 1975.

121. S. Glasstone and P. J. Dolan, *The Effects of Nuclear Weapons,* 3rd ed., 1977. A recent overview can be found in "The effects of nuclear war," Office of Technology Assessment, Congress of the United States (Allenheld, Osmun & Company, Montclair, N.J., 1980).

122. See, for example, the Federation of American Scientists' press release of October 4, 1975, criticizing the incompetent National Academy of Sciences study.

123. *Ecoscience,* op. cit., pp. 690–691.

124. K. Lewis, "The prompt and delayed effects of nuclear war," *Scientific American,* July 1979.

125. The exact conditions for generating firestorms are not known—see Glasstone and Dolan, op. cit., pp. 299–300.

126. National Academy of Sciences, op. cit., pp. 39–45.

127. *Man's Impact on the Global Environment: Report of the Study of Critical Environmental Problems,* MIT Press, Cambridge, 1970, p. 116.

128. See Ehrlich, Holm and Brown, *Biology and Society,* McGraw-Hill, New York, 1976.

CHAPTER VIII

1. Paul Ehrlich, "The Strategy of Conservation, 1980–2000," in M. E. Soule and B. Wilcox, eds., *Conservation Biology: An Evolutionary-Ecological Perspective* (Sinauer Associates, Sunderland, Mass., 1980).

2. See P. R. Ehrlich, A. H. Ehrlich, and J. P. Holdren, *Ecoscience: Population,*

Resources, Environment, W. H. Freeman, San Francisco, 1977, Chapter 14, for some history of environmental law in the United States.

3. *Conservation Foundation Letter,* "The Endangered Species Law is under scrutiny," April 1978.

4. It has been poetically described by Peter Mathiessen in "My turn," *Newsweek,* December 17, 1979.

5. David A. Etnier, *"Percina (Imostoma) tanasi,* a new percid fish from the Little Tennessee River, Tennessee," *Proceedings of the Biological Society of Washington* 88:469–488, 1976.

6. Details of the Tellico–Snail Darter case are from many sources, but three of the fullest are John Dernbach, " 'Little fish' versus 'big dam'; The Snail Darter and the TVA—It's not a funny story," *The Progressive,* December 1978; *Conservation Foundation Letter,* op. cit.; and Philip Shabecoff, "Behold the tiny Snail-darter: An ominous legal symbol?" *New York Times,* October 7, 1979.

7. Dernbach, op. cit.

8. Luther J. Carter, "Lessons from the Snail Darter saga," *Science,* 203:730, 1979.

9. Bill Vogt, "Now, the list-makers are endangered," *National Wildlife,* December/January 1980.

10. Elizabeth Kaplan, "Cruel twist in Snail Darter saga," *Not Man Apart* (Friends of the Earth), September 1979.

11. "Services adopt new listing regulations," *Endangered Species Technical Bulletin,* March 1980, p. 3.

12. Brad Kennedy, "Protecting wildlife," *New York Times,* January 13, 1980.

13. For discussions of the history and economics of whaling, see *Ecoscience,* op. cit.; and Robert M. May, "Whaling: Past, Present, and Future," *Nature* 276:319ff, 1978.

14. *Whales and Whaling: Report of the Independent Inquiry* (2 vols.), Australian Government Publishing Service, Canberra, 1978.

15. "The IWC: Debates in brief," *IUCN Bulletin,* August/September 1979, p. 73. This journal is an excellent source of information on whale politics as well as other global conservation issues.

16. John Walsh, "Moratorium for the Bowhead: Eskimo whaling on ice?" *Science* 197:847–850, 1977. See also Mike Weber, "Bowhead Whale: A U.S. dilemma," *Whale Center Newsletter,* Fall 1979; John Bockstoce, "Battle of the Bowheads," *Natural History,* May 1980; and various issues of *Not Man Apart,* 1977–1980.

17. "The IWC: Quotas could have been cut still more," *IUCN Bulletin,* August/September 1979; Christine Stevens, "Victory for whales," *Defenders,* October 1979; and Dr. Jeremy Cherfas, "The great white wash," *New Scientist,* July 19, 1979. Cherfas described the discussion surrounding the separation of pelagic from coastal whaling as "procedural wrangling reminiscent of undergraduate imitations of parliament."

18. D. Phillips and E. Kaplan in a preliminary report for staff of Friends of the Earth, June 11, 1980; Liz Kaplan, personal communication.

19. Reprinted in *IUCN Bulletin,* June 1979.

20. Paula Westdahl, "The nefarious pirate whalers," *Not Man Apart,* July 1980.

21. The story of the *Sea Shepherd*'s encounter with the *Sierra* is by Paul Watson, "Pirate whalers rammed out of business," *Greenpeace Chronicles,* September 1979 (Greenpeace, Vancouver, B.C., Canada). See also his "Pirate whaler smashed," *Defenders* (Defenders of Wildlife), December 1979.

22. Watson, *Greenpeace Chronicles,* op. cit.

23. *IUCN Bulletin,* October 1979.

24. "Creatures," *Audubon,* May 1980; Westdahl, op. cit.

25. E. Stokey and R. Zeckhauser, *A Primer for Policy Analysis,* W. W. Norton, New York, 1978.

26. Constance Holden, "Cracking down on illegal wildlife trade," *Science* 206: 801–802, 1979; Sam Iker, "The crackdown on animal smuggling," *National Wildlife,* October–November 1979, pp. 33–40.

27. Jon Tinker, "Controlling the global wildlife trade," *Atlas World Press Review,* July 1979; *IUCN Bulletin,* "Elephant numbers are falling heavily," January/February 1980.

28. Iker, op. cit.

29. Holden, op. cit.

30. Iker, op. cit.

31. Mike Lipske, "Trafficking in rare reptiles," *Defenders,* April 1980.

32. *IUCN Bulletin,* "Rhinos are no Dodos—yet," January/February 1980.

33. Lipske, op. cit.

34. J. A. Burton and T. Inskipp, "The zoo connection," *New Scientist,* January 5, 1978.

35. Holden, op. cit.

36. Tinker, op. cit.

37. Lipske, op. cit.

38. Gunnison National Forest Report, "Description of the Mount Emmons Mining Project," 1979.

39. Tom Huth, "Crested Butte: A town fights for its heritage," *Historic Preservation,* March/April 1979; Susan Cottingham, "Crested Butte takes on a mining giant," *Living Wilderness,* January/March 1979; and David Sumner, "AMAX comes to Crested Butte," *Sierra,* September/October 1979. All three articles feature color photos that will give you some idea what is at stake. *Historic Preservation* shows the charm of the town itself; the others show the beauty of its setting.

40. The mayor, who prefers to be called simply "Mitchell," was twice a victim of devastating accidents. As a consequence, he is confined to a wheelchair, a circumstance that has scarcely slowed him up. Mitchell has not hesitated to take his case to the national level, including visits to Congress and, on at least two occasions, the White House. He cheerily says that once there were 10,000 things he wanted to do in his life, but he has been forced by his accidents to reduce the list to 9,000. Friends of the Earth recently gave him a 9,001st by appointing him to its board of directors. For a profile of Mitchell, see Kenneth Brower, "Phoenix of Crested Butte," *Omni,* June 1979, pp. 127–128.

41. Charles H. Callison, "It's high time to scuttle the giveaway mining law," *Living Wilderness,* January/March 1979, pp. 4–9. A detailed examination of the congressional debates pertinent to the 1872 Mining Law has been done by attorneys associated with the town of Crested Butte. See also Heather Noble, "Environmental regulation of hardrock mining on public lands: Bringing the 1872 law up to date," *Harvard Environmental Law Review* 4:1, 145–163, 1980.

42. Tom Huth, op. cit.; John Hooper, "High alpine valley coveted by mining corporation," *Wilderness Report* (The Wilderness Society), May 1979.

43. Robert J. Regan, "U.S. walks tightrope over molybdenum supply," *Iron Age,* September 17, 1979. The sales of moly make a positive contribution to the U.S. balance of payments, a benefit AMAX is fond of pointing out. But the contribution is trivial

compared to what even a small increase in the efficiency of American cars would supply in the form of foreign exchange benefits.

44. A former student of Paul's, Ray White, now residing in Virginia, wrote to Senator John Warner of that state in 1979 about this matter. Our information comes from a letter to Senator Warner from R. B. Schwartzman, acting deputy assistant secretary for trade regulation in the Department of Commerce, in response to the senator's inquiries.

45. Quoted in Sumner, op. cit.

46. Letter of July 9, 1980.

CHAPTER IX

1. J. D. Hughes, *Ecology in Ancient Civilizations,* University of New Mexico Press, Albuquerque, 1975.

2. The story of Père David's Deer is based primarily on the account in James Fisher et al., *Wildlife in Danger,* Viking, New York, 1969. See also the volume edited by M. G. Soulé and B. A. Wilcox, *Conservation Biology: An Evolutionary-Ecological Perspective,* Sinauer Associates, Sunderland, Mass., 1980, from which the Conway quote comes. This book is the best broad technical treatment of the subject.

3. Unless, of course, one defines "horse" broadly to include all members of the genus *Equus,* which includes zebras and asses.

4. Fisher, op. cit., pp. 101–102.

5. J. Perry et al., "Captive propagation: A progress report," pp. 361–372, in R. D. Martin, ed., *Breeding Endangered Species in Captivity,* Academic Press, New York, 1975.

6. J. Volf, "Breeding of Przewalski Wild Horses," pp. 263–270 in R. D. Martin, op. cit.

7. 33:17.

8. Numbers of animals in captivity and the overall story of the oryx are from J. M. Dolan, "The Arabian Oryx: Its destruction, captive history, and propagation," *International Zoo Yearbook* 16:230–239, 1976.

9. Information on the return of the oryx and the status of the Tahr from Ray Vicker, "The sultan and the oryx," *International Wildlife,* May–June 1980. Recruitment of Bedouins into the oryx program is from *IUCN Bulletin,* April 1980.

10. "Pandaring," *Time,* June 2, 1980.

11. C. R. Schmidt, "Captive breeding of the Vicuña," in Martin, op. cit., p. 283.

12. See, for example, J. Perry and P. B. Kibber, "The capacity of American zoos," *International Zoo Yearbook* 14:240–247, 1974.

13. J. Perry et al., "Captive propagation," op. cit.

14. Random changes occur in the gene frequencies of all populations because the genes of each generation are a sample of those present in the previous generation and sampling error is an inevitable effect of the sampling process. Technically, the random changes are called genetic drift. What sampling error is, its relationship to population size, and the role of genetic drift in evolution are discussed in relatively simple terms in P. Ehrlich, R. Holm and D. Parnell, *The Process of Evolution,* McGraw-Hill, New York, 1974, pp. 97–102.

15. See the chapters by I. Franklin, M. Soulé, W. Conway, and J. Senner in M. Soulé and B. Wilcox, op. cit.

16. C. Hillman and J. Carpenter, "Masked mustelid," *Nature Conservancy News,* March/April 1980.

17. "Evolutionary change in small populations," pp. 135–149 in Soulé and Wilcox, op. cit., quote on p. 136.

18. A. Berger, "Reintroduction of Hawaiian Geese," in S. Temple, ed., *Endangered Birds: Management Techniques for Preserving Threatened Species,* University of Wisconsin Press, Madison, 1978, pp. 339–344.

19. J. Kear, "Returning the Hawaiian Goose to the wild," in Martin, op. cit., pp. 115–123.

20. Tim Halliday, *Vanishing Birds: Their Natural History and Conservation,* Holt, Rinehart and Winston, New York, 1975, pp. 184–186.

21. A. Berger, op. cit.

22. J. Fisher et al., *Wildlife in Danger,* Viking, New York, 1969, pp. 223–225.

23. C. Kepler, "Captive propagation of Whooping Cranes: A behavior approach," in S. Temple, op. cit., pp. 231–241.

24. S. Campbell, "Is reintroduction a realistic goal?" in Soulé and Wilcox, op. cit., pp. 263–269; T. Scherman, "Day of the falcon," *New York Times Magazine,* June 22, 1980.

25. Fisher et al., op. cit.

26. The story of killing the chick was a personal communication from David Brower, chairman, Friends of the Earth. The event was also described in the Science Section of the *New York Times,* July 15, 1980.

27. C. Koford, "Naturalistic condor plan outlined," *Condor Call,* June 1979.

28. See, for example, D. Mertz, "The mathematical demography of the California Condor population," *American Naturalist* 105:437–453, 1971.

29. Comments on Fish and Wildlife Service Plan, circulated in a Friends of the Earth news release, Spring 1980 (undated).

30. Comments on Fish and Wildlife Service Environmental Impact Assessment, October 12, 1979.

31. Letter from Congressman R. L. Lagomarsino, 19th District, California, to Friends of the Earth, March 25, 1980.

32. "Night of the condor," first published in *Omni* in 1979 and reprinted in *Not Man Apart,* February 1980.

33. See, for example, S. Walters, "The role of European botanic gardens in the conservation of rare and threatened plant species," *Gartnerisch-Botanischer Brief* 51:-2–21; A. Synge, "Botanic gardens and island plant conservation," in D. Bramwell, *Plants and Islands,* Academic Press, New York, 1980, pp. 379–390.

34. P. Raven, "Ethics and attitudes," in J. Simmons et al., *Conservation of Threatened Plants,* Plenum Press, New York, 1976, pp. 155–179.

35. E. Duffy, "The reestablishment of the Large Copper butterfly, *Lycaena dispar batava* Obth. on Woodwalton Fen Natural Nature Reserve, Cambridgeshire, England, 1969–73," *Biological Conservation* 12:143–157, 1977.

36. C. McFarland et al., "The Galápagos Giant Tortoises *(Geochelone elephantopus).* Part II: Conservation methods," *Biological Conservation* 6:198–212, 1974.

37. The theory as it exists today has been developed primarily to explain the bird faunas of islands, and traces directly to a classic paper, later expanded into a book, by the late ecologist Robert H. MacArthur and sociobiologist Edward O. Wilson, *The Theory of Island Biogeography,* Princeton University Press, Princeton, 1967.

38. For an introduction, see B. Wilcox, "Insular ecology and conservation," in Soulé and Wilcox, op. cit., pp. 95–117.

39. Also called Krakatau. For a summary, see B. Bolt et al., *Geological Hazards,* 2nd ed., Springer-Verlag, New York, 1977.

40. E. Willis, "Populations and local extinctions of birds on Barro Colorado Island, Panama," *Ecological Monographs* 44:153–161, 1974.

41. L. Gilbert, "Food web organization and the conservation of neotropical diversity," in Soulé and Wilcox, op. cit., pp. 11–33.

42. Unpublished data of D. Hooijer, from J. Terborgh, "Preservation of natural diversity: The problem of extinction prone species," *BioScience* 24:715–722, 1974.

43. "Benign neglect: A model of faunal collapse in the game reserves of East Africa," *Biological Conservation* 15:259–272, 1979.

44. See, for example, B. Wilcox, "Supersaturated island fauna: A species-age relationship for lizards on post-Pleistocene land-bridge islands," *Science* 199:996–998, 1978.

45. Wilcox in Soulé and Wilcox, op. cit.

46. J. Diamond, "Assembly of species communities," in M. Cody and J. Diamond, eds., *Ecology and Evolution of Communities,* Harvard University Press, Cambridge, 1975, pp. 342–444.

47. J. Diamond, "Patchy distributions of tropical birds," in Soulé and Wilcox, op. cit., pp. 57–74.

48. R. Foster, "Heterogeneity in disturbance in tropical vegetation," in Soulé and Wilcox, op. cit., pp. 75–92.

49. L. Gilbert, "Food web organization . . . ," op. cit., pp. 11–33.

50. For a brief discussion of succession, see P. R. Ehrlich. A. H. Ehrlich, and J. P. Holdren, *Ecoscience: Population, Resources, Environment,* W. H. Freeman, San Francisco, 1977, Chapter 4.

51. Halliday, op. cit., p. 72.

52. See, for example, E. O. Wilson, "The conservation of life," *Harvard Magazine,* January 1980, pp. 28–37; M. S. Gilpin and J. M. Diamond, "Subdivision of nature reserves and the maintenance of species diversity," *Nature* 285:567–568, 1980; A. J. Higgs and M. B. Usher, "Should nature reserves be large or small?" *Nature* 285:-568–569, 1980. See also T. E. Lovejoy and D. C. Oren, "Minimum critical size of ecosystems," paper presented at the *Symposium on Forest Habitat Islands in Man-dominated Landscapes,* American Institute of Biological Sciences (AIBS) annual meeting, Michigan State University, August 25, 1977 (to be published). An excellent overview of the problems of reserve design and management can be found in O. H. Frankel and M. E. Soulé, *Conservation and Evolution,* Cambridge University Press, New York, 1981, Chapter 5.

53. Sites for reserves in Amazonia are now being planned to coincide with the position of Pleistocene refugia—areas that remained tropical forest even at the height of the glacial period. For a discussion of this and access to the literature, see T. E. Lovejoy, "Designing refugia for tomorrow," chapter for *Proceedings of the Fifth Symposium of the Association for Tropical Biology,* to be published by Columbia University Press (mimeo preprint, 1980).

54. "Resolutions for the Eighties," *Harvard Magazine,* January–February 1980.

55. N. Myers, *Conversion of Tropical Moist Forests,* National Academy of Sciences, Washington, D.C., 1980, p. 133.

56. Brazil information is from Peter Eisner, "Brazil acts to keep green hell from

turning into a red desert," *San Francisco Sunday Examiner and Chronicle,* May 25, 1980.

57. L. S. Hamilton, *Tropical Rainforest Use and Preservation,* Sierra Club Office of International Environmental Affairs, New York, 1976.

58. E. Duffey, "The management of Woodwalton Fen: A multidisciplinary approach," in E. Duffey and A. Watt, *The Scientific Management of Animal Communities for Conservation,* Blackwell Scientific Publications, Oxford, 1971, pp. 581–597. Many other articles in this volume are pertinent.

59. " 'Prescribed fire' to aid birds finally stopped," *Peninsula Times Tribune,* May 7, 1980.

60. J. Hart, "Parks for the people: The national debate," *Sierra,* September–October 1979.

61. Cathy Smith, "Alaska bill becomes law," *Not Man Apart,* January 1981.

62. Robert Cahn, "The state of the parks," *Audubon,* May 1980.

63. Joseph Stocker, "Battle of the burro," *National Wildlife,* August/September, 1980.

64. Robert Cahn, "The state of the parks," *Sierra,* May/June, 1980.

65. "A bird in the hand," *New Scientist,* July 17, 1980, p. 185.

66. Norman Boucher, "Whose eye is on the Sparrow?" *New York Times Magazine,* April 13, 1980.

67. Gilbert, "Food web organization . . . ," op. cit.

68. Ibid.

69. R. Pyle, "Conservation of Lepidoptera in the United States," *Biological Conservation* 9:55–75, 1976.

70. Ibid., pp. 71–72.

71. *The Outline of Human History,* vol. 2, Macmillan, New York, 1921, p. 594.

72. R. Forman, H. Galli, and C. Leck, "Forest size and avian diversity in New Jersey woodlots with some land use implications," *Oecologia* 26:1–8, 1976.

73. One commonly hears the claim that Europe is not overpopulated; the uninitiated often cite the prosperous Netherlands, with a population density of some 900 people per square mile, to "prove" this point. In the technical literature, this is known as the "Netherlands Fallacy" (P. Ehrlich and J. Holdren, "Impact of population growth," *Science* 171:1212–1217, 1971). The point is that Europe as a whole and the Netherlands in particular can maintain such high population densities only because the rest of the world does not. Holland ranks second behind Denmark in its per capita imports of protein and gathers from the rest of the world many other materials crucial for its existence. If an impermeable barrier were constructed around the Netherlands (or Europe), the true degree of overpopulation would quickly become apparent. Similarly, the notion that much of the United States is "empty" or "sparsely settled," as is frequently heard on the mass media (for example, George Wills on *Agronsky and Company,* PBS, May 11, 1980; Tom Brokaw, *Today,* NBC, May 12, 1980) ignores the central point that it is *resource availability* and *environmental constraints,* not people per unit area, against which overpopulation must be measured.

74. J. R. Karr and J. G. Schlosser, "Water resources and the land-water interface," *Science* 201:229–234, 1978.

75. See, for example, J. Pires and G. Prance, "The Amazon forest: A natural heritage to be preserved," in G. Prance and T. Elias, *Extinction Is Forever,* New York Botanical Garden, New York, 1977, pp. 158–194. This article contains maps of Brazil's present and

proposed forest reserves. See also H. Irwin, "Coming to terms with the rainforest," *Gardens* (New York Botanical Garden) 1:29–33, 1977.

76. D. Janzen, "Tropical agroecosystems," *Science* 182:1212–1214, 1973. Other aspects of tropical moist forest agricultural difficulties are covered in A. Gómez-Pompa, C. Vázquez-Yanes, and S. Guevara, "The tropical rainforest: A nonrenewable resource," *Science* 177:762–765, 1972; and J. Parsons, "Forest to pasture: Development or destruction?" *Rev. Biol. Trop.* 24 (suppl. 1): 121–138, 1976.

77. G. Budowski, "A strategy for saving wild plants: Experience from Central America," in Prance and Elias, op. cit., pp. 368–373.

78. W. Clarke, "Maintenance of agriculture and human habitats within the tropical forest ecosystem," *Human Ecology* 4:247–259, 1976. This article also contains an interesting discussion of shifting agriculture.

79. "A new look at the plight of tropical rain-forests," *Environmental Conservation* 7:203–206, 1980. See also E. G. Benya and M. Zuliani, "Saving tropical forests," *BioScience* 30:724–725, 1980.

80. R. Aiken and M. Moss, "Man's impact on the tropical rainforests of peninsular Malaysia: A review," *Biological Conservation* 8:213–229, 1975.

81. A. Hammond, "Remote sensing (II): Brazil explores its Amazon wilderness," *Science* 196:513–515, 1977.

82. J. Cairns, Jr., J. Stauffer, Jr., and C. Hocult, "Opportunities for maintenance and rehabilitation of riparian habitats: Eastern United States," in R. Johnson and J. McCormick, eds., *Strategies for Protection and Management of Floodplain Wetlands and Other Riparian Ecosystems,* U.S. Department of Agriculture, Forest Service, Washington, D.C., GTR-WO-12, 1979.

83. W. Edmondson, "Recovery of Lake Washington from eutrophication," in J. Cairns, Jr., K. L. Dickson, and E. Herricks, eds., *The Recovery and Restoration of Damaged Ecosystems,* University Press of Virginia, Charlottesville, 1977, pp. 102–109.

84. Tobias O. Smollett in *Humphry Clinker,* quoted in A. Gameson and A. Wheeler, "Restoration and recovery of the Thames estuary," in Cairns et al., op. cit., p. 73.

85. Gameson and Wheeler, loc. cit.; T. Holloway, "Back from the dead: The restoration of the River Thames," *Environment,* June 1978.

86. P. Opler, H. Baker, and G. Frankie, "Recovery of tropical lowland forest ecosystems," in Cairns et al., op. cit.

87. "Waterway recovery," *Water Spectrum,* Fall 1978, p. 28.

88. See, for example, J. E. Randall, "An analysis of the fish populations of artificial and natural reefs in the Virgin Islands," *Caribbean Journal of Science* 3:31–47, 1963; F. H. Talbot, B. C. Russell and G. R. V. Anderson, "Coral reef fish communities: Unstable, high-diversity systems," *Ecological Monographs* 48:425–440, 1978.

89. See S. Temple, op. cit., Part II.

90. Artificial reefs, however, may serve as anchor points for settling corals, and they form a nucleus around which a natural reef can develop.

91. This section is based in part on P. Ehrlich, "Diversity and the steady state," in J. Comer, ed., *Quest for a Sustainable Society,* in press. A modified version of this article was published as "Variety is the key to life," *Technology Review,* March/April 1980.

92. George N. Appell, "The pernicious effects of development," *Fields Within Fields* . . . no. 14, Winter 1975.

93. From a letter written in 1855 to President Franklin Pierce concerning the purchase of his tribe's land; reprinted in *Greenpeace Chronicles,* September 1979. The name

of Seattle, built in the middle of the Duwamish's land, is a corruption of Chief Sealth's name.

94. C. Holden, "Park is sought to save Indian tribe in Brazil," *Science* 206:1160–1162, 1979.

95. The quote and the numbers of tribes are from G. Hawrylyshyn, "No match for progress," *International Wildlife,* March–April, 1976.

96. This section is based in part on P. Ehrlich, "The strategy of conservation, 1980–2000," in Soulé and Wilcox, op. cit.

97. Based on ibid., p. 338.

CHAPTER X

1. An overview and introduction to the literature can be found in P. R. Ehrlich, A. H. Ehrlich, and J. P. Holdren, *Ecoscience: Population, Resources, Environment,* W. H. Freeman, San Francisco, 1977; see also footnote 9.

2. P. R. Ehrlich and J. P. Holdren, "Impact of population growth," *Science* 171:-1212–1217, 1971.

3. See Ehrlich, Ehrlich, and Holdren, *Ecoscience,* Chapter 14. On relative deprivation, see L. C. Thurow, *The Zero-Sum Society,* Basic Books, New York, 1980, pp. 198ff.

4. R. L. Heilbroner, *The Worldly Philosophers,* 5th ed., Simon and Schuster, New York, 1980.

5. These costs are called externalities (or, more technically, external diseconomies) by economists, because they are costs of doing business that fall on others than the firm involved. A standard approach of neoclassical economics to "pollution" problems is to "internalize" externalities—to make the firm bear *all* the costs of doing business. Unfortunately, this is impossible in the real world because it would drive many prices to infinity, a fact lost on most neoclassical economists.

6. The costs now being paid for the extinction of the Passenger Pigeon can be estimated as a "shadow price." The sum of the prices each human being would be willing to pay to see those enormous swarms of birds and to be able to dine on the squabs—or be willing to pay to know the option still existed—would be the shadow cost to those living today. Each generation from now until the time no human being would pay a penny will incur such a cost.

7. See Ehrlich, Ehrlich and Holdren, op. cit., especially pp. 716–717.

8. Economists often confuse a period of stagnation in a growth-oriented economic system with what a steady-state economic (SSE) system would be like—see, for example, L. C. Thurow, op. cit., pp. 115ff. As Herman Daly, the leading steady-state economist, wrote recently, "The failures of a growth economy should not be used as arguments against a SSE! The fact that airplanes fall to the ground if they try to remain stationary in the air merely reflects the fact that airplanes are designed for forward motion. It does not constitute a proof that helicopters cannot remain stationary." "Postscript: some common misunderstandings and further issues concerning a steady-state economy," in H. E. Daly, ed., *Economics, Ecology and Ethics: Essays Toward a Steady-State Economy,* W. H. Freeman, San Francisco, 1980.

9. A start at creating a design for a steady-state or sustainable economic system has been made by Herman E. Daly of Louisiana State University. If civilization persists, Daly's name will take a place in the history of economics as the greatest of twentieth-century economists. The essence of his ideas may be found in his beautifully written *Steady-State Economics: The Economics of Biophysical Equilibrium and Moral Growth,*

W. H. Freeman, San Francisco, 1977. The ideas in this slim volume are easily accessible to the layperson. His more recent collection (note 8) is also highly recommended.

10. L. C. Thurow, op. cit., pp. 104–105. This is an interesting book in spite of Thurow's preposterous views on resource-environment issues (which simply reflect the state of economics today).

11. Ibid., pp. 114–115.

12. Not all treatments of the environment by main-line economists miss this point entirely. See, for example, J. V. Krutilla and A. C. Fisher, *The Economics of Natural Environments*, Johns Hopkins University Press, Baltimore, 1975.

13. H. O. Barnett and C. Morse, *Scarcity and Growth: The Economics of Natural Resource Availability*, Johns Hopkins University Press, Baltimore, 1963. This view is still absolutely standard in economics. See, for example, G. Anders, W. Gramson, and S. Maurice, "Does resource conservation pay?" *IIER Original Paper* no. 14, International Institute of Economic Research, Los Angeles, 1978.

14. J. L. Simon, "Resources, population, environment: An oversupply of false bad news," *Science* 208:1431–1435. That making an error that would shame a high school science student was not a slip of the pen is made clear by the entire article, which would have been a fine centerpiece for an April Fools' issue of any scientific journal.

15. Ibid.; see also R. Zeckhauser, "The risks of growth," *Daedalus*, Fall 1973, pp. 103–178. A well-known economist and author of a good book on policy analysis, Zeckhauser cannot analyze the risks of growth because he does not understand the constraints mandated by the second law of thermodynamics. He states: "Recycling is not the solution for oil, because the alternate technology of nuclear power generation is cheaper." The second law says that there are *no* circumstances in which oil, as a source of energy, can be recycled.

16. The law tells us that all real-world activities generate heat. No matter how successful people are at limiting their assault on natural ecosystems, the heat assault will always remain and, everything else being equal, will be proportional to the level of economic activity. No technological innovation will permit *Homo sapiens* to escape the consequences of the second law. Eventually all species, including humanity, will fall victim to *overheating* if economic growth continues—even if every other human and environmental problem is solved. See Ehrlich, Ehrlich, and Holdren, op. cit., for more details. For a fantasy on reaching the heat limit, see physicist John Fremlin's "How many people can the world support?" *New Scientist,* October 29, 1964.

17. Daly, *Steady-State Economics,* op. cit.

18. For example, the International Union for the Conservation of Nature and Natural Resources has recently developed a "World Conservation Strategy." For a description, see R. Allen, *How to Save the World,* Barnes and Noble, Totowa, N.J., 1980.

19. Poor people classically discount the future (Chapter 6, note 3) at a much higher rate than the rich.

20. Ehrlich, Ehrlich and Holdren, op. cit., Chapter 15.

21. See, for example, ibid., pp. 690–691.

22. For example, see the similar recommendation made in 1968 by two scientists, Lord Snow of the United Kingdom and Andrei D. Sakharov of the U.S.S.R. ("Father of the Russian hydrogen bomb"), ibid., p. 925.

23. Energy statistics from World Bank, *World Development Report,* 1979.

24. *Playboy* Magazine, 1979.

25. Available from the U.S. Government Printing Office, Washington, D.C.

26. *Denver Post,* July 25, 1980.

27. *Atlas Word Press Review,* April 1979, p. 13.

28. From the *New Republic,* 1978, reprinted in *High Country News,* vol. II, no. 22, November 16, 1979.

29. On the whole question of one-way transfers of wealth, see Kenneth Boulding, *The Economy of Love and Fear: A Preface to Grants Economics,* Wadsworth, Belmont, Calif., 1973. In his fine book, *The Sinking Ark* (Pergamon Press, New York, 1979), Norman Myers gives excellent examples of how grants economics could be used to retard the rate of extinction.

30. *Sand County Almanac,* Oxford University Press, New York, 1949 (reprinted 1970).

31. See note 18.

32. Ehrlich, Ehrlich and Holdren, op. cit., Chapter 5; Population Reference Bureau, *1980 World Population Data Sheet.* See also Parker Mauldin, "Population trends and prospects," *Science* 209:148–157, 1980, for a recent assessment.

33. For example, L. C. Thurow, op. cit.

34. "Night of the Condor," first published in *Omni* and reprinted in *Not Man Apart,* February 1980.

Acknowledgments

Over the years, our concerns about extinction have been shared with, and reinforced by, many of our fellow biologists. This book, in a sense, grew out of discussions with them over a decade or more. To list them all would be difficult indeed—and would surely lead to important omissions. But we must mention in this context Richard W. Holm and John B. Thomas (both Department of Biological Sciences, Stanford), Hugh Iltis (Department of Botany, University of Wisconsin), Peter H. Raven (Director, Missouri Botanical Garden), Michael E. Soulé (Institute for Transcultural Studies, Los Angeles), E. O. Wilson (Department of Biology, Harvard), and George M. Woodwell (Ecosystems Center, Woods Hole).

Dick Holm and Peter Raven also were kind enough to read the first draft of this book and to comment on it extensively. Other colleagues who have also criticized all or part of the manuscript are Loy Bilderback (Department of History, State University of California, Fresno), Thomas Eisner (Division of Biological Sciences, Cornell University), A. C. Fisher and John P. Holdren (Energy and Resources Program, University of California, Berkeley), John Harte (Lawrence Berkeley Laboratory), David E. Lincoln (Department of Biology, University of South Carolina), Kirk Smith (East-West Center), and Cheryl E. Holdren, Harold A. Mooney, and Bruce A. Wilcox (Department of Biological Sciences, Stanford).

Various people were helpful with the project in diverse ways—sending information, checking facts, and so on. These include Gordon R. V. Anderson (National Parks and Wildlife Service, Canberra), L. Charles Birch (Department of Zoology, University of Sydney), Keith S. Brown, Jr. (Instituto de Biologia, Universidade Estadual de Campinas), Richard C. Cassin, Dennis A. Murphy, David Regnery, and Ward B. Watt (Department of Biological Sciences, Stanford University), Anthony V. Hall (Bolus Herbarium, University of Cape Town), Thomas E. Lovejoy (World Wildlife Fund), John McCosker (California Academy of Sciences), Shirley McGreal (International Primate Protection League), Norman Myers (Nairobi, Kenya), David Brower, Elizabeth Kaplan, and David Phillips (Friends of the Earth), A. Hugh Synge (Royal Botanic Gardens, Kew), and Judith Wagner (Department of Biology, California State University, Hayward).

At Stanford, Jane Lawson Bavelas not only struggled with the conversion of handwritten material into first-draft typed manuscript, but made many helpful editorial suggestions. Siu Ling Chen also helped a great deal with these typing chores, and Mrs. Mary Johnson did a fine job of producing the final manuscript. Once again, Falconer Biology Library, under the direction of Michael Sullivan, did a splendid job of helping us deal with a massive literature. Claire Shoens was enormously helpful as usual, and Zoe Chandik, Sarah Gilman, and Judy Levitt always pitched in cheerfully whenever we had a problem. Don Biggs did more photocopying than we'd care to remember, always promptly and intelligently.

Sara M. Hiebert at the Rocky Mountain Biological Laboratory also provided secretarial assistance at several crucial points in the genesis of the manuscript.

Laura Burtness and Jill Holdren competently assisted in the task of proofreading. Those who have been foolhardy enough to write a book will appreciate what an enormous benefit the support of such people is.

Charlotte Mayerson (Random House) and Ginger Barber (Virginia Barber Literary Agency) labored diligently over the manuscript, and have earned our deep gratitude in the process. Their probing questions and editorial skills were enormously valuable to us. Like the many others who have helped us with this book, they should share in any credit that accrues to it; any blame must rest solely with us.

Finally we want once again to express our love and thanks to LuEsther, whose help has been so crucial to our work.

Index